SCIENTIFIC AND MEDICAL COMMUNICATION

Scientific and Medical Communication: A Guide for Effective Practice prepares readers to effectively communicate in professional scientific communities. The material in this book is firmly grounded in more than 500 published research findings and editorials by scientific writers, authors, and journal editors. Thus, this text provides the broadest and most comprehensive analysis of scientific writing. In addition, carefully selected and thoroughly annotated examples from the scientific and medical literature demonstrate the recommendations covered in the text. These real-world examples were carefully selected so that the scientific content can be understood by those without a detailed background in any particular scientific or medical field—thus clearly illustrating the content organization and writing style. This text will prepare individuals to write and edit scientific manuscripts, conference abstracts, posters, and press releases according to journal and professional standards. Readers will also learn to conduct effective searches of the scientific and medical literature, as well as proper citation practices.

Scott A. Mogull is an Associate Professor of scientific and technical communication in the Department of English at Texas State University in San Marcos, TX. Previously he was an Assistant Professor at Clemson University where he focused on teaching scientific writing to science majors. His primary research focuses on scientific and medical communication and scientific communication ethics. He has published research in the fields of microbiology, technical communication, medical writing, and medical rhetoric. For nearly a decade, Mogull has worked in the biotechnology, biodefense, and molecular diagnostics industry as a scientific communicator, product manager, marketing manager, and coordinator of global technical information. Since 2008, he has been on the editorial board of *Technical Communication Quarterly*, the journal of the Association of Teachers of Technical Writing (ATTW). He has a Ph.D. in Technical Communication and Rhetoric from Texas Tech University, an M.A. in Microbiology from The University of Texas at Austin, and an M.S. in Human Centered Design and Engineering from the University of Washington.

ATTW Book Series in Technical and Professional Communication

Tharon W. Howard, Series Editor

Scientific and Medical Communication: A Guide for Effective Practice
Scott A. Mogull

Plain Language and Ethical Action: A Dialogic Approach to Technical Content in the 21st Century
Russell Willerton

Rhetoric in the Flesh: Trained Vision, Technical Expertise, and the Gross Anatomy Lab
T. Kenny Fountain

Social Media in Disaster Response: How Experience Can Build for Participation
Liza Potts

For additional information on this series please visit www.attw.org/publications/book-series, and for information on other Routledge titles visit www.routledge.com.

SCIENTIFIC AND MEDICAL COMMUNICATION

A Guide for Effective Practice

Scott A. Mogull

Routledge
Taylor & Francis Group
NEW YORK AND LONDON

CONTENTS

SERIES EDITOR FOREWORD

Scott A. Mogull's book *Scientific and Medical Communication: A Guide for Effective Practice* is the fourth in the ATTW Book Series in Technical and Professional Communication (TPC), and as its subtitle suggests, this book joins the other titles in the Series through the focus on providing readers with "useful and usable" information about the practice of TPC. Mogull's book also joins Liza Potts's *Social Media in Disaster Response: How Experience Architects Can Build for Participation*, T. Kenny Fountain's *Rhetoric in the Flesh: Trained Vision, Technical Expertise, and the Gross Anatomy Lab*, and Russell Willerton's *Plain English and Ethical Action* in its emphasis on demonstrating the social and cultural impact that the field of TPC is having on our world. Indeed, there are few areas in our society that are having a more direct and significant impact on people's lives than are the practices of scientific and medical communication. By the middle of this century, the US Census Bureau is projecting that more than 1 of 5 Americans alone will be over the age of 65, nearly doubling today's rates. And as populations around the world age, the desperate need for clear, accessible, and above all *actionable* scientific and medical communication will almost surely increase in order to help medical professionals make informed decisions about how best to address the healthcare needs of an aging population.

Yet, as Mogull points out in the opening of his book, "For more than 100 years, readers of scientific and medical journals have been criticizing the poor quality of writing" (p. 3). At a time when socio-economic forces are in urgent need of clear, logical, and focused prose, Mogull shows that "learning to communicate in science and medicine can be particularly challenging" because "novice scientists and writers often mimic the poor writing examples that they find in the literature and thus perpetuate poor writing throughout SMC [scientific and medical communication]" (p. 5). This book is, therefore, an important effort to step across

Writing clearly does not necessitate dry and boring prose. My colleagues and I often bemoan the fact that science and scientific research are not sufficiently appreciated by the general public and that people often make decisions about their health or other aspects of their lives that would benefit from scientific thinking absent any knowledge of scientific data. Yet, we only have ourselves to blame. We speak and write in the specific lingo of our disciplines, using terminology and acronyms that only others in our narrow field can understand. While this language can help us be more precise and provide a shorthand for describing complex ideas, it can also serve to obscure what we are talking about. What we write is only meaningful when we consider to whom we're writing.

In *Scientific and Medical Communication: A Guide for Effective Practice*, Scott Mogull has provided a guide to mastering the craft of scientific writing and communication. He explains the structures of the various forms of communication and provides clear guidelines for practice. Using these tools will allow scientists to explain clearly their discoveries to others in the field and larger world and to give the conference presentation that everyone talks about the next day. Following Mogull's guidelines does not require strictly following a formula and abandoning a personal style. With practice, a good communicator develops a style that is distinct, is comfortable, and fits her personality without sacrificing all the rules. Therein lies the art of good writing.

Shelley M. Payne, Ph.D.
University Distinguished Teaching Professor
Professor, Department of Molecular Biosciences
Associate Dean, College of Natural Sciences
The University of Texas at Austin

SECTION I

Foundations

1

EFFECTIVE COMMUNICATION IN SCIENCE AND MEDICINE

In this chapter, you will learn:

- Characteristics of effective (and ineffective) scientific and medical communication
- Expectations of effective communicators
- Criteria for authorship in science
- Role of professional communicators in science and medicine

For more than 100 years, readers of scientific and medical journals have been criticizing the poor quality of writing. In a 1903 editorial published in the prominent journal *Science*, P. C. Warman pleaded with scientists "to devote to their writing some of the energy they manifest in the field and some of the patience they exercise in the laboratory" (567). Warman emphasized that "effective composition" (565) is more than simply writing clear sentences and paragraphs. Writers must consider the "sequence and unity, symmetry and proportion" (565) of statements in order to craft effective prose.

In the editorial, Warman (1903) stated that the underlying assumption is "an author's style is an exact expression of his mode of thought" (566). In other words, one who presents information clearly and logically must clearly and masterfully understand the material as well. Moreover, Antoine Lavoisier, the eighteenth-century scientist who is often considered the father of modern chemistry, stated that

> the impossibility of isolating language from science and science from language comes from the fact that every natural science is necessarily made

managed by the US National Library of Medicine and the US National Institutes of Health. In Chapter 3 we will cover a complete list of scientific and medical databases.) At this point, what is important is that you should not be intimidated by your unfamiliarity with a scientific or medical topic—everybody (even the top researcher or physician in a field) was unfamiliar with the particular topic at one point. At first your writing progress will be slow as you learn the specialized field, but the learning phase is only temporary. Soon your own knowledge and expertise of the topic will reach that of others on your research team.

2

SCOPE, GENRES, AND AUDIENCES OF SMC

In this chapter, you will learn:

- The role of communication in science and medicine
- The scope of scientific and medical communication
- Genres, audiences, and discourse communities in science and medicine
- Limitations of the scientific research method and the resulting influence on communication

I first began working in a research laboratory in the late 1990s as a graduate student studying microbiology at The University of Texas at Austin. Within weeks, I learned what separated the most respected scientists from the rest. The other graduate students and faculty never seemed to talk much about other people's research. The focus of hallway conversations always seemed to distill into some form of scientific communication: a published article in a scientific journal, a funded grant proposal, or, the pinnacle of prestige, a patent. When discussing scientific communication, everybody knew the stats of one another. For example, the number of scientific articles a researcher published within the previous few years (and particularly articles in prestigious journals) indicated somebody who was successful.

To use an analogy, scientific research follows the rules of natural selection: only the fittest individuals survive. In this case, fitness is measured by outcome of successful scientific communication. Within the laboratory, scientists often use the phrase "publish or perish" to stress the importance of publishing research. If scientists do not publish and receive a steady stream of grant money, their careers perish. As you see, a successful scientific career is subjected to a form of natural selection.

"Too bad about old Ainsworth. Published and published, but perished all the same."

Connecting scientific communication to one of Charles Darwin's theories is a useful analogy but another connection can be established as well. In fact, the foundations of our current scientific and medical communication enterprise began during the mid-1600s with a group of Darwin's colleagues forming the first scientific community, The Royal Society of London. In 1665, this group began publishing the first scientific journal, called *Philosophical Transactions* (Royal Society 2012). Over the next 350 years, scientific communication has developed and, to keep with the analogy, evolved under new environmental pressures (such as electronic media). But the overall goal has remained the same—the sharing and archiving of scientific and medical knowledge.

The primary repository of accumulated scientific knowledge is scientific and medical journals. Publishing original research in scientific and medical journals enables researchers to build on previous work and therefore avoid spending time and resources unnecessarily rediscovering previous findings (American Psychological Association 2010). Moreover, this sharing of research findings is not restricted by physical proximity of research groups nor left to restrictions of overlapping time frames (Garfield 1980; Montgomery 1996; Committee on Responsibilities of Authorship in the Biological Sciences National Research Council 2003). Publication of research also disseminates the latest advancements to medical practitioners, who are then able to implement the new findings in clinical settings (Derish et al. 2007).

Prominent scientists, physicians, and Nobel laureates regularly stress the importance of communication. In a commentary published in the journal *BioScience*, Phil Clapham (2005), who has published nearly 100 journal articles, stated that "publications are indeed everything in science" and that "failure to publish is a scientific crime" (390). As Clapham explained, scientific information that is not communicated is lost. Such information cannot be used to further new research nor contribute to making fully informed decisions about human health or other natural phenomena.

A classic example that illustrates the role of communication in science and medicine is the discovery of penicillin. The initial discovery of penicillin is credited to Sir Alexander Fleming, who in 1928 observed a strange phenomenon in which a mold colony of *Penicillium notatum* appeared to inhibit the growth of the *Staphylococci* bacteria that he was studying. During the course of his research, a petri dish was accidentally contaminated with mold spores, and Fleming observed that the *Staphylococci* bacteria were not growing in the region surrounding the perimeter of the mold colony. Since this region followed the shape of the mold colony, Fleming hypothesized that the mold colony was secreting a substance that inhibited growth of the bacteria. After additional investigation, Fleming published his findings in the *British Journal of Experimental Pathology* to share it with the scientific community. The story of penicillin resumed about a decade later when Sir Howard Florey and Ernst Chain at Oxford University, who had read Fleming's research article, successfully purified penicillin and demonstrated its therapeutic potential. Following this later work, penicillin was mass-produced and used to treat thousands of soldiers with bacterial infections during World War II. Penicillin began the antibiotic revolution, a revolution that saved the lives of countless individuals. In his Nobel Lecture, Fleming (2013) stated, "My publication in 1929 was the starting-point of the work of others who developed penicillin" (83). As you can see, the communication of research findings is as important as the work itself. By publishing, other researchers, medical practitioners, policy decision makers, and others can find and make use of the knowledge.

By using the information presented in this text, you will contribute to the advancement of science and medicine by effectively communicating research findings so others may make use of the research. As a personal benefit, you will also enhance your own career by developing a strong publication record, which serves as the foundation for obtaining funding of competitive research grants and protecting the intellectual property of your research by writing patents.

Scope of SMC

SCIENTIFIC COMMUNICATION is defined as the communication of scientific information to a highly technical expert audience of researchers studying closely related topics (Applewhite 1979; Derish et al. 2007; Harmon 1989). By definition, scientific information is knowledge that arises from systematic observation or experimentation, which is referred to as the SCIENTIFIC METHOD. MEDICAL COMMUNICATION is a subset of scientific communication that communicates clinically relevant research findings to technical audiences for research or application to therapy. In contrast, SCIENCE COMMUNICATION and HEALTH COMMUNICATION are the dissemination of scientific or medical research findings, respectively, to general audiences. Since the scope of this text is scientific and medical communication (SMC), you will find that the communications, or genres, that we emphasize are targeted to expert (technical) audiences. In the following section, we will examine the genres and audiences of SMC in further detail.

Genres

SMC consists of well-defined genres with specific objectives. GENRES are standardized types or categories of communication that have specific content, organization, and style, and are accepted, even expected, by the audience as the appropriate form of particular communications (Bargiela-Chiappini and Nickerson 1999; Bhatia 1993; Fahnestock 2005; Luzon 2005; Piqué-Angordans and Posteguillo 2006; Swales 1990). For example, the prototypical journal article is the genre that is used to summarize original research findings from a scientific investigation, and a grant proposal is the genre that is used to solicit funding for research (see Table 2.1). Even without reading these documents, the readers (the audience) will have a general idea of the type of information that is included in each document. As we will cover in this text, the organization and writing style of SMC genres is so formalized that readers expect to find specific information at the paragraph and even sentence level.

The major genres in SMC usually fall into one of the following categories based on the phase of the research investigation (see Figure 2.1):

- Preresearch phase: grant proposals are used to solicit funding to conduct a specific research investigation;
- Research phase: posters (and other types of oral presentations) are used to share preliminary findings with other researchers at a conference;
- Communication phase: journal articles and the other genres of original research (see Chapter 4) are used to formally share and archive research findings with other experts in the field;
- Commercialization phase: patents are used to protect intellectual property and enable application of research findings.

TABLE 2.1 Major genres of SMC.

Major Purpose	Genre	Content and Specific Objectives
Sharing Original Research Findings	Publications	
	Journal article (or research article or scientific paper)	Report of original data and scientific theories (or laboratory methods)
	Brief report (or letter, note)	Brief report of original data and scientific theories (similar to a research article but shorter and of more limited scope)
	Case study (or clinical observations)	Brief report of interesting observations to stimulate further research
	Conferences	
	Conference Abstract (*Note*: This is a stand-alone genre and is not referring to the summary of a research article.)	Summary of content being presented at a scientific conference
	Scientific Poster	Short summary of ongoing laboratory study; provides original data and tentative theories
	Seminar presentation (or "a talk," which is slang used in research laboratories)	Oral presentation that is a more extensive summary of ongoing laboratory study; provides original data and tentative theories
Summarizing and Synthesizing Original Research	Review article (or clinical review, literature review, narrative review)	Summary of the important research theories from multiple scientific or medical papers
Opinion	Editorial (or commentary)	Expert opinions about an important and relevant topic or research article
Research Funding	Grant proposal (or "grant" or "proposal")	Request that solicits funding for a particular research topic
Commercialization	Patent	Description of new technology to protect intellectual property and enable commercialization

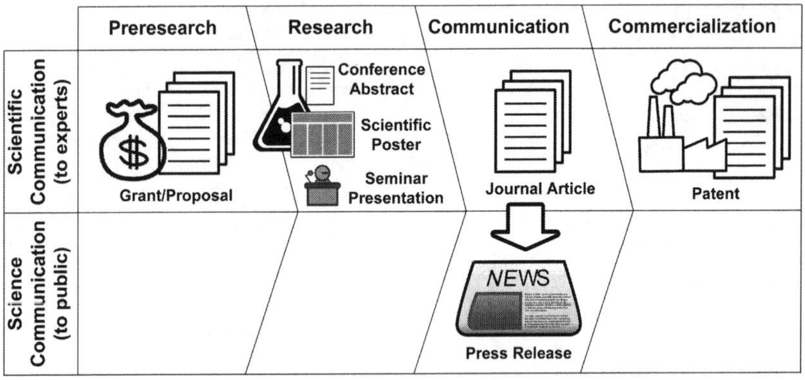

FIGURE 2.1 Major genres for communicating scientific and medical research.

Journal articles and the other genres of original research are classified as PRIMARY SOURCES, which are the original reports of research findings (American Psychological Association 2010). In contrast, SECONDARY SOURCES are genres that are not the first reports of scientific findings, such as literature review articles that synthesize the findings from many primary sources. Although not technically a genre of SMC, press releases often coincide with publication of major research findings in a scientific or medical journal to announce the research to the public (a general audience). In this text, we will cover the genres of the research and communication phases.

Audiences

The AUDIENCE is defined as the intended recipients of a communication. Notably, audiences are not passive recipients of information but develop meaning and understanding as they read (Flower 1981). They try to fit new information into their existing understanding of a topic—their existing "schema" (Huckin 1983)—and often selectively attend to different parts of a communication based on their needs (Alred, Oliu, and Brusaw 1992).

Genres are designed to meet the needs of the PRIMARY AUDIENCE in order to achieve a particular purpose. The primary audience is the main group of individuals who are the intended recipients or target of a particular communication genre. For example, the primary audience of scientific journal articles is other researchers who are conducting research on a similar topic. But other individuals may also read the journal article to accomplish other goals. For example, journalists who are writing a news story may also read an article to get some technical details to report to the general public. Such other groups who might also access a genre are SECONDARY AUDIENCES and often have different background education and needs or requirements for reading a communication. Unfortunately, to be an effective communicator to the primary audience, you are often unable to adequately accommodate every possible audience in a single genre. Rather, you should focus on the needs of the primary audience for each genre and use different genres to effectively communicate the same information to another group (such as a press release that contains the relevant technical information for journalists to use in news stories). One other audience type that is also important to consider when writing SMC is GATEKEEPERS, or individuals who may prevent a communication from reaching the primary audience. In SMC, journal editors and peer reviewers serve as gatekeepers by evaluating journal articles prior to publication. Conference posters and presentations also have gatekeepers but undergo a less stringent peer-review process often based solely on a summary of the research presented in an abstract. Since the only genre undergoing the complete peer-review process is journal articles, that is the only genre that should be credited and cited as the primary source of original research findings. As we will discuss later in the text, conference presentations are considered preliminary findings.

The point of this introduction to audience analysis is that each genre has a primary audience and effective communicators focus on the needs of a single audience when developing each genre. Certainly you must fulfill the expectations of the gatekeepers, as your success at having research accepted for presentation or publication depends on successfully addressing the requirements established by this group.

As Elizabeth Wyckoff (pers. comm.), a research scientist at The University of Texas at Austin, emphasizes, effective communicators must "know thy audience." A common mistake that novice scientist-authors often make, and one that professional communicators may inadvertently adopt, is overestimating the level of familiarity that an audience has with a particular research topic. Scientists spend years studying specific phenomena—becoming extremely familiar with the topic of investigation. When scientists transition to the writing or communication phase, they often overestimate the audience's knowledge and familiarity with the technical details. Even scientists who ask colleagues to review a manuscript before submission to a journal fall into this trap: colleagues within the same laboratory or those who socialize regularly with the researcher are often more familiar with the specific details of the research than the target audience. Professional communicators are in a good position to evaluate the relevant content for a communication because they are more distanced from the material and often approach the material as a reader (Reidenbach 2008). In the following sections, we examine the audiences of SMC separately because each has different needs and reading patterns.

Audience Analysis of Scientific Communication

The two important variables to consider when communicating to scientific audiences include:

- *Specific subject-matter knowledge*: Focused subject-matter knowledge containing detailed information of a specific topic (as specific and detailed as those studying the same few genes or proteins within a single organism) is particularly relevant because most written genres of scientific communication are targeted to other researchers studying similar phenomena. Seminars, or oral presentations, often have a slightly broader scientific audience.
- *Research methods experience*: Laboratory or field methods training and understanding of research methods are important for understanding the details of an experimental design and its implementation. Audiences unfamiliar with particular laboratory methods may have difficulty understanding the data. As a result, such audiences are more dependent on an explanation of *why* a laboratory method was selected, *how* the method addresses the research question, and *what* the data mean.

These variables influence the detail and depth that you need to provide in a communication genre of SMC.

Scientific communication audiences can be plotted along a continuum of specific subject-matter knowledge and research methods experience. Since scientific communication is the communication between scientists, the first time one would be considered a member of this audience is as a graduate student in a scientific field. Scientific training incorporates both theoretical instruction and laboratory training, so scientists advance linearly throughout the typical course of formal education. Individuals may fall within a general range of education and experience, as is illustrated in Figure 2.2. Although the figure provides a general model, individual scientists may be exceptions.

Audiences of scientific communication do not necessarily read a journal article in order from beginning to end, but rather they often search and scan for specific information (Bazerman 1985; Burrough-Boenisch 1999; Charney 1993; Harmon and Gross 1996). For example, scientists familiar with a topic routinely skip reading the Introduction and Methods sections and jump to the new research data and implications of the findings. Alternatively, I met one scientist who said he rarely reads the discussion and conclusion sections because he prefers to construct his own interpretation of the data. As you talk to different researchers, you will likely hear other unique reading patterns. In contrast to the reading patterns of scientists familiar with the content in their field, graduate students and others new to a research field often progress linearly through a journal article because they have less subject-specific knowledge and laboratory experience and therefore are more dependent on all sections of the article.

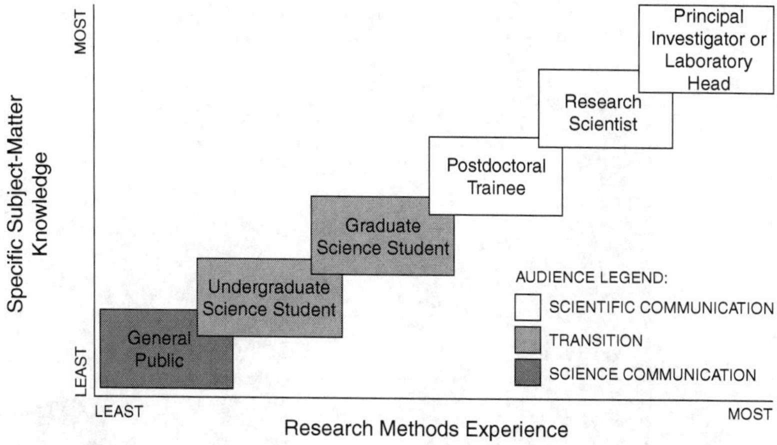

FIGURE 2.2 Continuum of scientific audiences. Typical scientific communication audiences advance linearly through specific subject-matter education and research methods experience, except for laboratory technicians, who focus on research methods.

Audience Analysis of Medical Communication

Audiences of medical communication differ from those of scientific communication. Medical audiences are predominantly medical doctors, medical students, and healthcare professionals or students (see Table 2.2). Additionally, these readers have a greater interest in literature review articles than journal articles of original research (see Figure 2.3). Academic researchers constitute a relatively small percentage of medical journal readers (13%) and have slightly different reading patterns than other healthcare professionals (*BMJ* 2007). Specifically, academic researchers in medicine are most likely to read primary research journal articles. The remaining secondary audiences represent very small percentages of the total readership. These groups include patients and patient-support groups (2%), other members of the public (4%), and the press (1%).

TABLE 2.2 Summary of visitors to medical journal website. Percentages of visitors remained relatively consistent from 2001–7, when data are available (*BMJ* 2007).

Category of Visitor	Percent
Medically qualified doctor	39
Medical student	11
Other healthcare professional/student	16
Academic researcher	13
Pharmaceutical employee	3
Librarian	2
Patient or patient-support group	2
Member of the public	4
Press	1
Other	9
TOTAL	**100**

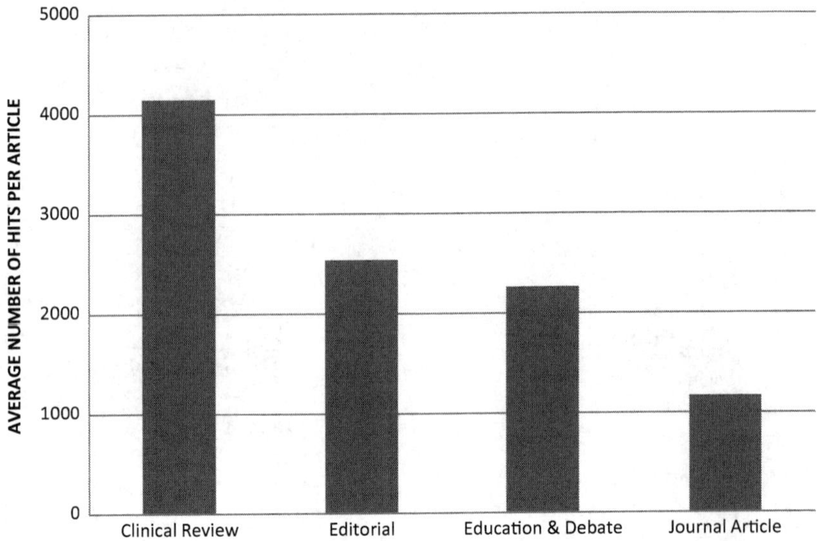

FIGURE 2.3 Average number of articles accessed in the *BMJ* medical journal by category. Data were calculated from a total of 5.38 million articles accessed in 2001 (Loke and Derry 2003).

In general, clinical practitioners have limited time for reading and are most interested in the clinical relevance of research. Review articles both provide a review of the existing primary research and, in the case of narrative reviews, recommend ways to apply these findings to clinical settings (Loke and Derry 2003). Clinicians tend to read the editorials more often than the research journal articles in medical journals (see Figure 2.3), which is important because editorials are opinion pieces that provide a particular slant or interpretation of the research published. Without examining the primary research data in the original article, clinical practitioners may come away with an incomplete view of the research findings.

Discourse Communities

The specialization of researchers leads us to a key feature of SMC. Since the purpose of journal articles is to share the findings from research studies among experts, articles in scientific and medical journals are not written for broad audiences. Instead, original research journal articles and journals are targeted to experts within a narrowly focused DISCOURSE COMMUNITY studying similar topics (Applewhite 1979; Lipson 1988; Patterson 1973; Reeves 2005; Zappen 1989). By definition, a discourse community is a group of researchers who study the same, highly specific research topic and formally share research findings within the community (by publishing in a limited number of appropriate journals and presenting research at relevant conferences). The size of a discourse community varies significantly by field. Factors that influence the size of a community usually reflect the relative importance of the research topic, amount of grant research funding, or potential for commercialization.

When you are working with a research team, you are joining part of an established discourse community. Discourse communities are important because they have a defined research specialty and established communication practices, which may serve as models for effective communication within the group. These communities generally share

- A common understanding of the process, function, or model of a particular scientific or medical issue (known as a PARADIGM);
- A certain level of knowledge with various scientific or medical topics as they relate to the field; and
- An understanding and expectation of a core set of research methods and approaches.

To be an effective communicator within a discourse community, your communication needs to follow the norms and expectations of the community. As we examine different genres and the sections within those genres, you will see how the discourse community influences how you will present information and determine the amount of detail to provide. The main reason to identify a discourse community is so you can gather the key publications from within that community and use these as models for your own writing. Therefore, the first step in the process to becoming an effective communicator in any discourse community is to collect the prominent and recent publications in the literature of that community, which we will cover in the next chapter.

Communication and the Scientific Method

Collecting and sharing knowledge are activities that are central to scientific advancement. SMC is the communication of information (or better yet, the "understanding") that arises from a systematic process, such as observation or experimentation, called the SCIENTIFIC METHOD. As you can see, this definition focuses on knowledge that arises following a formal process or method, which is designed to be objective. Simply put, the scientific method is a way of gathering knowledge, a general approach to research investigations that includes observation, hypothesis, experimentation, and conclusions (Schulte 2003). The scientific method is relevant to the communication process because the general organization of content in scientific and medical genres follows the same organization as the discovery process of the scientific method (Bazerman 1988; Gross 1985; Kelly 2010; Markel 1993; Schulte 2003; Swales 1990; Van Way 2007). Thus, SMC becomes a way of developing, refining, and formalizing a scientific argument that arises from data that accumulate over several months to years of research investigation (Bazerman 1983; Holmes 1987; Latour and Woolgar 1986; Norris and Phillips 2003; Woodford 1967; Zhang 1989).

In a larger context, the scientific method and formal communication process create a cycle of investigation and communication (see Figure 2.4). In this iterative cycle, the findings from published research papers, or the THEORIES, lead to more advanced research questions, or HYPOTHESES, which are subsequently investigated. As Clapham (2005) explained:

> Published knowledge [in science] is assimilated by colleagues and leads to more research: hypotheses are modified, rebutted, or confirmed, new paradigms are developed or old ones discarded. In a very real sense, publications *are* the scientific method.
>
> *(390)*

The scientific method and the scientific communication process are important for communicators because these processes influence the organization, style, and terminology (such as "theory" and "hypothesis") of all SMC. As you will see throughout this text, SMC differs significantly from the general discourse

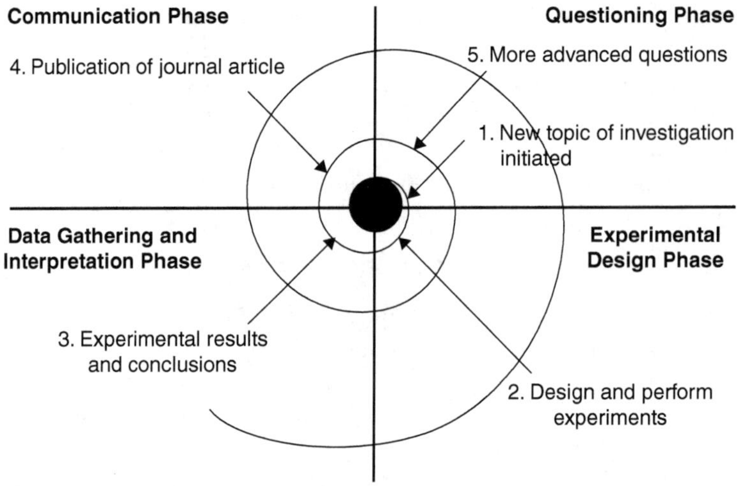

FIGURE 2.4 The scientific communication process is part of the scientific method. Published data and conclusions from research experiments become public knowledge, and this knowledge in turn stimulates new research questions and experimentation. These experimental data and conclusions must be shared through the formal publishing process to enter public knowledge. During each cycle, the understanding of a specific topic expands.

of science that you see published in popular and news media. In the rest of this section, we examine the limitations of the scientific method and see how that influences the degree of certainty of scientific findings.

In formal SMC, the certainty of research findings is conveyed cautiously. The language used in SMC differs from the general discourse of science, which you read in the news media. Most important is the philosophical distinction that influences the language. In popular discourse, the term "science" (or label "scientific") is imprecisely substituted for "fact" or "truth." This use of "science" is part of the lexicon of general audiences and introduces a false accuracy regarding the certainty of information. General audiences tend to think of scientific findings as certain, relying on the outdated theory of scientific positivism that science uncovers facts, which are undisputed (Cummings 1998; Miller 1992; Rude 1992). Although researchers systematically collect scientific data and use the scientific method to incorporate procedures for objectivity, scientists perform the collection, analysis, interpretation, and communication of data. Therefore, scientist's conclusions are limited by restrictions inherent in the methods and instruments used to collect the data and in the interpretation and analysis of those data. The current theoretical perspective of science is that scientific findings are temporary explanations that best fit the existing evidence and current understanding of the real world within the limitations of people's sensory and intellectual abilities (Yore et al. 2006). Therefore, the findings reported in scientific and medical journals represent the current data and understanding, but—critically—this information may be revised following the collection of new data (often arising from the advent and use of more sophisticated technology during the research process, new approaches to answer a question, or observations that conflict with conventional theory).

The search for the genetic material illustrates the refinement of scientific theories over time. In the mid-twentieth century, most scientists thought that genetic material was protein, not DNA, because DNA was too chemically simple to account for the diversity of life. But novel experiments and theories emerged that countered the prevailing paradigm. In a classic experiment, Oswald Avery, Colin MacLeod, and Maclyn McCarty (1944) showed that DNA, and not protein, changed the characteristics of organisms. Thus, Avery et al. showed that DNA, and not protein, was likely the genetic material. Later, James Watson and Francis Crick (1953), with Rosalind Franklin's electron microscope image of DNA, proposed a chemical structure for DNA and suggested the complex mechanism for its functioning as the molecule of heredity (Miller 1992; Shea 2008). This contemporary example illustrates that scientific findings (or "conclusions" from experimental research) are always subject to future research and technological advancements, such as the electron microscope that enabled imaging of smaller objects. For communicators, this example shows that you should present data and the theories arising from the data appropriately given the contingent findings of scientific investigation.

In general, when scientists talk about "science," they are referring to the scientific inquiry process—or, more specifically, the scientific method. Scientists generally use these terms to refer to the process of doing science inquiry, the verb, and not the resulting science knowledge, the noun (Yore et al. 2006, 126). Scientists may use terms such as "fact" or "proves" when talking about science in less formal contexts, but such terms of certainty are uncommon in formal SMC because of the tentative nature of scientific findings. As you will see in formal SMC, articulate scientists use the terms "evidence" or "observations" rather than "fact," and "supports" rather than "proves."

In SMC, communicators carefully use terms and present data that communicate the degree of certainty of the research findings. Most scientists believe that knowledge claims are not absolute, only *supported* or *falsified*. In addition, formal SMC that presents theories, particularly new theories, is saturated with careful presentation strategies (Reeves 2005) and hedging statements (Hyland 1996) that restrict and qualify the certainty of claims. For example, we see these strategies used in the first published account arguing that DNA was the molecule of heredity. Oswald Avery, Colin MacLeod, and Maclyn McCarty (1944) stated that "the experimental data presented in [their] paper *strongly suggest* [emphasis added] that nucleic acid, at least those of the desoyribose type, possess different specificities *as evidenced* by the selective action of the transforming principle" (152). Yet, over time as scientific theories mature, such careful presentation strategies are abandoned in lieu of more direct prose. For example in their paper presenting the structure of DNA, Watson and Crick (1953) stated, "It has not escaped our notice that the specific pairing we have postulated immediately suggest a possible copying mechanism for *the genetic material* [emphasis added]" (737). In this example, the new theory presented, the structure of DNA, is qualified but previous findings regarding the function of DNA are accepted without hedging terms and "the genetic material" is substituted for "DNA." Despite the shift to an objective style for more accepted scientific theories, these careful strategies are usually not replaced by terms of absolute certainty, such as "fact." Rather, according to the principles of science, scientific and medical findings are always subject to change or refinement—being contingent on data from future research. This is a particular problem in medical studies in which the validity of conclusions has been criticized for being overstated (Boutron et al. 2010; Ioannidis 2008).

In summary, statements of certainty are an important distinction between the communication of scientific and medical research to technical and to general audiences. General audiences expect definitive conclusions from scientific investigations, often interchange the terms "hypothesis" and "theory" without a consistent definition of either term, and are confused by scientists' hedging of claims. In contrast, scientists and physicians have a similar discomfort with scientific claims that are too definitive or bold. For these experts, such communication appears absolute and naive with regard to the scientific approach. Thus, throughout this text we will look at ways to adhere to a scientific writing style and appropriately qualify claims, as well as all of the other skills necessary, in order to communicate as professional members in scientific and medical discourse communities.

Concluding Points

In this chapter, we examined the role of communication in science and medicine. The major goal was to introduce you to the roles and expectations of effective communicators in science and medicine and to examine the theoretical foundations of SMC. We examined the major genres of SMC, distinguished between primary and secondary sources, and discussed the needs and reading patterns of the audiences. At this end of this chapter, we returned to the role of communication as part of the scientific method.

This text examines the communication of scientific and medical research, which includes writing journal articles (Section II), communicating research at conferences (Section III), and sharing research results with the public and news media (Section IV). In each section, we will examine the prototypical genre structure and discuss the approach that you should take to be an effective communicator within a discourse community. Within each chapter, we will also examine the issues of the scientific writing style that are relevant to the genre and practices of professional and ethical communication. I have organized this text so that you can read the introduction chapter to each genre and then focus on writing one section at a time. This text will guide you through the writing process so that you will be able to effectively communicate scientific and medical research and make a formal contribution to the advancement of science.

3

SEARCHING AND CITING THE SCIENTIFIC AND MEDICAL LITERATURE

In this chapter, you will learn to:

- Conduct a comprehensive search of the scientific and medical literature
- Effectively use databases and other resources for a comprehensive search
- Manage a library of journal articles
- Properly cite research

Overview

At the International Medical Congress in 1881, John Billings, from the Office of the Surgeon General of the United States Army, described many of the challenges, processes, and guidelines for referencing the scientific and medical literature. Many of the principles presented at that meeting have since become standards of SMC. To a notable list of attendees, which included Thomas Henry Huxley, Sir William Jenner, Baron Joseph Lister, and Louis Pasteur, Billings gave an address, entitled "*Our* [emphasis added] Medical Literature," in which he stressed the egalitarian nature of the literature and the responsibility of those involved in the archival and exchange of scientific and medical knowledge. Yet, Billings remarked, "If, perchance, among the dusty folios there are stray golden grains yet un-gleaned, remember that just in front are whole fields waiting the reaper" (61). The problem is finding the right piece of information when you need it.

Information overload is (and has been) a significant problem in SMC. We like to think of information overload as a recent problem (Bernstam et al. 2006), but finding necessary information from an excessive corpus has long been an issue. For example at the time of Darwin, Lister, and Pasteur, the written scientific and medical information accounted for one thirteenth of the entire world's literature and was increasing at 1,500 journal volumes and 2,500 unbound articles or essays annually (Billings 1881). To keep up with the pace of literature in the late 1880s, you would have to read more than four complete journal issues and almost seven additional articles every day. But the rate of publication has also increased exponentially. For example, the total MEDLINE database is approaching nearly 20 million references (see Figure 3.1). In 2014, you would have to read nearly 3,200 articles each day to keep up with the rate of publication, and this does not take into account all of the other articles that were published in journals not included in the MEDLINE database as well as all of the other genres (such as scientific presentations) that are not included in that database. To search such a vast and growing information archive, you certainly need a systematic approach. The goal of this chapter is to provide you with such an approach.

Writers of SMC must locate the latest REFEREED (or PEER-REVIEWED) journal articles on a topic to collect content for their own SMC. As mentioned in Chapter 1, one of the primary expectations of communicators in SMC is the ability to research the scientific and medical literature. The literature-search process is not simply a one-time keyword search of a single database. As you will see, properly searching

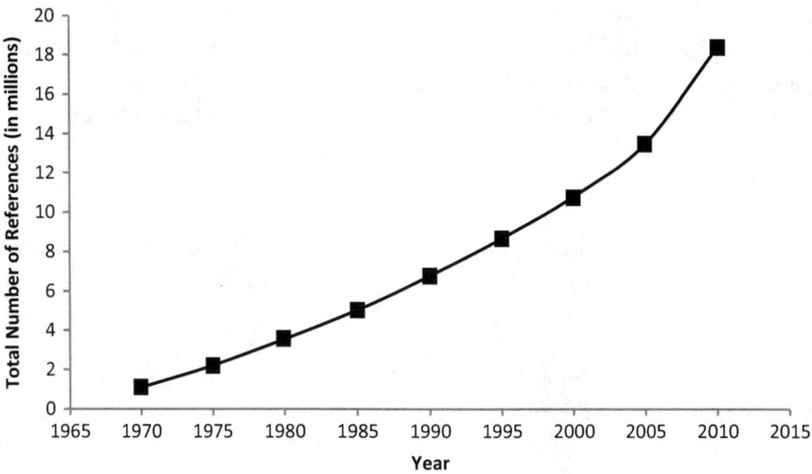

FIGURE 3.1 Total number of references indexed in MEDLINE by year.

Source of data: US National Library of Medicine (2014).

the scientific and medical literature requires significant time and effort. A systematic search of the SMC literature is a meticulous, multistep process that involves:

1. Building a library of relevant journal articles,
2. Updating the library regularly for the latest research, and
3. Monitoring for retractions of published scientific findings.

By conducting a systematic literature search, you might even become your research team's information manager and librarian—adding additional value to the entire team.

How to Prepare

To begin a literature search, you need to know the purpose or objective of the research—in concrete terms so you can use as keywords for searching databases. Additionally, you should ask other researchers on the team to recommend a few key articles that are closely related to current investigation that can supplement your keyword search and provide you with leads to other related articles.

What to Expect

In literature searches, most individuals use an INFORMATION FORAGING approach, which is a series of database searches that provides new leads and possible paths to follow to find the necessary information (Pirolli and Card 1999). Rather than a straightforward search to a single best resource, one wanders through information fields and follows certain paths while discarding others. Therefore, you will modify your search query and search terms continuously as you get closer to the information you need. Although parts of the search process are automated, you will direct much of the process as well (Bates 1989). The approach develops as you identify key articles, authors, and journals—and use these factors to focus your future searches.

In addition to the topic of the search results, you must also attend to the genre (see Table 2.1). Typically only refereed JOURNAL ARTICLES of original research are accepted citations in SMC. Original research journal articles have various names in different journals, such as brief reports, letters, notes, case studies, or clinical observations. The important feature is to distinguish original research from other genres, such as editorials or commentaries, which are opinion pieces rather than original research. REVIEW ARTICLES or LITERATURE REVIEWS are comprehensive summaries of original research journal articles by experts in the field and are generally acceptable sources for broader claims that encompass or synthesize a field of research. You should never rely solely on secondary reports of information as references, such as review articles. Additionally, even though the sources you find are legitimate sources for identifying references to other journal articles on the same topic, each paper has a limited and biased list of sources (Awrey et al. 2011). Therefore, you need to conduct your own database search to ensure that you have examined all of the research from a more comprehensive resource. The GREY LITERATURE, a category of technical and semitechnical work that was not evaluated by a comprehensive peer-review process, usually should not be cited. The grey literature that you might come across includes unpublished manuscripts, theses (and dissertations), conference proceedings, scientific posters, conference abstracts, presentations, patents, grants, technical reports, newsletters, product catalogs, textbooks, and academic lectures. Additionally, genres of popular and news media should not be referenced in SMC. In summary, all of your sources must have been through the complete peer-review process and have the content scrutinized by experts in the field prior to publication.

The literature-search process evolves as you uncover new sources and is iterative through refined keyword searches of major databases and targeted searching of references from key articles, authors, and journals. We will examine each of these issues in this chapter. To be an effective communicator, you require quality content that you collect by repeating your literature search before each major revision of a paper and immediately before submission of the manuscript (Taylor 2002). Later in this chapter, we will discuss maintaining a log of literature searches to make each subsequent search more efficient that the initial search.

Designing Effective Database and Keyword Searches

A keyword search of databases is often the first step to locating published research on a topic (Aoki, Enticott, and Phillips 2013; Bartels 2009; Chapman, Morgan, and Gartlehner 2009; Kurup and Hersey 2010; Webster et al. 2010). Although you have likely previously performed casual keyword searches of databases cataloging academic literature, you will find that effective communication in SMC requires that you take a more systematic approach to the literature search. A comprehensive keyword search should include the following steps:

1. Selecting appropriate databases,
2. Selecting effective keywords to yield a manageable number of database records,
3. Accessing the full-text articles, and
4. Recording database search parameters for future searches.

Although I organize the literature-search process into four steps for our discussion, the process is iterative rather than linear. For example, you are likely to move through steps two to four multiple times before exhausting all citation records available in a particular database. At that point, you should move to a second (or third) database to ensure that you have thoroughly searched the topic. At the point of saturation, you will no longer uncover new references and your comprehensive search will be complete. From that point forward, you will simply need to repeat the searches to identify the latest published research.

In this section we will discuss each step in detail. In our discussion, I show examples from MEDLINE/PubMed, the most popular database and host interface of biomedical literature. All database HOST INTERFACES (the interfaces that a user inputs search parameters) provide similar features and options although the location and ease of use vary. You will likely have to spend time with any DATABASE HOST (the particular interface provided by one company to access a database) to familiarize yourself with the interface and to become proficient in locating the sources that you need.

"What burns me up is that the answer is right here somewhere, staring us in the face."

Step 1: Selecting Appropriate Databases

When selecting databases, your goal is to identify the top two to three databases for each research topic. A TOP DATABASE is defined as one with the most records, or sources, for a particular topic. By using multiple databases, you minimize the bias and limited INDEX, which is the collection of records of any single database. When you search databases, make sure that they are actually databases of different records (such as MEDLINE and Google Scholar) and not simply different database interfaces. Different database hosts, or the interface you use, may search the same database (index). For example, PubMed and Ovid Medline both search the US National Library of Medicine's MEDLINE database. If you used both PubMed (the US National Library of Medicine's host or interface) and OVID MEDLINE (a commercial database host), you would be searching the same collection of records, so you would not uncover any unique articles. But, you might prefer the interface or features of one database host over another.

The databases that you use depend on the discipline and the type of information that you need. Table 3.1 lists popular databases in SMC. Additionally, large institutions may subscribe to other databases services, so you should check your library for a list of the largest, most relevant, and nonoverlapping databases.

Unfortunately you cannot rely on any single database (or even a completely automated search) for a comprehensive search of the scientific and medical literature. Each database varies by several critical features, so you should use multiple databases and try different keyword combinations to conduct a comprehensive literature search. You will find that search results can vary significantly between databases because different databases index different collections of SMC journals, use different search algorithms, and search different content fields (e.g., full text or title and abstract only). To illustrate the differences between database results, a search for "ORS in the treatment of childhood diarrhea" returned nearly 16,000 results in Google Scholar, but the same search in PubMed (except with removal of "stop" words) returned 89 results (Ramji 2011). Your initial reaction to these results might be to conclude that Google Scholar is a "better" search engine. Certainly some databases provide more records on some topics than others, but each database identifies unique results and provides advantages and disadvantages (Nourbakhsh et al. 2012; Shultz 2007; Steinbrook 2006). Two important differences between PubMed/MEDLINE and Google Scholar is that PubMed prioritizes recently published papers and provides a more focused set of results (e.g., 89 results would be much easier to examine individually than 16,000). Certainly the results depend significantly on the selection of keywords, which is the next step.

Step 2: Selecting Effective Keywords to Yield a Manageable Number of Database Records

The goal of a keyword search is to uncover a manageable list of citations that you can analyze in greater detail for relevance to find the specific information

TABLE 3.1 Popular databases of refereed literature in SMC.

Database[1]	URL	Discipline	Open Access (Free)
AGRICOLA	http://agricola.nal.usda.gov	Agriculture and related sciences	Yes
Agricultural Sciences and Technology Database (AGRIS)	http://www.ntis.gov/products/agris.aspx	Agriculture	Yes
arXiv	http://arxiv.org	Physics, mathematics, computer science, quantitative biology, quantitative finance and statistics	Yes
Astrophysics Data System	http://adswww.harvard.edu	Physics and astronomy	Yes
Biological Abstracts	http://www.ebscohost.com/academic/biological-abstracts	Biological sciences	No
BioOne	http://www.bioone.org	Biological sciences (specializing in small professional societies)	Yes
CAB Abstracts	http://www.cabdirect.org	Applied health and agriculture	No
Chemical Abstracts Service (CAS)/SciFinder	http://scifinder.cas.org	Chemistry and biochemistry	No
$Chem_x Seer$	http://chemxseer.ist.psu.edu	Chemistry	Yes
Cochrane Library	http://www.thecochranelibrary.com	Medicine (clinical trials)	No
Cumulative Index to Nursing and Allied Health Literature (CINAHL)	http://www.ebscohost.com/academic/cinahl-plus-with-full-text	Nursing and allied health	No

EMBASE	http://www.elsevier.com/online-tools/embase	Biomedicine and pharmacology	No
Google Scholar	http://scholar.google.com	Multidisciplinary	Yes
HighWire	http://highwire.stanford.edu	Multidisciplinary	Yes
JSTOR	http://www.jstor.org	Multidisciplinary	No
MEDLINE (via PubMed)	http://www.ncbi.nlm.nih.gov/pubmed	Biomedicine	Yes
PubChem	http://pubchem.ncbi.nlm.nih.gov	Chemistry	Yes
PsycINFO	http://www.apa.org/pubs/databases/psycinfo/index.aspx	Psychology	No
Scopus	http://www.scopus.com	Multidisciplinary	No
TOXNET	http://toxnet.nlm.nih.gov	Toxicology and environmental health	Yes
Web of Science	http://thomsonreuters.com/web-of-science	Multidisciplinary	No
WorldCat	http://www.worldcat.org	Multidisciplinary	Yes

[1] *Sources*: Aoki, Enticott, and Phillips (2013); Bartels (2009); Benson and Silver (2013); Kurup and Hersey (2010); Matthews, Bowen, and Matthews (2000); McMillan (2012); Ramji (2011); Steinbrook (2006); and Vanhecke et al. (2007).

that you need. In your initial keyword search, you will need to select two to four keywords based on the research question and then use those keywords to search multiple databases. KEYWORDS or terms that are useful for searching databases include some combination of the following:

- **Population, patient group, or organism** being studied (if humans, then you do not need to specify the organism, but you may refine your search to relevant populations or patient groups; alternatively, if not human subjects, then use the full genus and species scientific name, such as *Escherichia coli*, and also try any common names);
- **Disease, process, or phenomenon** (be sure to spell the term accurately, particularly features that might influence a database search such as American versus British spelling, spaces or no spaces, hyphens, or any other special characters);
- **Specific test variables** (particularly the intervention or treatment conditions); and
- **Location** (if important, as would be the case in field research or examining the disease of populations in a particular geographical location).

Generally, you should begin searching a database with the name of the organism (or population) and one or two key terms. Then you can revise your search strategy based on the success of the search. In Examples 3.1 and 3.2, I have identified the research goal from two journal articles. In these examples, I illustrate how to find and select concrete keywords that are useful for searching databases. During the process, you are likely to uncover unfamiliar terms. Often these unfamiliar terms are useful keywords. (You may use such terms for a keyword search even if you do not know what the terms mean at this point. After you locate and read some recent articles, you will begin to understand the terminology.) At this point in the process, the important thing is to make sure that you spell each term correctly to use in the database searches.

EXAMPLE 3.1 SELECTING KEYWORDS FROM A TYPICAL MEDICAL RESEARCH GOAL OR QUESTION.

Research Goal: Examine melatonin secretion and the prevalence of type 2 diabetes.

Procedure:

- **Highlight the names of organisms or populations**: type 2 diabetics
- **Highlight any term with special formatting** (which is a good way to find key terms): type 2 diabetes (note use of a numeral rather than the spelled word)

- **Highlight nouns and verbs** (prioritizing concrete nouns and specific verbs): melatonin secretion and incidence of type 2 diabetes (as two separate terms)
- **Identify the top two to four key terms** (although there is not a set number of key terms that you can search in databases, limiting your identification to the top few terms will help you to organize your search): melatonin secretion and type 2 diabetes

Source: McMullan, Ciaran J., Eva S. Schernhammer, Eric B. Rimm, Frank B. Hu, and John P. Forman. 2013. "Melatonin Secretion and the Incidence of Type 2 Diabetes." *JAMA* 309 (13): 1388–96.

EXAMPLE 3.2 SELECTING KEYWORDS FROM A SCIENTIFIC RESEARCH GOAL OR QUESTION.

In this slightly more challenging example, you will select keywords from a research goal with unfamiliar terms.

Research Goal: Examine the effect of *hfq* on gene expression and virulence in *Neisseria gonorrhoeae*.

Procedure:

- **Highlight the names of organisms or populations**: *Neisseria gonorrhoeae*
- **Highlight any unusual term** (particularly terms with special formatting): *hfq* and *Neisseria gonorrhoeae* (note that these terms are italicized and have unique capitalization patterns; see Chapter 5 for more information on terminology)
- **Highlight nouns and verbs** (prioritizing concrete nouns and specific verbs): gene expression and virulence (as two separate terms)
- **Identify the top two to four key terms** (although there is not a set number of key terms that you can search in databases, limiting your identification to the top few terms will help you to organize your search): *Neisseria gonorrhoeae*, *hfq*, gene expression, and virulence

Source: Dietrich, Manuela, Rebekka Munke, Marion Gottschald, Elke Ziska, Jan Peter Boettcher, Hans Mollenkopf, and Alexandra Friedrich. 2009. "The effect of *hfq* on global gene expression and virulence in *Neisseria gonorrhoeae*." *The FEBS Journal* 276 (19): 5507–20.

The size of the research field and selection of database will influence the number of results that you obtain through each database search (see Figure 3.2). Often you may not accurately predict the success of each search or number of results that you will receive until you try a few database searches (Bartels 2009). The scope of a manageable list of citations to inspect is somewhat of an individual preference, but I usually target an upper limit of 200 to 300 citations that I then examine individually for relevance by title and then by abstract.

If your search yields too many results, you can reduce the number of records by increasing the number of keywords that you use in each individual search and using BOOLEAN OPERATORS (such as AND, OR, or NOT) to refine your search. Complex nesting patterns of keywords with these operators increase the precision of database searches (see Table 3.2). Additionally, you might try limiting the date of publication to the last 5 to 10 years since most content that you reference in SMC should be within this time frame (Price 1989; Grant 2001). In PubMed, for example, you can filter the search results by 5 or 10 years with a single click directly on the results page (see Figure 3.2, left-hand column). Alternatively if your search yields too few results (fewer than 10 to 20 records), you might try reducing the number of keywords used in each search or reducing the specificity of the target organism or population by using a broader or more inclusive term. For example, you might change the specific name of the organism, "*Neisseria gonorrhoeae*," to the type of organism, "bacteria."

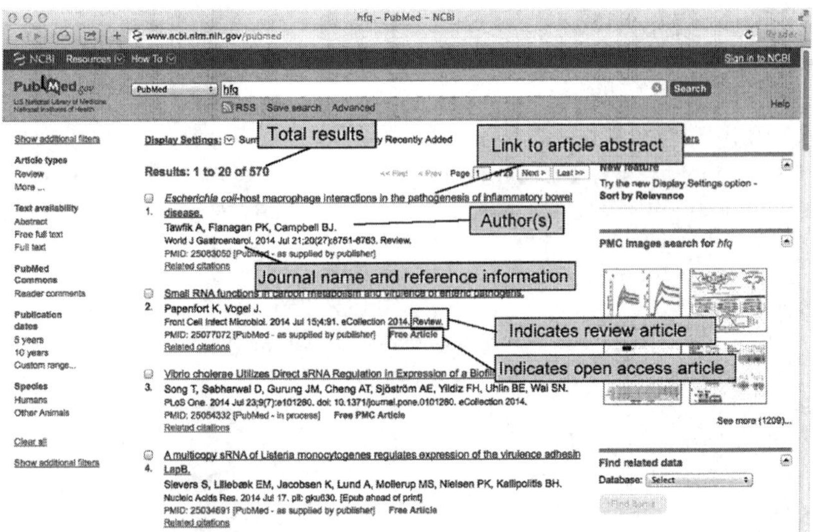

FIGURE 3.2 PubMed/MEDLINE search results for "*hfq*." Journal article titles are hyperlinked to the article record. Authors are listed below the article title followed by the journal name and reference information. Review articles are indicated to distinguish from original research journal articles.

TABLE 3.2 Boolean operators as used to focus database search results.

Boolean Operators	Description	Sample Search	Representation of Search Results
AND	Returns results with both keywords	*Neisseria gonorrhoeae* **AND** *hfq*	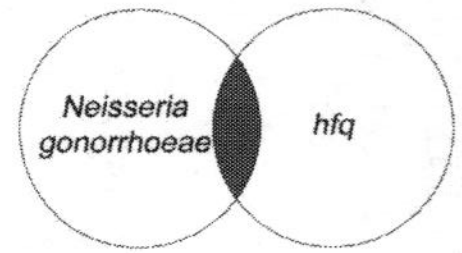
Complex AND	Returns more specific results—records must contain all three keywords	*Neisseria gonorrhoeae* **AND** *hfq* **AND** virulence	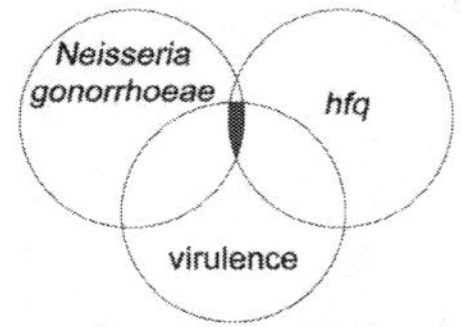
OR	Returns results with either keyword	*Neisseria gonorrhoeae* **OR** *hfq*	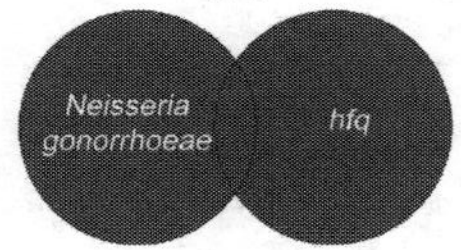
NOT	Excludes results of one keyword	*hfq* **NOT** *Neisseria gonorrhoeae*	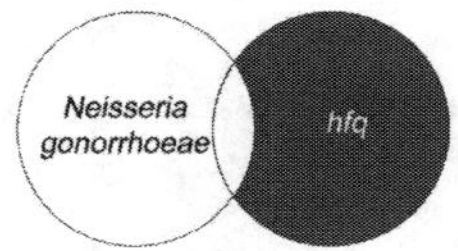
Nesting mixed operators (with brackets)*	Returns results with primary keyword and either of two other terms	virulence **AND** [*Neisseria gonorrhoeae* **OR** *hfq*]	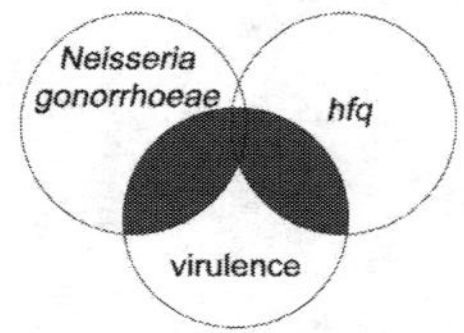

*Contents within brackets are processed first.

Note: Adapted from Webster et al. (2010).

If you need guidance determining broader search terms or narrower search parameters, you might try the MEDICAL SUBJECT HEADINGS (MeSH), which is a controlled list of INDEX TERMS or labels used to categorize biomedical articles in databases with a controlled vocabulary. Although you can search the list directly through the US National Library of Medicine interface (http://www.nlm.nih.gov/mesh/MBrowser.html), I find the results provided by the US National Center for Biotechnology Information (http://www.ncbi.nlm.nih.gov/mesh) to be more useful because the terms are displayed in a hierarchal tree structure (see Figure 3.3).

You can also refer to the list of terms that each journal uses to index articles. The editorial staff of a journal manages these lists of terms, so you will find these terms are less systematic but may be useful for identifying search terms in some research fields.

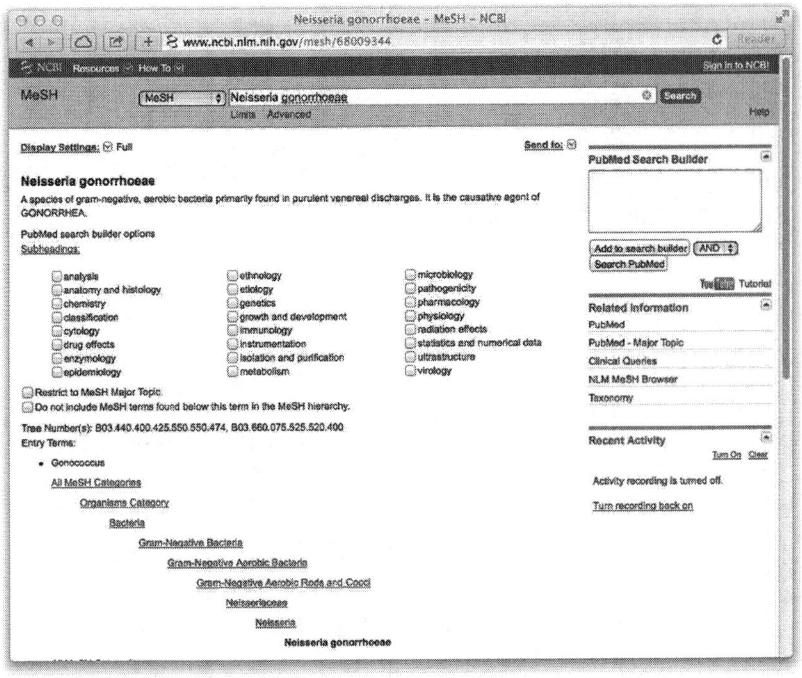

FIGURE 3.3 Medical Subject Headings (MeSH; http://www.ncbi.nlm.nih.gov/mesh) organize keywords in a hierarchal tree structure (bottom of image). The more inclusive terms (those higher in the list) are useful for a broader database search. In this example, search of "*Neisseria gonorrhoeae*" with the tag "[MeSH Terms]" provides the taxonomic lineage. The terms above "*Neisseria gonorrhoeae*" (underlined hyperlinks) would broaden a database search. Alternatively, the subheadings (top of image) can be used to narrow a search.

Finally, if broadening the terms used in a database does not yield a sufficient number of results, you should consider changing databases. In such situations, you might not be using the best database for the particular topic.

Step 3: Accessing the Full Text of Articles

In SMC, you should never rely on the information from Abstracts of journal articles for the material that you reference (International Committee of Medical Journal Editors 2014; Taylor 2002). As we will cover later in this text, Abstracts are often poorly structured and lack essential information (Ad Hoc Working Group for Critical Appraisal of the Medical Literature 1987; Falagas and Vergidis 2004; Rennie and Glass 1991; Salager-Meyer 1990, 1994, 2006). The Abstract of the research provided in databases (see Figure 3.4), will help you decide if the information is relevant and therefore if you should access the full text of the journal article. In addition to retrieving the full text of a journal article, you need to distinguish between actual observations (found in the Results section) and researcher interpretation of the data (found in the Discussion section and summary statements in the Abstract and Introduction). Notably, you should cite data as evidence but treat the interpretation or conclusions as a putative explanation that may be revised with future evidence (see Chapter 2).

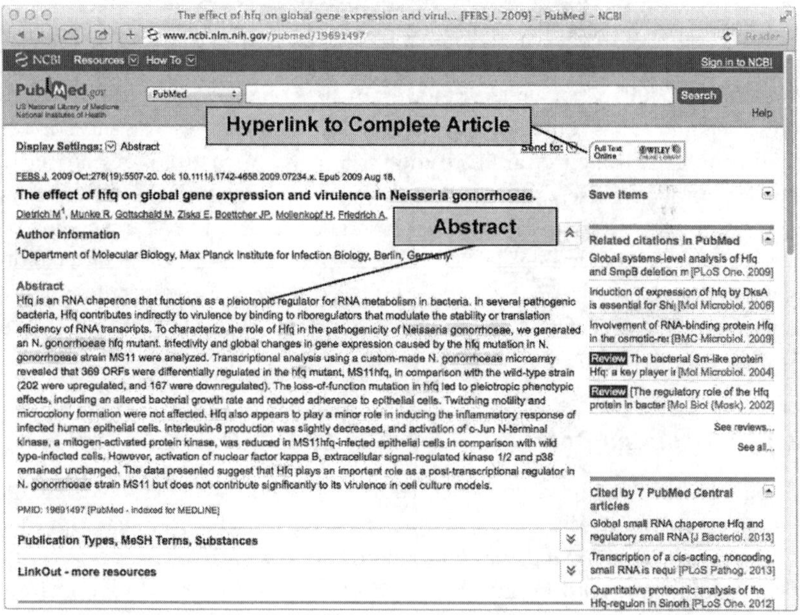

FIGURE 3.4 Detailed article record from PubMed provides the Abstract and hyperlink to the complete article.

Ideally you need to cite one to four of the most relevant and recent original research journal articles to corroborate a statement (American Psychological Association 2010; Nair 2005; Taylor 2002). In total, the literature review in the Introduction section of a journal article contains about 15 to 20 citations of previous research. Most of the original research that you cite should be published within the past 5 to 10 years, although important references that are older may also be included (Grant 2001).

Many scientific and medical journals are OPEN ACCESS, or provide free access to content online. Other articles are open access, even in journals that are not, because the authors have paid publication fees to make the content available online. Open access reflects modernization of the publishing model of SMC (Albert 2006; Laakso and Bjork 2012). A major change occurred in 2005 when the US National Institutes of Health (NIH) recommended grant recipients publish open access for NIH-funded research. At that time, approximately 27% of biomedical articles were freely available online (Matsubayashi et al. 2009). Since 2008, the NIH has required that investigators upload the final version of journal articles to PubMed Central (PMC) within 12 months of publication (US National Institutes of Health 2008). (Although it has a similar name as PubMed, PMC is a different database. PMC is a free archive of full-text journal articles in the biomedical sciences. PubMed automatically searches the content in PMC as well as other archives, such as the MEDLINE database.) As a result, greater than 50% of biomedical articles published in 2010 were available as open access through PMC, Google, or both (Kurata et al. 2013). As you might expect, open-access articles are cited more often than nonopen-access articles because they are more readily available (Eysenbach 2006). But even if the full text of a relevant journal article is not available online for free, you still should access content and reference it. Often you may access subscription-based journals through your institution's library. Alternatively, researchers in your laboratory are likely to be members of professional societies and therefore have online access to many journals. You should check with researchers in your workplace before purchasing articles online from the publisher or using a citation retrieval service. Citrome, Moss, and Graf (2009) suggest that you could request a digital copy of an article from the author through email, although this depends on the responsiveness of the author and may violate copyright.

Step 4: Recording Database Search Parameters for Future Searches

As you search the literature, you should document your search parameters so that you can repeat productive searches in the future and avoid repeating nonproductive searches. I recommend that you record the date, database, type of search, search parameters, and the number of results in a record sheet (see Example 3.3). When you record the search parameters, you should record your entry in the search box as well as the translation of your search into the database search string (if available from the database host as was done in Example 3.3). Recording the precise search parameters ensures exact duplication of the search. Such translated search strings are increasingly important as your searches become more complex (see Figure 3.5). Additionally, recording the number of results will help you remember the effectiveness of each search and quickly determine the number of new articles published since your last search. For example, the search for "hfq[All Fields]" in MEDLINE increased by 37 citations between February 24, 2014, and August 4, 2014 (see Example 3.3).

EXAMPLE 3.3 RECORDING OF PUBMED/MEDLINE DATABASE SEARCHES.

In the search parameters column, Boolean operators are in caps and bold indicates the search parameters provided by the database (which will facilitate future searches).

Date	Database	Type of Database Search			Search Parameter Database String[1]	Number of Results[2]
		Keyword	Forward Tracking	Related Articles		
2-24-14	PubMed/ MEDLINE	✓			hfq: **hfq[All Fields]**	533
2-24-14	PubMed/ MEDLINE	✓			Neisseria gonorrhoeae AND hfq: **("neisseria gonorrhoeae" [MeSH Terms] OR ("neisseria"[All Fields] AND "gonorrhoeae" [All Fields]) OR "neisseria gonorrhoeae" [All Fields]) AND hfq[All Fields]**	1
2-24-14	PubMed/ MEDLINE	✓			hfq (filtered article types by "review"): **hfq[All Fields] AND Review[ptyp]**	51
8-4-14	PubMed/ MEDLINE	✓			hfq: **hfq[All Fields]**	570

[1] Search terms are not italicized because text formatting is unavailable in database searches.

[2] The number of results reflects the effectiveness of each database search and is useful for comparing the number of new articles in searches at different times.

(a)

(b)

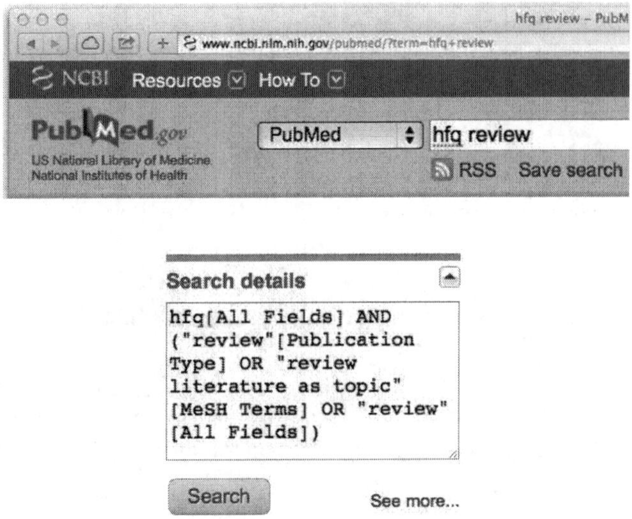

FIGURE 3.5 Comparison of keywords entered into PubMed (a) search box and (b) translation of search string included on the results page.

Using Key Articles to Find Additional References

As you identify one or more KEY ARTICLES, which are highly relevant journal articles of original research or literature reviews, your literature search will become more targeted and efficient (Bernstam et al. 2006; Chapman, Morgan, and Gartlehner 2009; Greenhalgh and Peacock 2005; Hu, Rousseau, and Chen 2011). A key article allows you to

1. Examine the list of references for other highly relevant articles (called "CITATION TRACKING"),
2. Search databases for newer articles that cite the key article (called "FORWARD TRACKING"), and
3. Search other databases for similar articles based on the database algorithm.

The benefit of citation tracking and forward tracking is that researchers in the field have purposely selected these articles and made informed, conscious connections between the content of different articles. This benefit of these two approaches differs from the keyword searches and database searches for similar articles because software programming of database algorithms drives those searches (Aoki, Enticott, and Phillips 2013). Every approach has advantages and disadvantages (see Table 3.3). The point is to use different approaches to compensate for the limitations, such as publication time lag (see Figure 3.6), of other approaches. Example 3.4 shows how to record each of the methods described in this section.

TABLE 3.3 Major advantages and disadvantages of literature-search approaches.

	Advantages	Disadvantages
Keyword search	• Potentially provides the greatest number of results • Typically spans the entire publication time frame of indexed records • Does not require a key article	• Results depend on quality of keywords selected and other search parameters selected by user (who may not be as familiar with the topic as an expert) • Search limited to journals and articles indexed in each database searched
Citation tracking	• Sources identified and evaluated by experts in the field	• Requires a key article • Likely incomplete or biased list • Does not contain articles published around the same time of the key article or more recently
Forward tracking	• Sources identified and evaluated by experts in the field	• Requires a key article • Requires authors of recent articles to have cited the key article • Only identifies articles after a key article is published • Likely incomplete or biased list
Related article searches	• Does not require keyword selection or determining of search parameters • Potentially covers the citation gap caused by the publication time lag	• Requires a key article • Depends on database algorithm • Limited to journals and articles indexed by the database

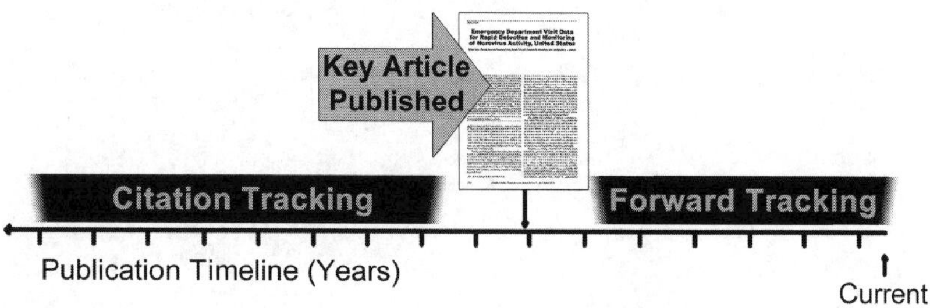

FIGURE 3.6 Key articles are useful for identifying references to highly related research. Checking the reference section in key articles and conducting database searches for newer articles citing the key article are complementary approaches in the literature-search process. But a gap of a few years exists immediately prior to and after the publication of a key article (a publication time lag as illustrated above the publication timeline).

EXAMPLE 3.4 RECORDING OF PUBMED/MEDLINE DATABASE SEARCHES.

The results vary because each database has a different corpus or collection of articles.

Date	Database	Type of Database Search			Search Parameter Database String	Number of Results
		Keyword	Forward Tracking	Related Articles		
2-24-14	PubMed/ MEDLINE		✓		The effect of hfq on global gene expression and virulence in Neisseria gonorrhoeae (key journal article by Dietrich et al. 2009)	7
2-24-14	PubMed/ MEDLINE			✓	The effect of hfq on global gene expression and virulence in Neisseria gonorrhoeae (key journal article by Dietrich et al. 2009)	24
2-25-14	Google Scholar		✓		The effect of hfq on global gene expression and virulence in Neisseria gonorrhoeae (key journal article by Dietrich et al. 2009)	17
2-25-14	Google Scholar			✓	The effect of hfq on global gene expression and virulence in Neisseria gonorrhoeae (key journal article by Dietrich et al. 2009)	101
3-1-14	PubMed/ MEDLINE	✓			"Dietrich M"[Author]	347

When using the automated searches to find articles that cite a key article or related articles, you first need to locate the key article in the database. To maximize the search accuracy when searching for a specific article, you can use the article's Title or the DIGITAL OBJECT IDENTIFIER (DOI), which is a unique identification number assigned to digital publications such as individual journal articles. Searches for newer articles that cite an article (forward tracking) or for related articles are often prominent options in databases. For example, PubMed/MEDLINE lists related citations and articles citing the article on the publication record page (see Figure 3.4, right-hand column). In another example, Google Scholar provides hyperlinks to articles that cite each article and to related articles beneath each entry on the results page (see Figure 3.7). Remember that for any database search, the results are limited to the articles indexed in its collection and the results of any single database search are often unique (Kraft 2014). For example, PubMed/MEDLINE listed 24 articles citing the sample article in Figure 3.4, whereas Google Scholar listed 17 citing the same article (see Figure 3.7).

Finally, you should search specifically for additional articles written by the authors of key articles. Since authors often specialize in particular topics, this is a useful method to find related research. I recommend that you conduct a search for the research published by the first author, last author, and corresponding

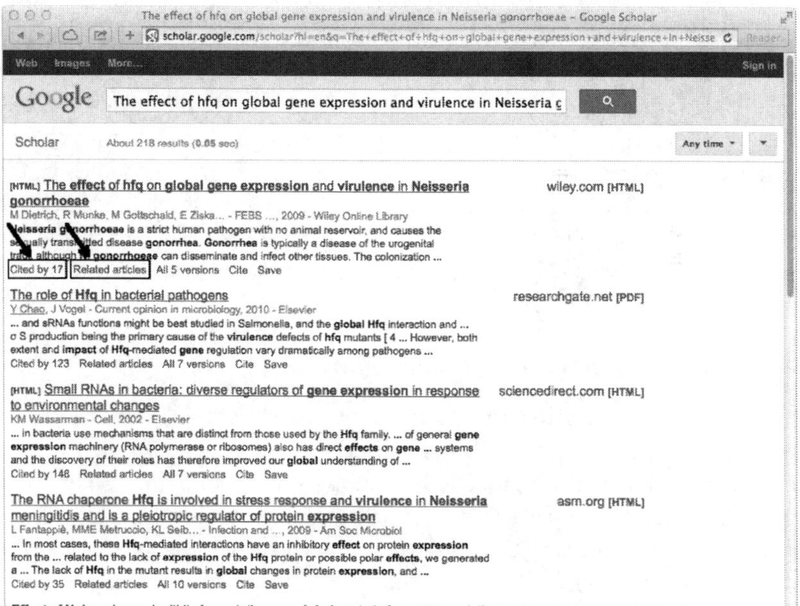

FIGURE 3.7 Google Scholar provides hyperlinks to other articles in the database that cite an article (forward tracking) and to similar (related) articles.

author—all of whom are the major authors of a journal article (see Chapter 1). When searching by author, you should use the author's last name and initials as the keyword in the display format of the database host. Authors generally publish all their scholarly work under the same name style (with or without the middle initial or name), much like a brand name. For example, you can search PubMed for other articles written by the first author in the record shown in Figure 3.4 by conducting a keyword search for "Dietrich M" (or simply click on the hyperlink of the author's name from the article record). In contrast, Google Scholar formats the name beginning with the first initial as "M Dietrich," so you should use that structure as the keyword (see Figure 3.7).

Updating and Managing a Library of Journal Articles

Ideally, writers and researchers should repeat database searches every six months to keep current with the latest research (Bartels 2009). But the search frequency depends on the pace of publication in a particular research field and should be done more often when writing research manuscripts. As a guide, you should repeat the most productive database searches every six months, before each major revision of a paper, and then again before you submit a manuscript for publication. In addition to repeating these manual searches, you might also register for the electronic table of contents (eTOCs) updates and keyword alerts from journal websites—especially from journals that you consider the most appropriate for the field and those to which you are likely to submit your research manuscript for publication. Many databases also provide email notifications that will notify subscribers when new research is published on a particular topic—both by selecting keyword parameters or by establishing alerts for key articles (Citrome, Moss, and Graf 2009; Luzon 2007). Establishing automated notifications may take some time initially to set up, but doing so will help ensure that you are aware of new research and may save time later if you plan to be involved in a research field for the long term.

Checking for Retractions and Corrections

You should always check to ensure that you are citing the latest, correct information by checking for corrections or retractions that have been published after the original journal article. A complete retraction of an original journal article in the medical literature is rare—a rate of 0.014% to 0.016% for original research journal articles—but corrections are somewhat common (Friedman 1990; Luo et al. 2013; Steen 2011).

To check for fully retracted journal articles, the International Committee of Medical Journal Editors (2014) recommends that writers and researchers conduct an advanced search in PubMed and use "retracted publication" as the publication type. The limitation of this method is that it will only provide completely retracted journal articles from the journals indexed in the database.

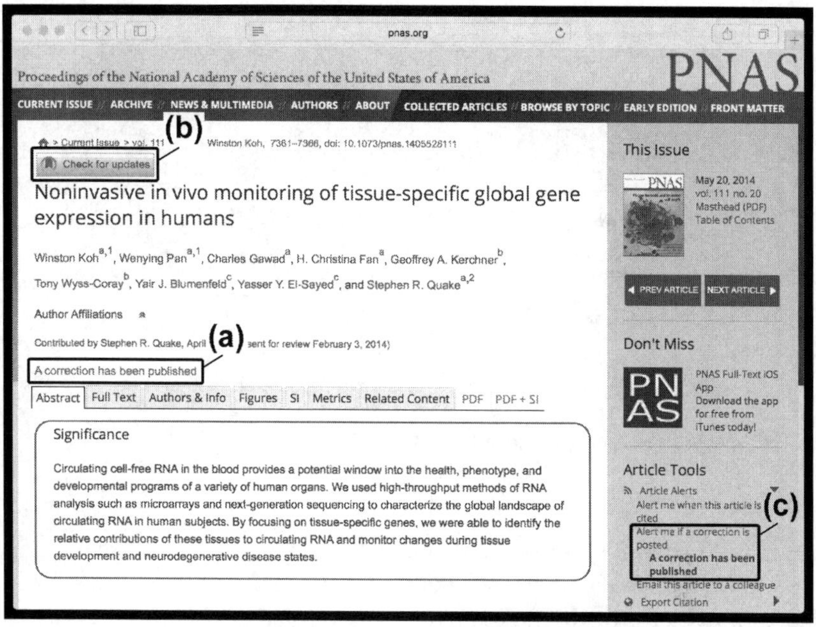

FIGURE 3.8 Web page from *PNAS* for a journal article that has a correction. These corrections are indicated by (a) prominent statements that "A correction has been published." Readers can also (b) check the latest edition of the article by clicking the CrossMark logo or (c) subscribe to article alerts if a correction is posted.

Rather to verify that you are citing correct information, you should visit the journal's website for the articles that you cite because the journal editors are responsible for issuing corrections and notifying readers. Some easier methods are being implemented, such as the CrossMark system that allows you to click on the logo to hyperlink directly to the article updates. For example, the *Proceedings of the National Academy of Sciences* provides three different options to notify or check for the corrections to a journal article (see Figure 3.8). The most prominent of these options is the notification statement ("A correction has been published") on the webpage for the article (see Figure 3.8a). If you do not see such a statement, you may click the CrossMark logo to verify if the article is current or if it has been updated (see Figure 3.8b). Finally, to avoid revisiting the webpage for articles periodically, you could subscribe to receive article alerts (see Figure 3.8c), which will provide you with updates of revisions to journal articles that you reference in your own work.

Managing Files and Using Reference-Management Software

As you find useful sources, you should begin downloading and archiving a copy of each complete journal article. Researching scientific and medical articles rapidly becomes an information-management issue of organizing all of the sources.

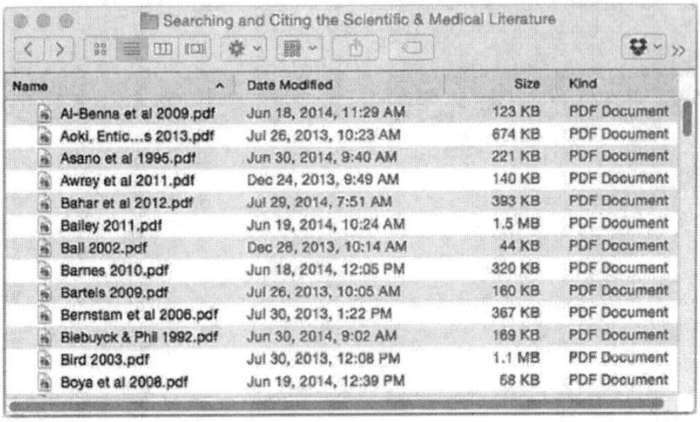

FIGURE 3.9 Articles organized by author and date.

A simple approach is to name the files by the author's name and date, saving them in a topical folder, and then automatically backing up the folder to a cloud-based service (see Figure 3.9). This is an effective strategy because many citation styles use the same information (such as author-date format for in-text citations and the author names and publication dates are used to organize the reference list). Some people prefer to include the titles of articles in the file name, although long file names are difficult to search in file folders. Alternatively, you might prefer using reference-management software such as BibTeX, EndNote, RefMan, or RefWorks. These software packages provide an interface that includes all of the citation information and also include other useful tools such as automatically creating a list of references, which is an important way to reduce bibliographic errors, as we will discuss in the next section.

Avoiding Faulty Citation Practices

In the late 1970s, June Goodrich and Charles Roland (1977) of the Mayo Foundation and Mayo Clinic published the article "Accuracy of Published Medical Reference Citations." In true experimental design, the authors stated, "Our null hypothesis may be stated as thus: that reference citations published in major North American medical journals provide accurate information for readers" (15). Their question that referenced information is accurate reflects a concern of the integrity of the SMC literature—at least we hope and believe that articles in medical journals accurately cite information. Also, we might think that this question should be true when examining the list of 10 journals that the authors examined (which included *JAMA* and *The New England Journal of Medicine*). But of the 2,195 references examined, 634 (29%) contained errors. As Goodrich and Roland (1977) explain, such errors have harmful consequences:

> Quite aside from these disturbing specific consequences of erroneous reference citations, there is the general question of credibility. What reliance can readers have in authors and in journals which, despite their combined efforts, permit the publication of such slipshod work? The reference list is part of the article. The article is the literary embodiment of the research. If the reference list is significantly inaccurate, what about the data in the Results section?
>
> The question of the relative responsibility of author and of journal is moot. Both parties should have vested interests in ensuring optimal accuracy. Yet our results show that the efforts are insufficient, at least in the journals studied during this period.
>
> *(19)*

Although arrogant or careless authors may dismiss errors in the reference list as not being important, the point is that such errors, particularly such simple ones to correct, reflect the overall credibility of the author and cast doubt on the entire work. Pervasive errors undermine the ethos of the author and readers may question where else in the experimental design, recording, or reporting of data the author discounted proper attention to detail. Such concern is not without merit, as 27.5% of scientists in a survey admitted to inadequate record keeping in research projects, which was—by far—the top reported behavior of scientific misconduct (Martinson, Anderson, and de Vries 2005).

Unfortunately, inaccurate information and sloppy citation practices extend beyond the list of references and into the text of the scientific and medical literature—where the effects can be much more impactful. In the following section, we will examine the problem of inaccurate citation practices in more detail and what you should do to ensure the accuracy and quality of your own work.

Miscitation

MISCITATION is a general term for the improper quotation of information from other sources or inaccurate bibliographic information (see Figure 3.10). Within the literature of those who study error rates and patterns, different authors use the term "miscitation" differently. To keep the issues clear, I use the term MISQUOTATION to refer to an inaccurate or misleading statement in the text of an article that fails to properly convey the information from the original source (de Lacey, Record, and Wade 1985; Evans, Nadjari, and Burchell 1990). Misquotations are further subdivided into two categories—as major or minor errors—based on the severity of the mistake. A MAJOR ERROR is when a claim is unsubstantiated by, unrelated to, or contradicted by the referenced article (Evans, Nadjari, and Burchell 1990). A MINOR ERROR is when a claim is inaccurate but does not change the overall meaning from the referenced article (Awrey et al. 2011).

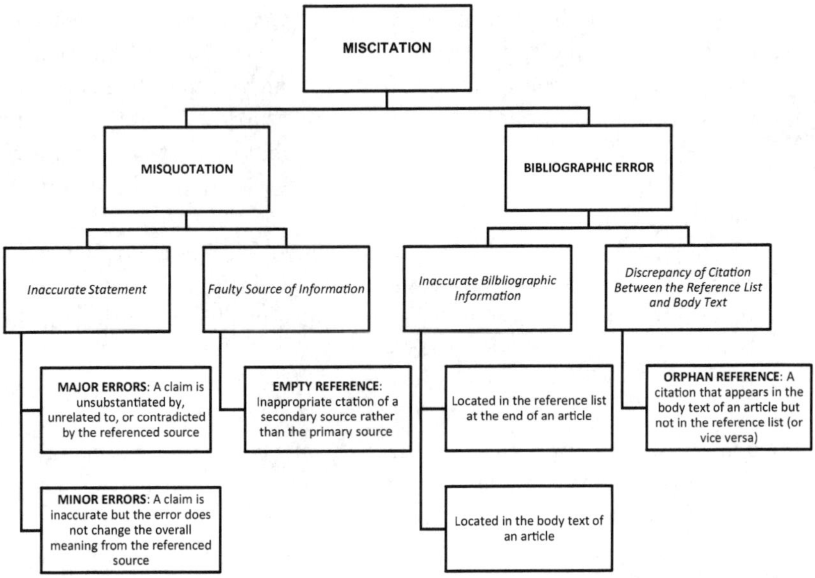

FIGURE 3.10 Categories of miscitations. Quotation errors (or misquotations) are mistakes in claims whereas bibliographic errors are mistakes in reference information.

Misquotations errors often arise from oversimplifying a summary statement (de Lacey, Record, and Wade 1985) or inappropriately expanding a generalization beyond the scope warranted in the original journal article (Evans, Nadjari, and Burchell 1990). But misquotations also arise from careless attention to detail. Table 3.4 illustrates several of the common misquotations found in the literature.

In a comprehensive literature review of misquotations in medical journal articles, I found approximately 21% of in-text quotations randomly checked were inaccurate (see Table 3.5). Of these misquotations, nearly two-thirds were categorized as "major" and one-third as "minor" errors. Several studies in scientific fields reported similar rates of misquotations. For example, a study in ecology reported that only 76% of referenced claims were clearly supported (Todd et al. 2007). A study in marine biology reported that approximately 76% of referenced claims were clearly supported (Todd et al. 2010). And a study in physical geography found that approximately 81% of referenced claims were clearly supported (Haussmann et al. 2013). Consistently in SMC, approximately 20% of referenced claims were erroneous. Indeed major errors, which represent the majority of miscitations, distort the audience's understanding of the material and lead us to question the author's work, understanding of the material, and possibly intention.

Bibliographic errors, or mistakes made in the reference information, frequently appear in reference lists. Most bibliographic errors are mistakes (such as

TABLE 3.4 Careless attention to detail may result in major and minor quotation errors. These examples are representative of some errors found in the medical literature. Errors and their corrections are highlighted in bold.

Category	Sample Errors[1]	Correct Information (Original Source)	Description
Major Errors	In the study by Jones et al., **89.4% of patients** did not respond to treatment.	In the study by Jones et al., **8.94% of patients** did not respond to treatment.	The decimal point is in the wrong place, making the percentage much different than the original data.
	In the study by Jones et al., the **majority of patients** did not respond to treatment.	In the study by Jones et al., the **minority of patients** did not respond to treatment.	The results were improperly summarized and the error significantly changes the intended meaning.
Minor Errors	Jones et al. studied **893 patients** with abscesses.	Jones et al. studied **983 patients** with abscesses.	The 9 and 8 were transposed, although this error does not affect the intended meaning.
	Jones et al. studied **42 patients**.	Jones et al. studied **42 abscesses** in 40 patients.	The actual object studied was incorrectly reported, although this error does not affect the intended meaning.
	A 24-hour observation period was sufficient **in all patients** to rule out a hollow-viscus injury.	A 24-hour observation period was sufficient **in *almost* all patients** to rule out a hollow-viscus injury.	Incorrectly reporting modifiers or qualifiers incorrectly conveys the certainty in the statement.

[1] Adapted from Awrey et al. 2011 and de Lacey, Record, and Wade 1985.

spelling) or omissions of author names, article titles, or journal names, but also include incorrect volumes, issues, publication years, or page numbers. Bibliographic errors may also appear within the text of an article when including the name or year of publication in a name-date format. ORPHAN REFERENCES are also a type of bibliographic error in which the citation appears in a paper but is not included in the reference list, or vice versa (Lukic et al. 2004).

TABLE 3.5 Prevalence of miscitation in the medical literature.

Field	Publication Year of Sample	Misquotations			Total Bibliographic Errors in the Reference List[2]	Source
		Total	Major[1]	Minor[1]		
Anatomy	2001	19% (52/272)	94% (49/52)	6% (3/52)	36% (71/199)	Lukic et al. 2004
Burn	2006	14% (16/117)	50% (8/16)	50% (8/16)	14% (16/117)[3]	Al-Benna et al. 2009
Emergency medicine	1991	35% (51/145)	82% (42/51)	18% (9/51)	28% (40/145)	Goldberg et al. 1993
General medicine	1975	–	–	–	29% (634/2195)	Goodrich and Roland 1977
	1984	15% (45/300)	42% (19/45)	58% (26/45)	24% (71/300)[3]	de Lacey, Record, and Wade 1985
	1999	–	–	–	27% (413/1557)	Siebers and Holt 2000
Orthopedic	2007–8	38% (152/398)	72% (110/152)	28% (42/152)	32% (64/200)	Davids et al. 2010
Ophthalmology	2003	25% (50/200)	60% (30/50)	40% (20/50)	18% (35/200)	Buchan, Norris, and Kuper 2005
Public health	1986	30% (45/150)	51% (23/45)	49% (22/45)	31% (46/150)[3]	Eichorn and Yankauer 1987

Radiology	1993	9% (9/95)	78% (7/9)	22% (2/9)	51% (48/95)	Hansen and McIntire 1994
Surgery	1987	29% (40/137)	93% (37/40)	8% (3/40)	48% (72/150)[3]	Evans, Nadjari, and Burchell 1990
	1997	17% (26/153)	65% (17/26)	35% (9/26)	38% (63/168)[3]	Fenton et al. 2000
	2004	8% (20/258)	80% (16/20)	20% (4/20)	14% (37/261)	Reddy et al. 2008
	2007	19% (168/900)	52% (87/168)	49% (82/168)	8% (68/900)	Awrey et al. 2011
	2009	18% (73/408)	49% (36/73)	51% (37/73)	73% (183/249)	Luo et al. 2013
Errors/Total Citations		747/3533	481/747	267/747	1861/6886	
Normalized % (Based on total number of citations)		**21%**	**64%**	**36%**	**27%**	

[1] Percent of total errors reported.

[2] Total errors reported unless indicated. Note that some citations in the reference list may contain more than one error.

[3] These articles provided only the number of citations containing errors (not total errors).

Bibliographic errors are somewhat more prevalent than quotation errors. As summarized in Table 3.5, approximately 27% of citations included mistakes. Additionally, Lukic et al. (2004) found 80 orphan references in 38 journal articles in the anatomy field. Although bibliographic errors are arguably less critical than quotation errors, bibliographic errors may make locating the original source of information difficult, time consuming, or occasionally impossible for readers. Additionally, bibliographic errors reduce accurate citation counts and thus do not give proper credit to the original authors, which may affect the careers of those improperly cited (Taylor 2002). So as a professional courtesy to readers and researchers, you should ensure the accuracy of bibliographic references—and expect that they will do the same for you. One of the easiest methods to ensure the integrity of bibliographic data is to use reference-management software as mentioned previously. Indeed, the International Committee of Medical Journal Editors (2014) recommends verifying bibliographic information from electronic bibliographic sources (such as MEDLINE) to minimize errors. This information can be directly imported into your reference-management software, thus streamlining the referencing task (Ghai, Saxena, and Makkar 2007).

Empty References (or the "Lazy Author Syndrome")

Citation of a secondary source, rather than the original research article, is an unethical and misleading practice that is generally inconsistent with the purpose of referencing. This practice of citing a secondary source, or EMPTY REFERENCING, is problematic because it does not provide the original source of data for the readers to directly investigate the claim. Additionally, inappropriately citing a secondary source does not provide proper credit to the original researchers, which can negatively affect their careers.

The practice of copying citations from secondary reports rather than the original paper (for example, from the Introduction section of an article or a literature review article) is a pervasive problem in scientific literature. I became sensitized to this problem when writing a journal article on the infection of *Shigella* in the intestine. The story was somewhat of a laboratory legend among our research team. As the story goes, the principal investigator, Shelley Payne (pers. comm., August 12, 2012), Associate Dean of the College of Natural Sciences and Professor of Molecular Genetics and Microbiology at The University of Texas at Austin, noticed that a literature review paper incorrectly reported the growth rate of *Shigella* in the intestine and this error was perpetuated in other papers that cited the review. Since our research team investigated the infectious process of *Shigella*, the growth rate of the organism was an important piece of data that affected our own research. We designed and performed experiments that were based on the time it took the organism to grow.

Unfortunately, anecdotal reports of empty referencing, or the "LAZY AUTHOR SYNDROME," are common. Haralambos Gavras (2002) of Boston University School of Medicine reported that

> in recent years, there has been a disturbing trend in medical articles to casually, but wrongfully, attribute scientific findings to the author of the latest review in the field, rather than to the scientist(s) who originally described the findings. This is a disservice to the readers of the article, who are given inaccurate information, and an affront to the scientists, whose contributions is not properly recognized.
>
> *(831)*

In the editorial, Gavras suggests that a major source of the problem is that some authors—being "lazy"—conduct a simple keyword search in MEDLINE and cite the latest review article indexed by those words. A related problem, which might contribute to the "lazy author syndrome," is that writers may not read the sources that they cite. For example, Simkin and Roychowdhury (2003, 2005) estimated that authors only read about 20% of the references that they cite.

Drake et al. (2013) noted that such improper practices reflect poorly on the authors and may distort information:

> Several of the authors of this article have noted disturbingly inaccurate reporting of our study findings when cited by others. We have also noted copying and pasting of phrases and citations from published articles in a way that distorts or even completely changes the meaning of the source. For us, this calls into question the quality and meaning of scientific advancement.
>
> *(1)*

If unidentified, this distortion of referenced information throughout the scientific and medical literature could, at best, confuse other researchers or, at worst, undermine future research investigations. As Steel (1996) stated in a commentary published in *The Lancet*, "Many wheels have been re-invented, and scientists whose papers have been ignored rarely complain vociferously, mainly because 'it happens all the time'" (144).

Several researchers have investigated and traced the problem of empty references. Harzing (2002) noted that her interest "was borne out of sheer amazement and indignation that serious academics seemed to get away with something students at all levels were warned not to do" (127). Although the prevalence of empty references in SMC is difficult to estimate, Table 3.6 shows an average of 19% of inappropriate citations to secondary sources in the medical literature. In a study of articles in ecology journals, researchers found that review articles were cited significantly more often than original research articles in the same journal and that 22% of citations to review articles were "empty references" (Teixeira et al. 2013).

TABLE 3.6 Prevalence of empty references (or inappropriate citing of secondary sources) in the medical literature.

Field	Publication Year of Sample	Prevalence of Empty References	Source
Anatomy	2001	24% (65/272)	Lukic et al. 2004
Emergency medicine	1991	41% (60/145)	Goldberg et al. 1993
General medicine	1984	6% (17/300)	de Lacey, Record, and Wade 1985
Surgery	2007	19% (171/900)	Awrey et al. 2011
	2009	2% (8/408)	Luo et al. 2013
Errors/Total citations		313/1617	
Normalized % (Based on total number of citations)		**19%**	

Citing a secondary source often distorts the information (Greenberg 2009; Rastegar and Wolfe 2002; Teixeira et al. 2013). Furthermore, such faulty citation practice propagates errors and establishes false authorities (Greenberg 2009). The editors of *Nature Cell Biology* (2009) note that empty references, or inappropriate attributions of concepts, fall under plagiarism. To address this problem, *Nature Cell Biology* and other journals encourage authors to cite primary sources rather than secondary sources and spot check accepted manuscripts with semantic-text comparison tools that detect empty references. Additionally, journal editors are increasing the number of references allowed to enable authors to cite more original research rather than secondary sources.

Citation Manipulation (or "Citation Gaming")

CITATION MANIPULATION is the inappropriate citation of sources for the sole purpose of increasing the citation counts of those sources (Council of Science Editors Editorial Policy Committee 2012). The reason that some authors are inclined to inappropriately cite their own work (SELF-CITATION) or to exchange citations among colleagues (CITATION SWAPPING) is because many metrics of research productivity are based on citation rates. Citation manipulation also occurs among journal editors and reviewers, who may coerce authors during the peer-review process to inappropriately cite their own articles or other articles from the journal.

Citation manipulation can be somewhat difficult to distinguish from appropriate citation practices when examining the list of references because researchers, colleagues, and journals tend to publish research on the same topic and form a discourse community (see Chapter 2). The distinction between appropriate and inappropriate citations depends on the role of the information being referenced—if that information is necessary or appropriate. Although such dubious citations may pass through the peer-review process, since reviewers rarely question selected references unless there is an obvious omission, individuals engaging in citation manipulation are apparent to experts within a discourse community and thus develop a reputation for this form of unethical conduct. Moreover, performance metrics monitor unusually high self-citation. The Council of Science Editors Editorial Policy Committee (2012) notes that "although such behavior may result in a short-term gain, the strategy may not work in the long term" (13).

Random-Citing Scientists

The MODEL OF RANDOM-CITING SCIENTISTS (see Figure 3.11) illustrates how scientists are biased to citing articles that were cited in recent articles (Simkin and Roychowdhury 2007). In this model, a journal article of original research is initially cited by a few scientists who read the article. In subsequent publication generations, the likelihood or frequency of citation depends on the number of times the original journal article was cited in each previous publication cycle. In contrast, "unlucky" (1669) articles that were not cited in the first generation may never be cited. The point of the model is that the number of citations an article has received in the latest publication generation influences the likelihood that the article will be cited again.

The model of random-citing scientists has implications for the process of citation tracking discussed previously. Using the reference list of key articles as a source for finding other articles should be only one part of an overall search strategy (Greenhalgh and Peacock 2005). That is, you should recognize the limitations and bias of citation tracking. Unfortunately, many researchers may be relying too much on the previously cited journal articles that they have identified through citation tracking rather than conducting independent, comprehensive database searches. Furthermore, the model of random-citing scientists also illustrates how miscitation or errors can be amplified and perpetuated throughout the scientific literature if authors rely on the citation tracking method and do not read the original article. Thus, you should always conduct your own comprehensive keyword search of multiple databases and also read the complete text of

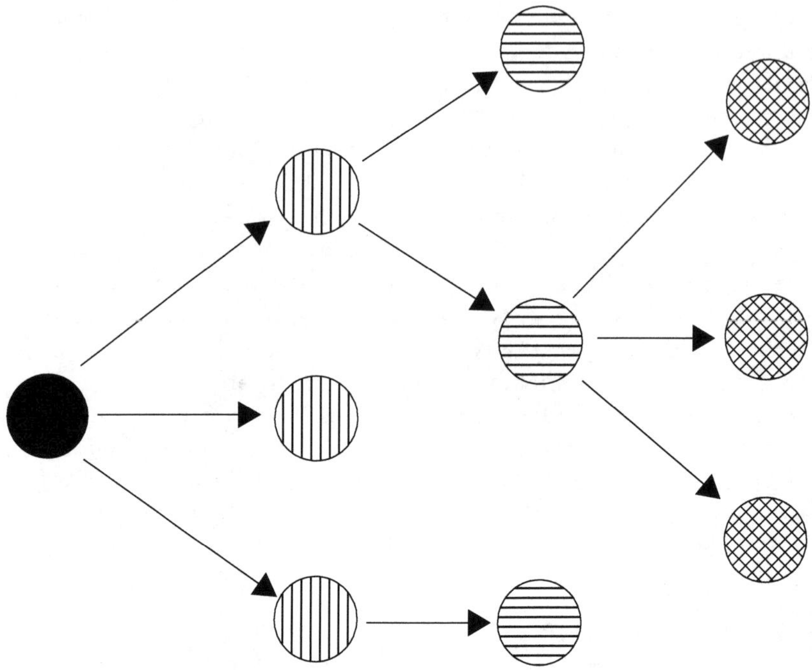

FIGURE 3.11 Model of random-citing scientists. In this model, an original research article (black circle) was cited in the first generation of articles by three other journal articles (vertical lines) written by the scientists who have read the original article. This process continues so that certain "lucky" (1669) papers that have been cited in the first publication generation continue to be cited in each subsequent publication generation cycle (continuing with horizontal lines and crosshatch). Some paths are fruitful and lead to more citations whereas others are dead ends.

Source: Simkin and Roychowdhury (2007).

any article that you cite so that you reference the most appropriate articles and accurately summarize the relevant content.

Proper Citation Practices

The purpose of referencing information and citing sources in SMC is to enable verification of any claim when the data or details are not present in the current communication. Often previous work is cited to provide the necessary background or context for the current research so that the audience will understand and appreciate the importance of the new information (Thompson 1995).

Yet authors must be discerning of which claims to cite. The general rule is to cite information that does not appear in the textbooks within the field. As

P.K Nair (2005), the editor in chief of *Agroforestry Systems* recommends, "Do not repeat well-known facts nor state the obvious." The historical basis for this guideline is that most scientists and physicians did not have ready access to the latest published research in journals—except for those living close to one of the world's largest libraries (Billings 1881). Most scientists and physicians had to rely on textbooks and possibly a limited number of journal subscriptions. Additionally, citations served as an efficient means to educate readers who do not necessarily have time to review all of the background detail.

The guiding principle in determining what statements to cite should be the purpose of the citation. In a letter published in *BMJ*, St. Leger (1985) of the Department of Community Medicine at the University of Manchester use the phrase "spurious scholasticism" to describe practice of excessive citation of well-established knowledge. As St. Leger stated, "For example, physicists do not refer to Newton's original writing every time they use his mechanics, whereas social scientists have a tendency to cite an original source for the seemingly most trivial insight" (1420).

Determining "common knowledge," or what to cite, is challenging for novice writers who are new to a scientific discourse community because most of the details about the topic are new and unfamiliar. Additionally, common knowledge is not static but evolves over time. For example, the double helical structure of DNA as reported by Watson and Crick in 1953 would have been cited throughout the mid-1950s but not as the structure became widely accepted and added to textbooks. Most articles have a peak of citations approximately 3 years after publication and then taper off and are rarely cited after 10 to 15 years (Simkin and Roychowdhury 2007). One notable exception are the rare "SLEEPING BEAUTY" (1668) journal articles, which are relatively unnoticed for a period of time and then gather significant citations because the findings were premature or ahead of their time (Garfield 1980; Van Raan 2004). A classic example of a sleeping beauty is the work by Gregor Mendel, who made significant discoveries in plant genetics, which took over 30 years for the scientific community to appreciate (Van Raan 2004). Most articles, however, contribute to the incremental advancement of scientific knowledge within the current period and therefore have a brief window of citations between 2 to 10 years after publication before the field advances and newer findings replace the old (see Figure 3.12).

I have found recent journal articles on similar research topics to be particularly useful when determining the appropriate information to cite. With a small collection of recent journal articles on the same research topic, you can analyze each section (such as the Introduction) for cited information and use those sources as models for identifying claims that you should cite in your own writing (although be sure to refer to the original publication to ensure the accuracy of your citation). In the case of Introductions, for example, recent journal articles on the same topic generally cover the same background information—with very similar claims of importance and summaries of research findings. Thus, writing

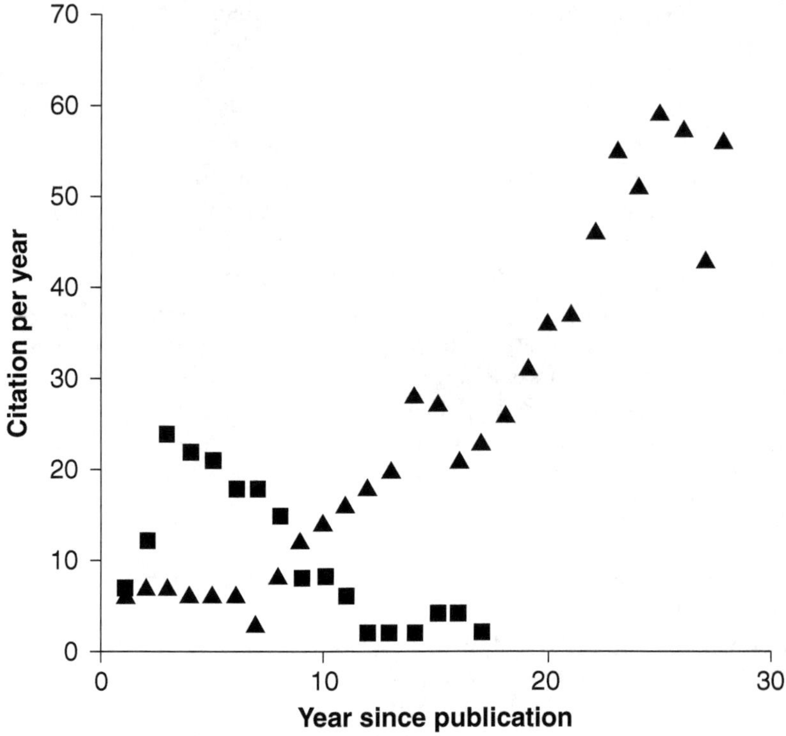

FIGURE 3.12 Citation frequencies of typical articles (squares) and "sleeping beauties" (triangles). Note that most articles have a brief window of citations between 2 to 10 years after publication.

Source: Simkin and Roychowdhury (2007).

an effective background is often a matter of synthesizing the background information presented in other papers with the most recent findings on the topic. Having recent articles on the topic are particularly useful because that content will have been authored by members of the specific discourse community and successfully passed through the peer-review process—so you will know which claims require support. Even though you can use the Introduction of published articles to guide your writing, you should not assume that this means that you do not need to do an exhaustive literature search. To establish your own credibility within your research team and in the field, you will need to include all of the recent and relevant research that leads logically and directly to the research question of your article.

In SMC, the purpose of a citation is to give the readers the source that will provide more information about claims that were investigated elsewhere—so that you do not have to repeat the evidence. In the absence of repeating the data, citations serve as your "evidence" for each claim (Taylor 2002). Proper citation practices in SMC are as follows:

- **Cite the original source of information from peer-reviewed literature.** Refer to the specific data or conclusions from original research journal articles and never cite information in Abstracts (International Committee of Medical Journal Editors 2014; Taylor 2002), but instead access the complete journal article and find relevant information to ensure that you are not misrepresenting the content (Harzing 2002). Occasionally you may cite a broad (or general) claim or an original insight from a literature review article or monograth (a book-length literature review; Harzing 2002). If you use a secondary source to support a broad claim, you should always qualify the source as a secondary source (such as referring to the work as a "review") so that readers do not assume that the claim is from original research (Taylor 2002). Furthermore, you should avoid grey literature sources, such as textbooks, or conference presentations and certainly popular media sources since these sources are not the primary (or original) source and have not been through a comprehensive peer-review process.
- **Limit the number of citations to the most appropriate and relevant source(s) for each claim.** Typically only a few (up to four) of the most relevant and recent sources need to be cited to corroborate a statement (American Psychological Association 2010; Nair 2005; Taylor 2002). But the number may vary somewhat based on the amount of research available on the topic.
- **Clearly indicate the statement that a citation supports** (Harzing 2002). Many referencing systems have authors insert the in-text reference at the end of a sentence. But if a single sentence is loaded with different claims from multiple sources, you should insert each reference immediately after each claim is stated. Only if all the sources corroborate the claim should you stack the citations at the end of the sentence.
- **Summarize the relevant information from the source in your own words.** In SMC, you should avoid direct quotations. The only reason to use a direct quote would be to present a particularly well-articulated interpretation from a leading scholar on the topic. Most SMC articles do not include quotations, particularly not of data, although I have seen some different quotation patterns within different discourse communities. Generally it is best to avoid direct quotations, but if you are inclined to use one, I recommend that you review recent articles from the target journal and discourse community to see if those articles include direct quotations. If not, then I discourage you from using one.

By following these guidelines, you will increase the accuracy and integrity of citations and add substantial value to SMC. By citing the most relevant and critical findings from the previously published research, you demonstrate to journal editors, peer reviewers, and readers that you (and all of the authors on the paper) are well informed (American Psychological Association 2010; Iverson et al. 2007). As Grant (2001) explains, "A reader can learn a lot about authors' perspectives, expertise, and possibly even bias from looking at the references cited" (63). Yet as Huth (1982) cautions, citations "should not be used to show off wide knowledge" (52) but rather to identify the most directly relevant information. In fact, an excessive and unfocused review of the literature is indicative of novice authors and undermines a professional ethos.

Concluding Points

When I cover researching the primary literature in class, students often want to know when they have completed or finished a literature search. Asking when a literature search is complete is somewhat of a loaded question because, as we have discussed, new research is always being published and part of your responsibility as an effective communicator is to stay informed of the latest articles. Practically, however, you will reach a point of saturation in your searches where additional searches of different keywords and databases, and tracking of citations from key articles, no longer reveal novel sources. When you reach that point of saturation, you can consider your literature search complete . . . at least for the moment. You will need to repeat a literature search during each major revision of an article and immediately prior to submission. Additionally, automated notifications that you set up on journal or database host websites should alert you to any new relevant articles.

SECTION II
Writing Journal Articles

4

COMMUNICATING RESEARCH IN PRIMARY JOURNALS

In this chapter, you will learn to:

- Plan articles of original research
- Identify a journal to submit research
- Obtain style and formatting instructions from a journal

Overview

When I began the writing phase of my molecular biology research, I remembered the advice from a senior scientist, Elizabeth Wyckoff (pers. comm.), who had decades of experience and a few dozen articles. We sat down in the back of the laboratory and she began by saying, "What you want to do is to *tell a story*." On scrap paper she began to scribble down the list of key points that I should present. For example, I had data showing that a specific mutant strain of bacteria was able to infect, but not spread, human cell-tissue cultures. In another experiment, I had findings that the mutant bacteria grew more slowly than normal bacteria in a culture media that mimicked intercellular conditions. In a yet another experiment, I had data showing that these mutant bacteria were much more sensitive to acidic conditions than normal bacteria. Each of these key points, and several more, was the outcome of a different experiment. At this point, I, like many researchers, had collected a handful of individual DATA SETS, or the results from discrete experiments, but I did not see how all of this data fit together. The *story* that I was to develop and convey in a journal article would connect each of these individual data sets and provide a logical flow leading from one experiment to the next and then fitting all of the results that I reported into the overall thesis of the journal article.

When Dr. Wyckoff and now I suggest that you "tell a story" in SMC, you should not confuse the term "story" with fiction writing. Writers of SMC often refer to the "story" or "narrative" of research as a post hoc rationalization that connects the outcomes of fruitful experiments but excludes accounts of tangential investigations and dead ends (Schriger 2005). As the noted scholars Bruno Latour and Steve Woolgar (1986) described in *Laboratory Life: The Construction of Scientific Facts*, the *story* that researchers construct is "widely regarded by outsiders as well organized, logical, and coherent, [but] in fact consists of a disordered array of observations with which scientists struggle to produce order" (36). Much of the organization and connections of the research occurs during the research phase, yet the entire story or narrative is usually not formally articulated until the writing process (Zhang 1989). Therefore, you should refine your view of a journal article in that it "is not a *report* of what went on in the laboratory, but rather the *product* of the laboratory effort" (Markel 1993, 17). This revised perspective emphasizes an important change from the naive perspective that journal articles merely summarize and detail a log of events, as in a laboratory notebook, to realizing that journal articles are actually a carefully constructed argument regarding a scientific phenomenon.

How to Prepare

Journal articles are written after the data have been collected and organized and the research team is ready to share the findings as a complete story (Phillips 2009). From a practical standpoint, this means that the researchers will have outlined, or at least thought about, the argument in the article. As you will see, each individual experiment from multifaceted investigation provides a different data set that becomes a piece of the larger argument, which ultimately forms the thesis of the journal article. Although the thesis has been conceptualized, you should be flexible and consider the thesis somewhat tentative because it usually evolves and is refined in each draft of the journal article manuscript. As you will see, with this perspective, the distinction between research and writing becomes less defined or partitioned.

What to Expect

As a writer, you role is to take the data from fruitful experiments and write a manuscript to submit to a scientific or medical journal. MANUSCRIPTS should be considered a specific genre that communicates research between a research team and the journal editors and reviewers. Although the content of the manuscript and journal article are the same, the design or appearance is much different. Manuscripts must adhere to specific format and style guidelines that look more similar to a double-spaced research paper submitted in a college course than the final design of an article in the printed journal. Proper manuscript formatting facilitates the peer-reviewing, copyediting, and typesetting processes. Unfortunately for writers, journals have different style and formatting requirements, and editors of scientific and medical journals often refuse to review manuscripts that do not follow these requirements.

This chapter is designed to guide you, as a writer, through manuscript planning and design. It serves as an overview of the communication of original research to scientific and medical journals. In following chapters, we will cover the specific content and scientific writing style of each section of a journal article.

Planning Articles of Original Research

Before you begin writing the manuscript for a journal article, you need to collect information and make decisions about the publication venue. Specifically, you need to gather technical content from the research team, select a TARGET JOURNAL (where you plan to submit the manuscript), and obtain style and submission guidelines from that journal (called the INSTRUCTIONS TO AUTHORS).

Step 1: Gathering Technical Content from the Research Team

Technical content is the core of SMC. As Peter Driscoll (1997) of Hope Hospital stated, "If you [as the author] are not clear on what is to be described, then it is likely that the reader will be equally confused" (65). As the writer you need to collect and organize the technical content, which should include the following:

1. The thesis statement or overall argument of the research article,
2. Data sets to present (either as raw data or in tables and figures, and preferably in order of presentation for the argument), and
3. Experimental protocol or procedures used to obtain the data (annotated with the specific detail that you need to write the Methods section).

In addition to the technical content for the journal article, you should gather any previously published journal articles, poster presentations, review articles, or grant proposals authored by colleagues that were related to the current research. The research genres (journal articles and poster presentations), in particular, will serve as useful models for constructing your manuscript.

In general, researchers are willing to share articles and other background information if they have the information readily available and they are providing it at the beginning of a project. More senior researchers expect that novice members of the research team will require assistance at the beginning of new writing projects and are usually generous with providing help and guidance. But you should also demonstrate your own initiative in collecting information that is available in databases (see Chapter 3). You should only request information that your colleagues have readily available or that you are unable to find on your own since these researchers will, at some point, expect you to be able to work independently and locate the information that you need.

Step 2: Selecting a Target Journal

Before you begin writing the manuscript, you should identify the target journal that is the one journal to which you plan to submit a manuscript for review and allow the opportunity to make the initial publishing decision. By convention, authors submit a research manuscript to only one journal at a time and wait for a publishing decision before submitting it to another journal. Therefore you should carefully select a target journal during the planning phase to avoid a delay in publishing the research that would occur by submitting it to a journal that is not an appropriate fit for the research (Boellstorff 2011; Gasparyan 2013; John 2009; Wachs 1996). The manuscript review process often takes several months, so a rejection by one journal will add several months to the process and, additionally, often requires revision in style and format before you can submit the research to another journal (Naylor and Muñoz-Viveros 2005). Ideally the entire manuscript should be revised for a different journal because each journal has a slightly different audience and scope. As Teijlingen and Hundley (2002) of the University of Aberdeen warned authors, "A common mistake made by novice researchers is to write a paper and then look for a journal in which to publish" (508). Rather, you need to know the exact audience, scope, style, and formatting of a journal before you begin to write the manuscript. Otherwise you will waste time writing a paper without a clear concept of the audience and the content will be too generic and therefore will need extensive revision later.

The target journal is something that the researchers, who are familiar with a discourse community, will have considered as they conduct the research. As the writer, you might ask the researchers to name a journal to which they would plan to submit the article initially. If you are new to a discourse community, you should be sure to accurately record the exact name of the target journal because many journals have subtle differences in titles (see Example 4.1) and you want to ensure that you follow the style and formatting guides of the right journal.

EXAMPLE 4.1 TITLES OF SEVERAL INFECTIOUS DISEASE JOURNALS.

Note that many journals have similar names, which may cause confusion when the journal is being discussed in informal conversations. This is especially challenging for those new to a discourse community.

American Journal of Infectious Diseases
BMC Infectious Diseases
Clinical Infectious Diseases

Current Opinion in Infectious Diseases
Emerging Infectious Diseases
Infectious Diseases in Clinical Practice
International Journal of Infectious Diseases
The Journal of Infectious Diseases
Journal of Infectious Diseases & Therapy
Journal of Microbiology and Infectious Diseases
Journal of Pediatric Infectious Diseases
Journal of the Pediatric Infectious Diseases Society
The Open Infectious Diseases Journal
The Pediatric Infectious Disease Journal
Transplant Infectious Disease

In some situations you may be asked to recommend a target journal for publication. If so, then you should consider the following most commonly considered criteria (listed in order of importance) in order to find the most appropriate target journal (Frank 1994; Gasparyan 2013; Özçakar et al. 2012; Søreide and Winter 2010):

1. Audience and journal type,
2. Prestige and impact factor,
3. Publishing model (including open access and fees), and
4. Publication turnaround time.

Audience and Journal Type

When selecting a target journal, you should match your research topic to the aims and scope of the journal as closely as possible. This process includes considering the audience's need or use for the information, such as basic research or practitioner focus (Roederer et al. 2013). Jeremy Boss and Susan Eckert (2003) of Emory University remind researchers and writers that "the most important thing is to get your work out there where people can see it by publishing regularly in journals that are widely respected, read, and cited by your peers."

If you are new to a field, one way to identify a target journal is to conduct a database search on the specific research topic (see Chapter 3) and identify the journals that appear most frequently. Taking this approach, you will identify journals that have regularly and recently published research on the topic. Often you will develop a short list of potential journals that need to be further evaluated. Although selecting one journal from a short list of relevant potential journals may be debated among different researchers, Michael John (2009) of the faculty of medicine at Vita-Salute San Raffaele University stated that a "wrong" (60) journal is "a journal where you will not have any interested readers" (60).

In addition to the audience of a potential journal, you should also consider the INDEXING, or listing of the research in databases. As discussed in Chapter 3, each database contains a unique list of journals, so other researchers may not find your research if it is not indexed in an appropriate database. For example, many biomedical researchers prefer to submit research to journals indexed in MEDLINE/PubMed. In practice, journal indexing in databases may be the single most important criteria in reaching an audience. Additionally, this also introduces one point of caution in that you must select the appropriate database for the research topic if you plan to use the method described above to identify a target journal. If you are unsure of the "best" database, then you should try multiple databases searches approaches as covered in Chapter 3.

Journals can be divided into the following four types, as defined by Jeremy Boss and Susan Eckert (2003) from Emory University School of Medicine and School of Nursing:

- BROAD TOP OF THE HEAP (e.g., *Science* and *Nature*)—publishes cutting-edge research on "hot" topics that defines or shifts the current paradigm in a field and is of interest to a broad scientific audience;
- NEAR THE TOP (e.g., *Cell* and *Nature Genetics*)—also publishes cutting-edge research within a more specific area and provides more depth in each article than broad top of the heap journals;
- SOCIETY LEVEL (e.g., *Proceedings of the National Academy of Sciences*)—publishes research on broader topics, usually by members of a professional society; and
- SPECIALTY or SUBSPECIALTY (e.g., *Immunology*)—publishes in-depth research on a niche topic.

Each type of journal publishes slightly different research scope and depth, and your research topic and findings are generally more appropriate for one type than another.

Broad top of the heap journals are extremely competitive and only publish research that is truly innovative or a "hot" topic and is relevant to a more general scientific community (Boss and Eckert 2003). For example, *Science* (n.d.) has an ACCEPTANCE RATE (the percent of papers accepted out of total submissions) of less than 7%. Similarly, *Nature* (n.d.) accepts approximately 8% and *JAMA* (n.d.) accepts approximately 9%. In contrast, specialty journals often have a moderate acceptance rate of 20% to 50% (*Journal of Neuroscience* n.d.). Although the acceptance rate should not be used as a primary reason for selecting a journal, you should carefully evaluate the pros and cons of submitting research to a top journal with a very low acceptance rate.

In most cases, you should select the most appropriate specialty journal rather than a broad top of the heap journal such as *Science* or *Nature*. Some authors spend a significant amount of time and effort submitting research to top journals when the research is more suited for a niche journal (Myers 1985). For example,

Greg Myers (1985) of the University of Lancaster observed a researcher who submitted a manuscript to *Nature* (twice) and then *Science* before it was accepted by the *Journal of Molecular Evolution*. Myers (1985) also followed another scientist who submitted a manuscript to *Science* (twice), then *Nature*, and then *Proceedings of the National Academy of Sciences* (*PNAS*) before it was accepted by *Hormones and Behavior*. In both examples, the researchers invested significant time and effort in rewriting their manuscripts prior to each submission. By accurately assessing the fit of your research at the onset, you can save yourself from such unnecessary investments. In fact, the most cited articles are from niche journals of a discourse community rather than top journals such as *Science* or *Nature* (Boss and Eckert 2003).

Prestige and Impact Factor

Along with journal type, researchers consider the prestige of a journal and the IMPACT FACTOR, a metric used to compare journals within a field (Boss and Eckert 2003). Journal prestige is the perception of a journal's reputation, impact factor, and breadth of distribution. In selecting journals, younger researchers (25 to 40 years old) tend to rely more on the impact factor whereas older researchers (over 40 years old) tend to rely more on the perceived reputation of the journal (Søreide and Winter 2010). A journal's reputation is partly determined by its longevity of publication and brand name, which are factors that researchers learn as they become immersed within a discourse community. Older researchers' longevity and familiarity within a discourse community enable them to have a better sense of an appropriate journal and be less reliant on metrics such as the impact factor.

The impact factor, or the average number of times that each article in a journal is cited over a two-year period (Garfield 1955, 1986, 2006), can be used to compare journals within a field. A journal with a higher impact factor has more average citations per article and is considered more influential than a journal with a lower impact factor. Although the impact factor is the most commonly used metric, it is not without controversy (Baethge 2012; Boss and Eckert 2003; *PLoS Medicine* Editors 2006). The major concerns of the impact factor are as follows:

- Most citations from a journal are from a limited set of articles (citation frequency is not evenly distributed among all the articles in a journal);
- Fields and specialties vary in size, so the impact factors cannot be used to compare journals in different fields;
- Review articles are cited more frequently than the original research, which distorts the metric;
- References are calculated from journals indexed in the Thomson Reuters (formerly ISI) Web of Knowledge, which does not index all journals; and
- The metric evaluates the influence of research within scientific discourse communities and not the quality of research or its influence beyond a group of researchers, such as on the public.

Despite these concerns, the impact factor remains the most widely accepted measure to compare journals. Therefore, you can compare and select journals within a discourse community by using the latest impact factor, which is often reported on a journal's web page.

Publishing Model (Including Open Access and Fees)

Since the widespread use of the internet beginning in the 1990s, journal publishers have been experimenting with different publishing models. Initially, free online access to research journal articles, called OPEN ACCESS, was unique to the newer, online-only, low-overhead journals. Researchers selected open-access journals (Swan and Brown 2004) for these primary reasons (in order):

1. Belief in the principle of free access to research,
2. Perception of faster publication,
3. Possibility of a larger audience, and
4. Belief that research will be cited more.

Yet researchers also expressed concerns of the quality of these journals and the higher author fees for publication when compared to traditional, print-based journals that distributed production costs through subscriptions (Schroter, Tite, and Smith 2005; Warlick and Vaughan 2007).

Over time, the differences between open access and subscription or traditional print-based journals have diminished (Björk, Roos, and Lauri 2009; Björk and Solomon 2012). Many society journals have adopted an open-access model and many other print-based journals allow researchers to select an open-access model for their article by paying additional fees (called ARTICLE PROCESSING CHARGES). Additionally, the more recent open-access adaptations allow authors to maintain copyright ownership of their article (rather than transfer to the journal) and permit uploading a version of the article to open-access repositories (such as PubMed) after an embargo period. Furthermore, major research funders such as the National Institutes of Health (NIH) and the Wellcome Trust, and some institutions, require that research supported by these institutions is published in open-access journals.

The access model and fees must be considered in the journal selection process, although these issues typically rank lower in importance than those reported previously (Gasparyan 2013; Özçakar et al. 2012; Søreide and Winter 2010). The fees, which may be as high as a few thousand dollars (USD), are usually included in research grants but may be waived for scientists from developing nations.

Publication Turnaround Time

In the competition to be first to publish research and be credited with the scientific finding (see Chapter 2), researchers often consider the time from manuscript submission to publication. This timeline usually ranges from 12 to 30 months, but varies by field (Masic 2011). Most journals provide information about the turnaround time from submission to print or alternatively PREPRESS PUBLICATION, which makes the article available on a website prior to release of the print journal (Benson and Silver 2013). Another option includes FAST TRACK, which can accelerate the review to days and publication to weeks for cutting-edge research that may be of important clinical relevance. For example, *BMJ* attempts to publish select articles within four weeks (*BMJ* n.d.) and *The Lancet* offers reviews of select manuscripts within 72 hours and publication in four to eight weeks (*Lancet* n.d.).

Step 3: Obtaining Instructions to Authors from the Target Journal

Once you have selected a target journal, the next step is to collect the manuscript requirements, which journals usually provide in the author resources section of their website (see Figure 4.1). Typically called the "Instructions to Authors" (see Figure 4.2), you will find that this resource answers most (if not all) of your questions about the

- Audience and scope of the journal,
- Types of articles published,
- Style guidelines, and
- Manuscript formatting and submission guidelines.

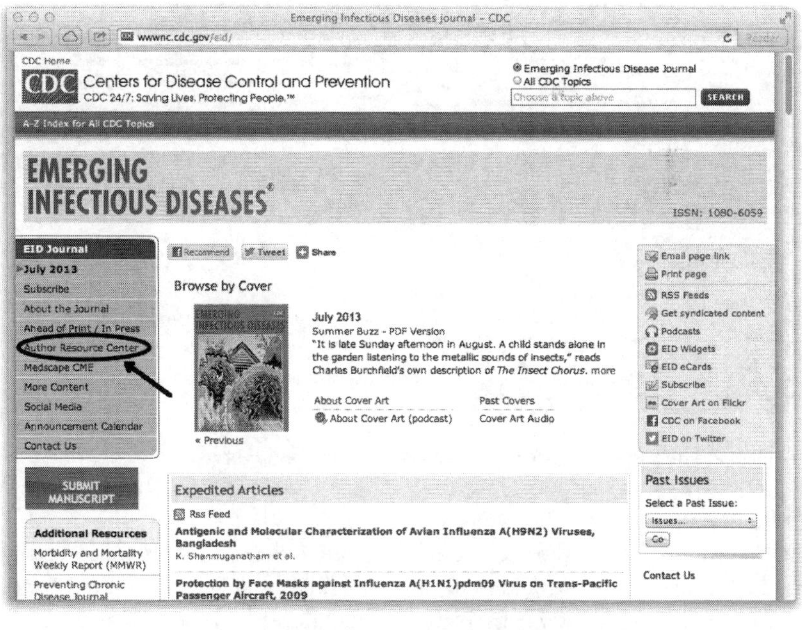

FIGURE 4.1 The author resources section for *Emerging Infectious Diseases* is found as a link from the journal's home page.

Emerging Infectious Diseases is a peer-reviewed journal established expressly to promote the recognition of new and reemerging infectious diseases around the world and improve the understanding of factors involved in disease emergence, prevention, and elimination.

The journal is intended for professionals in infectious diseases and related sciences. We welcome contributions from infectious disease specialists in academia, industry, clinical practice, and public health, as well as from specialists in economics, social sciences, and other disciplines. Manuscripts in all categories should explain the contents in public health terms. For information on manuscript categories and suitability of proposed articles, see below and visit www.cdc.gov/eid/ncidod/ EID/instruct.htm.

Emerging Infectious Diseases is published in English. To expedite publication, we post some articles online ahead of print. Partial translations of the journal are available in Japanese (print only), Chinese, French, and Spanish (www.cdc.gov/ncidod/EID/trans.htm).

Instructions to Authors

Manuscript Submission. To submit a manuscript, access Manuscript Central from the Emerging Infectious Diseases web page (www.cdc.gov/eid). Include a cover letter indicating the proposed category of the article (e.g., Research, Dispatch), verifying the word and reference counts, and confirming that the final manuscript has been seen and approved by all authors. Complete provided Authors Checklist.

Manuscript Preparation. For word processing, use MS Word. List the following information in this order: title page, article summary line, keywords, abstract, text, acknowledgments, biographical sketch, references, tables, and figure legends. Appendix materials and figures should be in separate files.

Title Page. Give complete information about each author (i.e., full name, graduate degree(s), affiliation, and the name of the institution in which the work was done). Clearly identify the corresponding author and provide that author's mailing address (include phone number, fax number, and email address). Include separate word counts for abstract and text.

Keywords. Use terms as listed in the National Library of Medicine Medical Subject Headings index (www.ncbi.nlm.nih.gov/mesh).

Text. Double-space everything, including the title page, abstract, references, tables, and figure legends. Indent paragraphs; leave no extra space between paragraphs. After a period, leave only one space before beginning the next sentence. Use 12-point Times New Roman font and format with ragged right margins (left align). Italicize (rather than underline) scientific names when needed.

Biographical Sketch. Include a short biographical sketch of the first author—both authors if only two. Include affiliations and the author's primary research interests.

References. Follow Uniform Requirements (www.icmje.org/index.html). Do not use endnotes for references. Place reference numbers in parentheses, not superscripts. Number citations in order of appearance (including in text, figures, and tables). Cite personal communications, unpublished data, and manuscripts in preparation or submitted for publication in parentheses in text. Consult List of Journals Indexed in Index Medicus for accepted journal abbreviations; if a journal is not listed, spell out the journal title. List the first six authors followed by "et al." Do not cite references in the abstract.

Tables. Provide tables within the manuscript file, not as separate files. Use the MS Word table tool, no columns, tabs, spaces, or other programs. Footnote any use of boldface. Tables should be no wider than 17 cm. Condense or divide larger tables. Extensive tables may be made available online only.

Figures. Submit figures as separate files, in the native format when possible (e.g., Microsoft Excel, PowerPoint). Photographs should be submitted as high-resolution (600 dpi) .jpg or .tif files. Other file formats may be acceptable; contact fue7@cdc.gov for guidance. Figures should not be embedded in the manuscript file. Use color only as needed. Use Arial font for figure lettering. Figures, symbols, lettering, and numbering should be clear and large enough to remain legible when reduced to print size. Large figures may be made available online only. Place figure keys within the figure; figure legends should be provided at the end of the manuscript file.

Videos. Submit as AVI, MOV, MPG, MPEG, or WMV. Videos should not exceed 5 minutes and should include an audio description and complete captioning. If audio is not available, provide a description of the action in the video as a separate Word file. Published or copyrighted material (e.g., music) is discouraged and must be accompanied by written release. If video is part of a manuscript, files must be uploaded with manuscript submission. When uploading, choose "Video" file. Include a brief video legend in the manuscript file.

Types of Articles

Perspectives. Articles should not exceed 3,500 words and 40 references. Use of subheadings in the main body of the text is recommended. Photographs and illustrations are encouraged. Provide a short abstract (150 words), 1-sentence summary, and biographical sketch. Articles should provide insightful analysis and commentary about new and reemerging infectious diseases and related issues. Perspectives may address factors known to influence the emergence of diseases, including microbial adaptation and change, human demographics and behavior, technology and industry, economic development and land use, international travel and commerce, and the breakdown of public health measures.

Synopses. Articles should not exceed 3,500 words and 40 references. Use of subheadings in the main body of the text is recommended. Photographs and illustrations are encouraged. Provide a short abstract (150 words), 1-sentence summary, and biographical sketch. This section comprises concise reviews of infectious diseases or closely related topics. Preference is given to reviews of new and emerging diseases; however, timely updates of other diseases or topics are also welcome.

Research. Articles should not exceed 3,500 words and 40 references. Use of subheadings in the main body of the text is recommended. Photographs and illustrations are encouraged. Provide a short abstract (150 words), 1-sentence summary, and biographical sketch. Report laboratory and epidemiologic results within a public health perspective. Explain the value of the research in public health terms and place the findings in a larger perspective (i.e., "Here is what we found, and here is what the findings mean").

Policy and Historical Reviews. Articles should not exceed 3,500 words and 40 references. Use of subheadings in the main body of the text is recommended. Photographs and illustrations are encouraged. Provide a short abstract (150 words), 1-sentence summary, and biographical sketch. Articles in this section include public health policy or historical reports that are based on research and analysis of emerging disease issues.

Dispatches. Articles should be no more than 1,200 words and need not be divided into sections. If subheadings are used, they should be general, e.g., "The Study" and "Conclusions." Provide a brief abstract (50 words); references (not to exceed 15); figures or illustrations (not to exceed 2); tables (not to exceed 2); and biographical sketch. Dispatches are updates on infectious disease trends and research that include descriptions of new methods for detecting, characterizing, or subtyping new or reemerging pathogens. Developments in antimicrobial drugs, vaccines, or infectious disease prevention or elimination programs are appropriate. Case reports are also welcome.

Photo Quiz. The photo quiz (1,200 words) highlights a person who made notable contributions to public health and medicine. Provide a photo of the subject, a brief clue to the person's identity, and five possible answers, followed by an essay describing the person's life and his or her significance to public health, science, and infectious disease.

Commentaries. Thoughtful discussions (500–1,000 words) of current topics. Commentaries may contain references but no abstract, figures, or tables. Include biographical sketch.

Another Dimension. Thoughtful essays, short stories, or poems on philosophical issues related to science, medical practice, and human health. Topics may include science and the human condition, the unanticipated side of epidemic investigations, or how people perceive and cope with infection and illness. This section is intended to evoke compassion for human suffering and to expand the science reader's literary scope. Manuscripts are selected for publication as much for their content (the experiences they describe) as for their literary merit. Include biographical sketch.

Letters. Letters commenting on recent articles as well as letters reporting cases, outbreaks, or original research, are welcome. Letters commenting on articles should contain no more than 300 words and 5 references; they are more likely to be published if submitted within 4 weeks of the original article's publication. Letters reporting cases, outbreaks, or original research should contain no more than 800 words and 10 references. They may have 1 figure or table and should not be divided into sections. No biographical sketch is needed.

Books, Other Media. Reviews (250–500 words) of new books or other media on emerging disease issues are welcome. Title, author(s), publisher, number of pages, and other pertinent details should be included.

Conference Summaries. Summaries of emerging infectious disease conference activities are published online only and should contain 500–1,000 words. They should focus on content rather than process and may provide illustrations, references, and links to full reports of conference activities.

Online Reports. Reports on consensus group meetings, workshops, and other activities in which suggestions for diagnostic, treatment, or reporting methods related to infectious disease topics are formulated may be published online only. These should not exceed 3,500 words and should be authored by the group. We do not publish official guidelines or policy recommendations.

Announcements. We welcome brief announcements of timely events of interest to our readers. Announcements may be posted online only, depending on the event date. Email to eideditor@cdc.gov.

FIGURE 4.2 The Instructions to Authors from the journal *Emerging Infectious Diseases*.

The Instructions to Authors is a critical resource that you *must* follow exactly (Audisio et al. 2009; Benson and Silver 2013; Korner 2008; Naylor and Muñoz-Viveros 2005; Welch 2007). Each journal has different requirements and the requirements for one journal are generally not accepted by other journals (John 2009; Schriger, Arora, and Altman 2006), so you can save yourself significant time and effort if you follow the Instructions to Authors for your target journal as you begin writing the manuscript. In the discussion that follows, we will examine the key information provided in these guidelines.

Audience and Scope of the Journal

As we discussed in Chapter 2, research journal articles are targeted to other researchers studying similar topics in a narrowly focused discourse community (Applewhite 1979; Knottnerus and Tugwell 2007; Lipson 1988; Masic 2012; Patterson 1973; Reeves 2005; Zappen 1989). In the Instructions to Authors, journal editors describe the primary audience and scope of research topics that are published in their journal (see Figure 4.2). As a writer, you should use this information to verify that a journal is appropriate for your research (Audisio et al. 2009; John 2009). This information should help you avoid the primary complaint of journal editors—the submission of manuscripts that do not fit the scope of the journal because the authors are unfamiliar with the publication (Worsham 2008). Additionally, this information will help you determine the amount of detail that you should include in your manuscript and whether you should focus more on theoretical or practical implications of the research.

"I see by the current issue of 'Lab News,' Ridgeway, that you've been working for the last twenty years on the same problem I've been working on for the last twenty years."

Types of Articles Published

The Instructions to Authors specifies the types of articles that a journal publishes. Although the names of each type of article vary between journals, the standard categories include the following:

- JOURNAL ARTICLES—Reports of original research from experimental, observational, or theoretical studies or the development of new laboratory methods. Journal articles are concise reports, typically ranging from 3,000–5,000 words (Brand and Huiskes 2001; Kurmis 2003).
- BRIEF REPORTS, LETTERS, OR NOTES—Shorter reports of original research, often ranging from 600–1,500 words. Letters, in particular, have become a forum for rapidly publishing innovative research that is often of interest to a larger scientific community (Blakeslee 1994; Zylke 2013). The molecular structure of DNA by Watson and Crick (1953) is an example of a letter, which is just over one page. As a genre, letters covey a greater sense of urgency and importance than do standard journal articles and are often written in first person, although the organization of content remains in the same inductive structure as that of most SMC. Note that the label "letter" is used differently by different journals and should be distinguished from editorials, which are opinion pieces.
- CASE STUDIES or CLINICAL OBSERVATIONS—Original research that provides a descriptive account of a single individual or example.
- EDITORIALS or COMMENTARIES—Opinion pieces written by researchers to comment on other research published in the journal.

The Instructions to Authors also specifies limits for each type of article (such as maximum number of words or characters, tables or figures, and references). You should try to keep manuscripts within the stated limits. Manuscripts that do not follow these guidelines are often returned for revision or even rejected (Audisio et al. 2009; Benson and Silver 2013). Journal editors are reluctant to consider or accommodate manuscripts that exceed the limits because they have a limited page budget (number of pages permitted to be published each year), and a significantly longer article would take space from other articles and affect the publication plan (Benson and Silver 2013; Welch 2007). As Stephen Welch, executive editor of the journal *CHEST*, stated, "Most editors prefer short, to-the-point articles" (1591).

Original research in SMC is presented in an inductive manner, in which research of a particular case leads to larger generalizations (Gross 1985). Journal articles and other genres of original research are usually divided into four major sections: INTRODUCTION, METHODS, RESULTS, and DISCUSSION (see Table 4.1). The mnemonic IMRAD (pronounced ɛm-ræd) is formed from the first letter of each section (plus the "a" from "and"). Occasionally you may see the acronym

TABLE 4.1 Summary of the IMRAD sections and their purposes.

Section	Purpose
Introduction	Provides context for the present study and summarizes the most relevant research so readers can understand the present study.
Methods	Describes the experiments performed and how data were collected to provide an argument that the approach and tools were an appropriate and valid design to answer the research question.
Results	Presents the data from the experiments listed in the Methods section as evidence for a scientific argument.
Discussion	Argument for how the experimental data answer the research question and how the research can be applied to a larger context.

"AIMRAD," which adds an "A" at the beginning to emphasize the importance of the Abstract for electronic database search and retrieval (Iverson et al. 2007; Piqué-Angordans and Posteguillo 2006). The formal organization of the IMRAD structure developed over centuries (Bazerman 1984; Gross, Harmon, and Reidy 2009; Piqué-Angordans and Posteguillo 2006; Swales 1990) and presents information in an order that reflects the broad scientific discovery process (Bazerman 1988; Gross 1985; International Committee of Medical Journal Editors 2014; Markel 1993; Woolgar 1981). This prototypical structure is followed in most journals, although some leading journals such as *Science* now post the Methods section online and only include sufficient details in the printed article so readers can understand the overall approach. What is important is that readers of SMC expect to find particular information in each section, so adhering to the IMRAD structure facilitates the communication process (Chiswick 2002; Piqué-Angordans Posteguillo 2006). As clinician and scientist Oscar Ratnoff (1981) noted, "The structure of a [scientific] paper is so stereotyped, it wouldn't be accepted for publication unless it is divided up in a rigid way" (96–7). Journals in SMC can receive hundreds of manuscripts each week (*Nature* n.d.), so manuscripts that do not follow these conventions are returned to the authors for revision.

As illustrated in Figure 4.3, the IMRAD organization resembles an hourglass in that it begins with general information to establish a context for the present research, rather rapidly narrows down to the specific details of the current study (which continue throughout the majority of the article), and eventually concludes by applying the specific research findings to a larger context (Hill, Soppelsa, and West 1982; Schulte 2003). Even if articles do not formally use the IMRAD headings, as is the case with shorter genres such as letters, nearly all reports of original research still follow this structure.

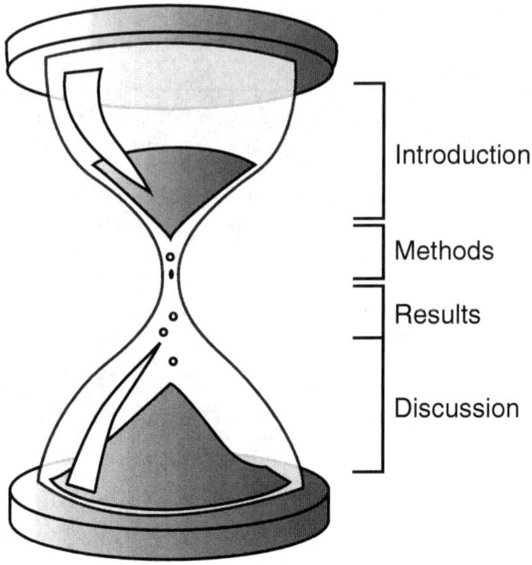

FIGURE 4.3 The content in IMRAD journal articles follows an hourglass shape from general (or broad) information that narrows to cover a specific case for the research investigation. This specific information is maintained throughout the body of the communication and then expands to cover broad implications at the end of the discussion.

Style Guidelines

The Instructions to Authors specifies the style guidelines for the prose, in-text citations, and reference list (Benson and Silver 2013; Schriger, Arora, and Altman 2006). It also specifies the accepted (and preferred) scientific nomenclature, abbreviations/acronyms, and units of measure. Additionally, for international journals, such guidelines specify spelling conventions (specifically if you should use American English or British English spelling).

Generally, journals take one of two approaches for style guidelines. In one approach, journals publish their own, unique style guidelines either in a comprehensive instructions document (for example, see the *Journal of Bacteriology* Instructions to Authors at http://jb.asm.org/site/misc/authors.xhtml) or as a supplement to the Instructions to Authors (for example, see the complete style guide for *Emerging Infectious Diseases* at http://wwwnc.cdc.gov/eid/). In the other approach, journals publish a brief Instructions to Authors and refer to a more comprehensive style manual that is published by an organization (such as those listed in Table 4.2). Such style manuals provide a set of disciplinary norms that are accepted by many journals within a particular field (Hagge 1997).

Recent articles on similar topics from the target journal will illustrate style decisions recently made by the editorial team. Although referring to such articles

TABLE 4.2 Major style manuals in SMC published by organizations.

Organization	Style Manual	URL
American Chemical Society	*The ACS Style Guide: Effective Communication of Scientific Information*	http://pubs.acs.org/series/styleguide
American Medical Association	*AMA Manual of Style: A Guide for Authors and Editors*	http://www.amamanualofstyle.com
American Psychological Association	*Publication Manual of the American Psychological Association*	http://www.apastyle.org
Council of Science Editors	*Scientific Style and Format: The CSE Manual for Authors, Editors, and Publishers*	http://www.scientificstyleandformat.org
International Committee of Medical Journal Editors	*Recommendations for the Conduct, Reporting, Editing and Publication of Scholarly Work in Medical Journals* (previously the *Uniform Requirements for Manuscripts*)	http://www.icmje.org/recommendations

is less reliable than referring to the style manual, you can see style decisions in context by doing so. Occasionally you may encounter discrepancies between the Instructions to Authors from a journal and the recommended style guide. If this occurs, you should usually follow the guidelines provided by the journal editors (in the Instructions to Authors) because editors occasionally revise the style guidelines. If the discrepancy remains unclear and is not clarified by examining recent articles, you should contact the editor for clarification.

Manuscript Formatting and Submission Guidelines

The manuscript formatting guidelines facilitate the peer-review and publishing processes (Overstreet 2010). These guidelines help minimize the workload of reviewers and editors, which is a concern of journal editors (Sanford 1991). Each journal has precise requirements that developed from the particular experiences of that group of publishers and editors (Harmon and Gross 1996). Therefore, an individual Instructions to Authors document or style manual may not specify a requirement that another journal emphasizes. Unfortunately, many journal editors will not consider manuscripts that do not adhere to their journal's formatting requirements.

As mentioned previously, manuscripts are formatted more like a college research paper rather than how the article will appear in the journal. Journal editors usually require manuscripts to be submitted as a Microsoft Word file (Boellstorff 2011; Coghill and Garson 2006). The manuscript formatting requirements are often very specific; in practice, they are similar to those for preparing a thesis or dissertation. For example, journal editors specify the exact typeface and formatting for the body text and each heading level, margins, location of page numbers, number of characters in the running head, and use of italics or other styles. Some journals provide Microsoft Word templates that include the required formatting, which can be downloaded from the journal's website. Refer to Appendix A for general formatting requirements for research manuscripts. This appendix may be used if the Instructions to Authors does not specify complete requirements or answer all your questions.

In addition to the formatting requirements, the Instructions to Authors usually specifies the content to include on the pages and sections of the manuscript. For example, the title page typically includes the article title, running head, authors, institutions, and keywords. Manuscripts often have a separate page for the Abstract and then begin the Introduction on a new page. Tables and figures, along with their captions, are usually inserted at the end of the manuscript, or uploaded as separate documents, rather than being integrated into the text. Since the Instructions to Authors includes many specific requirements, you will save a lot of time by accessing it early in the writing process rather having to reformat or revise your manuscript later.

Concluding Points

In this chapter, we covered the steps necessary to prepare writing an article of original research and introduced the IMRAD structure. Underlying the IMRAD structure of journal articles of original research is an internal microstructure that further dictates content and style. The director of a prominent viral research center described how the structure of scientific communication adheres to a certain pattern:

> If you take any manuscript in the primary literature, it's really highly structured and they [scientific writers] are asked to adhere to most of those structural elements. For example, if you just take a section of a results paragraph from a paper, the first, second, third, fourth, fifth sentence almost always follows a particular pattern. They [novice scientific writers] just can't see what those patterns are when they first start reading them, but this is what I'm asking them to adhere to in terms of structure of writing.
>
> *(pers. comm., 2008)*

Since each section of an article is quite different, you will find the article easiest to write one section at a time.

Researchers who are experienced authors of SMC often begin with the Methods or Results sections because those sections are easier for them to write since they are the ones conducting the experiments (Harmon and Gross 2010). But novice writers and professional communicators might try to write the Introduction section first because doing so will help you become familiar with the subject matter because you will need to read several published examples in order to write this section. Moreover, writing the Introduction section first will help you to construct the content into a linear and coherent argument that is similar to how readers will view the research.

In the chapters to follow, we will discuss the specific content and style of each section of the IMRAD organization as well as all the other parts of a journal article. When you are ready to write each section, I encourage you to read the corresponding chapter in this book before you begin writing because each section has a very precise structure and style.

5

THE INTRODUCTION SECTION

In this chapter, you will learn to:

- Select content to present in the Introduction section
- Organize content in the Introduction section following the "Create a Research Space" model
- Write in a scientific style and edit style problems that commonly arise in the Introduction section

Overview

The Introduction is the first content section of an IMRAD-structured journal article. The purpose of the Introduction is to familiarize the audience, who are subject-matter experts, with the most recent and relevant research so they can understand how the current research findings fit into the current understanding of the topic (Chiswick 2002; Day and Gastel 2006; International Committee of Medical Journal Editors 2014; Kurmis 2003; Piqué-Angordans and Posteguillo 2006). The Introduction should provide sufficient information so that the audience can understand and evaluate the results of the current research without having to refer to any other journal articles on the topic (American Society for Microbiology 2014; Vande Kopple 1985). A strong Introduction also establishes the value of the research topic to journal editors, reviewers, and readers, and creates continuity of a "story" from previously published research to the present study (Cals and Kotz 2013a).

The content presented in Introductions of journal articles often adheres to a standard organization and style (Harmon and Gross 1996; Piqué-Angordans and

Posteguillo 2006; Skelton 1994; Swales 1990, 2004). The conventions, covered in this chapter, will facilitate your writing as well as the reading by the audience who expect to find specific information in a particular order and style. By the end of this chapter, you should be able to write an Introduction section that is appropriate for a scientific or medical journal article.

How to Prepare

The first step in writing an Introduction is to clearly articulate the primary objective and conclusion of the research (ideally in one or two sentences). This purpose statement will help you to remain focused during the writing process. Additionally, you should gather the following:

- Instructions to Authors for the target journal (the journal that you plan to submit the article to for consideration);
- Style guide, if specified in the Instructions to Authors; and
- Two to three recent journal articles on similar research topics (probably articles that you will cite in the background section and preferably at least one article from the target journal).

(For more information on finding these materials, see Chapter 4.) Together, these documents will help you to write the content and identify the particular conventions for the target journal.

What to Expect

First, you should keep in mind that the Introduction provides a highly focused summary of the most relevant research on a topic—not a comprehensive literature review (American Psychological Association 2010). The end of the Introduction attracts particular attention from experienced readers because that is where authors often state the experimental objective of the study (Cals and Kotz 2013a). Most Introduction sections conclude with a summary of the research question or purpose, which transitions directly into the Methods section that follows.

The Introduction is a concise section of approximately 250 to 500 words, or a maximum of 10% to 15% of the total word count of the journal article (Brand and Huiskes 2001; Cals and Kotz 2013a; Chiswick 2002; Hofmann 2010; Nair 2005; Plaxco 2010; Thrower 2008a; Van Way 2007; Zeiger 2000). Longer Introduction sections are discouraged because they may exhaust, confuse, or mislead readers by including too much information (Zeiger 2000).

The formatting of the Introduction is simply a few paragraphs of the manuscript. Since the Introduction is the first body section of a journal article, it rarely has a heading (American Psychological Association 2010). Additionally, because the Introduction should be a short section, subheadings are also uncommon.

Organization of the Introduction Section: The "Create a Research Space" Model

Most Introduction sections of journal articles follow the "CREATE A RESEARCH SPACE" (or "CARS") model (Swales 1990, 2004, Swales and Najjar 1987). Across science, medicine, and engineering, researchers found 70% to 100% consistency of the model, which means this is a highly conserved structure (Kanoksilapatham 2005, 2012; Nwogu 1997; Skelton 1994; Swales 1990, 2004; Swales and Najjar 1987).

In the CARS model, content is organized from known research to the unknown, hypothesis, or question (Kanoksilapatham 2005; Skelton 1994). The literature cited in the Introduction section funnels from general information about the topic to specific information that directly leads to the current research (Cals and Kotz 2013a; Hofmann 2010, Zeiger 2000). Introduction sections typically contain the following content, which is organized into "moves" and "steps":

Move 1: Establish a research territory, *by*

Step 1: Establishing the importance of the research topic, and
Step 2: Summarizing previous research to contextualize the present study.

Move 2: State the gap or inconsistency in the previous research to establish a niche for the present study.

Move 3: Stating the objective of the present study (which is to fill the research niche stated in Move 2).

An example of an Introduction section following the CARS model is provided in Examples 5.1 and 5.2. In the following section, we will cover each Move in detail.

EXAMPLE 5.1 SAMPLE INTRODUCTION SECTION FROM A MEDICAL JOURNAL ARTICLE ANNOTATED WITH THE "CREATE A RESEARCH SPACE" (CARS) MODEL.

This example illustrates the typical amount of content of an Introduction section (which includes 17 references).

Coffee Drinking and Cutaneous Melanoma Risk in the NIH-AARP Diet and Health Study

Cutaneous melanoma is the fifth most common cancer and the leading cause of skin cancer death in the United States, with an estimated 77 000 new cases and 9500 deaths in 2013.[1]	Move 1, Step 1: Establish the importance of the research topic
Exposure to ultraviolet radiation (UVR), particularly UVB, is the only consistentlyassociated exogenous risk factor for melanoma.[2] Yet, other exposures are likely important.	Move 1, Step 2 (and broad justification of the study)

Experimental evidence lends biological plausibility to a possible protective role of coffee consumption in UVB-induced carcinogenesis. In vitro and animal studies have shown that coffee constituents suppress UVB-induced skin carcinogenesis,[3] induce cell apoptosis,[4] protect against oxidative stress and DNA damage,[5] reduce inflammation in epidermal cells,[6] and inhibit changes in DNA methylation.[7] The protective effects of coffee constituents, especially caffeine, on UVB-induced skin cancer demonstrated by murine and cell culture models have been corroborated by epidemiological studies of coffee consumption and risk of nonmelanoma skin cancers.[8–11]	Move 1, Step 2: Summarizing previous research to contextualize the present study
In contrast, the few existent epidemiological studies of coffee consumption and melanoma are marked by inconsistent results.[12–17]	Provide research findings that lead from Move 1, Step 2 to Move 2
With few lifestyle factors as viable targets for melanoma prevention and the worldwide popularity of coffee drinking, it is important to resolve these conflicting findings.	Move 2: State the gap or inconsistency in the previous research to establish a niche for the present study
In the current study, we analyzed data from the National Institutes of Health (NIH)–AARP Study, which include nearly four times as many cases of malignant melanoma as the largest prospective study to date, to better understand the association of coffee drinking with risk of malignant melanoma, as well as risk of melanoma in situ.	Move 3: State the objective of the present study, which occupies the research niche

Source: Loftfield, Erikka, Neal D. Freedman, Barry I. Graubard, Albert R. Hollenbeck, Fatma M. Shebl, Susan T. Mayne, and Rashmi Sinha. 2015. "Coffee Drinking and Cutaneous Melanoma Risk in the NIH-AARP Diet and Health Study." *Journal of the National Cancer Institute* 107 (2): djv013. Accessed February 24, 2015. doi: 10.1093/jnci/djv013.

EXAMPLE 5.2 SAMPLE INTRODUCTION SECTION FROM A SCIENTIFIC JOURNAL ARTICLE ANNOTATED WITH THE "CREATE A RESEARCH SPACE" (CARS) MODEL.

This example illustrates the typical amount of content of an Introduction section (which includes 12 references).

Outbreak-associated *Salmonella enterica* Serotypes and Food Commodities, United States, 1998–2008

Salmonella enterica is estimated to cause 1.2 million illnesses each year in the United States and to be the leading cause of hospitalizations and deaths from foodborne disease.[1] Because of the major public health	Move 1, Step 1: Establish the importance of the research topic

role of *Salmonella* infections, the US Department of Health and Human Services has made decreasing the nationwide incidence of these infections by 25% a Healthy People 2020 national goal.[2] Overall, salmonellosis incidence has not decreased in the past decade; the incidence has substantially increased for some serotypes and decreased for others.[2, 3] Focused attention on determining sources of *Salmonella* infections will be vital to reach the 25% target reduction in these infections.	Summarize Move 1, Step 2 (and broad justification of the study)
Salmonella serotypes differ in their natural reservoirs and ability to cause human infections;[4–6] only a small proportion of >2,500 serotypes cause most human infections.[4,7] In 2009, only 20 serotypes comprised >82% of the ≈36,000 serotyped human-derived *Salmonella* isolates in the United States that were reported to the Centers for Disease Control and Prevention.[3] A few serotypes have been associated with specific animal reservoirs. For example, serotype Dublin, which caused 103 laboratory-confirmed human infections in 2009,[3] is found predominantly in cattle.[5]	Move 1, Step 2: Summarize previous research to contextualize the present study
However, reservoir sampling alone has limited use in predicting the contribution of a reservoir to the incidence of human illness.[8]	Provide research findings that lead from Move 1, Step 2 to Move 2
Outbreak data and case-control studies have linked some serotypes to certain foods or exposures (e.g., serotype Enteritidis to eggs and chicken).[9–11] Information obtained during outbreak investigations is a key tool in understanding which foods are common sources of pathogens contributing to foodborne infections. During outbreak investigations, illnesses can be linked to a particular food by using epidemiologic or laboratory evidence.[12]	Summarize of any secondary line(s) of research that intersects with and helps explain/ justify the present study
To our knowledge, no systematic examination of *Salmonella* serotypes and food vehicles implicated in outbreaks has been reported.	Move 2: State the gap or inconsistency in the previous research to establish a niche for the present study
We analyzed foodborne disease outbreak data to determine associations between food commodities and serotypes to help inform future outbreak investigations, foodborne illness source attribution analyses, and control measures.	Move 3: State the objective of the present study, which occupies the research niche

Source: Jackson, Brendan R., Patricia M. Griffin, Dana Cole, Kelly A. Walsh, and Shua J. Chai. 2013. "Outbreak-Associated *Salmonella enterica* Serotypes and Food Commodities, United States, 1998–2008." *Emerging Infectious Diseases* 19: 1239–44.

Content of Introductions

Writers sometimes have difficulty beginning introductions of any genre because of a perception that there is an endless set of possible approaches and styles

(Arrington and Rose 1987). In SMC, however, the Introduction contains specific information and often follows the CARS model (Swales 1990, 2004; Swales and Najjar 1987), as we just discussed. In the following section, we will examine the content of this model. Although we will discuss each CARS Moves in order, you might find the writing process easier if you begin with Move 3 because doing so will help you to keep your Introduction focused on the purpose of your research study, especially when summarizing previous research in Move 1, Step 2 (Chiswick 2002).

Move 1: Establishing a Research Territory

Move 1, which establishes the research territory, accounts for most of the content in the Introduction. At the beginning of the Introduction (Step 1), authors state the importance or relevance of the broad research topic of the current study. This short lead segues into a brief, focused summary of recent and relevant research findings (Step 2). In the following subsections, we will examine Step 1 and Step 2 in detail.

Step 1: Establishing the Importance of the Research Topic

Most journal article Introductions begin with one to two sentences stating the importance of the research topic (see Example 5.1). Establishing the importance of a research topic may initially appear to be a difficult task, but doing so is one of the more conventional statements within a discourse community. After you have collected several other journal articles on the same topic, you will see that many of the articles often begin with relatively similar claims of the importance (compare the first sentences of Examples 5.2, 5.3, and 5.4). Since there is a relative consensus of the importance of a topic among researchers within a field, you should create an introductory statement that makes a similar assertion of importance.

EXAMPLE 5.3 ESTABLISHING THE IMPORTANCE OF THE RESEARCH TOPIC (MOVE 1, STEP 1) FROM REFERENCE 10 IN EXAMPLE 5.2.

The similarity of the first sentences (underlined) reflects a relative consensus of importance among the discourse community.

Salmonella organisms infect an estimated 1.4 million people annually in the United States, resulting in ~16,000 hospitalizations and >500 deaths.[1] The economic impact of salmonellosis in the United States is considerable, costing \$0.5 to \$2.3 billion per year.[2]

Source: Kimura, Akiko C., Vasudha Reddy, Ruthanne Marcus, Paul R. Cieslak, Janet C. Mohle-Boetani, Heidi D. Kassenborg, Suzanne D. Segler, Felicia P. Hardnett, Timothy Barrett, and David L. Swerdlow. 2004. "Chicken Consumption Is a Newly Identified Risk Factor for Sporadic *Salmonella enterica* Serotype Enteritidis Infections in the United States: A Case-Control Study in FoodNet Sites." *Clinical Infectious Diseases*, 38 (Suppl 3): S244–52.

EXAMPLE 5.4 ANOTHER SIMILAR OPENING STATEMENT ESTABLISHING THE IMPORTANCE OF THE RESEARCH TOPIC (MOVE 1, STEP 1) FROM REFERENCE 11 IN EXAMPLE 5.2.

(Note that the slightly different numbers of hospitalization and deaths between Examples 5.2 and 5.3 is due to reports from different years; the authors should have specified the year or time frame to contextualize these statistics and avoid confusion.)

Salmonella causes an estimated 1.4 million infections, 15,000 hospitalizations and 400 deaths each year in the United States.[1] *Salmonella* serotype Enteritidis (SE) accounts for approximately one-fifth of all human *Salmonella* isolates reported to the Centers for Disease Control and Prevention (CDC) each year.[2, 3]

Source: Marcus, R., J.K. Varma, C. Medus, E.J. Boothe, B.J. Anderson, T. Crume, K.E. Fullerton, M.R. Moore, P.L. White, E. Lyszkowicz, A.C. Voetsch, and F.J. Angulo. 2007. "Re-Assessment of Risk Factors for Sporadic *Salmonella* Serotype Enteritidis Infections: A Case-Control Study in Five FoodNet Sites, 2002–2003." *Epidemiology and Infection* 135: 84–92.

Although statements of importance are often similar, you need to evaluate and research the claims in order to avoid perpetuating incomplete statistics as illustrated in Examples 5.2, 5.3, and 5.4. When writing such problem statements, you should include both a tangible condition and cost, not necessarily financial, that the audience agrees are important (Williams 2011). When you present the cost, you should also provide the magnitude or scope to the real world by presenting recent statistics (Cals and Kotz 2013a; Samraj 2002, 2005). An example of a recent statistic might be the rate of skin cancer, such as "Cutaneous melanoma is the fifth most common cancer and the leading cause of skin cancer death in the United States, with an estimated 77,000 new cases and 9,500 deaths in 2013" (as stated in Example 5.1).

When writing the problem-cost statement, you should adhere to the following conventions:

- **Begin by stating the problem** (Peterson 1961) **and lead with the subject of the problem.** In Examples 5.2, 5.3, and 5.4, you should note that each paper leads with the name of the organism (i.e., *Salmonella*) and immediately states the problem. This informative beginning immediately tells readers the subject of the paper.
- **Present specific statistics or details that establish the scale of the "cost." Be sure to provide sufficient context for the statistics.** For example, "*Salmonella enterica* is estimated to cause 1.2 million illnesses each year in the United States." Avoid nonspecific terms such as "large," "many," or "few" and generic claims that something is a problem because you (as the author) said it was a problem. Rather, you should search and cite a recent statistic from a credible source, which you can find by reading the Introductions of several similar journal articles and examining the statistics and sources those sources reference. For example, in Examples 5.2, 5.3, and 5.4, the writers of each journal article presented similar statistics of the number of infections caused by *Salmonella* in the United States. However, statistics vary over time as illustrated by the different number of infections mentioned in these sources (e.g., 1.2 million stated in Example 5.2 and 1.4 stated in Examples 5.3 and 5.4). Therefore, when citing a statistic, you should also provide the relevant time frame and source of the data to fully contextualize the information and avoid confusion (see Example 5.5).
- **Avoid subjective and unfounded claims.** For example, avoid any statements that a research topic is "interesting" or "has not received sufficient attention" (Ortinau 2011). Such statements are vague and such empty claims waste space. Rather, establish the importance by providing relevant statistics and citing studies to substantiate these claims as described above.

In conclusion, the importance of the research topic introduces the broad subject area and leads directly into a brief and focused literature review (Move 1, Step 2).

EXAMPLE 5.5 ESTABLISHING THE IMPORTANCE OF THE RESEARCH TOPIC (MOVE 1, STEP 1; FROM REFERENCE 8 IN EXAMPLE 11.1).

Note that the statistic is contextualized (see underlined phrases) so that readers have the pertinent details to understand the statistic without having to access the reference article.

Salmonellosis remains a substantive case of morbidity and mortality. Using 1996 and 1997 data from the Centers for Disease Control and Prevention

(CDC) FoodNet sentinel site program, Mead et al. [1] estimated that ~1.4 million nontyphoidal *Salmonella* cases occur in the United States each year.

Source: Sarwari, Arif R., Laurence S. Magder, Priscilla Levine, Ann Marie McNamara, Susan Knower, Gregory L. Armstrong, Ruth Etzel, Jill Hollingsworth, and J. Glenn Morris Jr. 2001. "Serotype Distribution of *Salmonella* Isolates from Food Animals after Slaughter Differs from That of Isolates Found in Humans." *The Journal of Infectious Diseases* 183: 1295–9.

Step 2: Summary of Previous Research to Contextualize the Present Study

The bulk of the Introduction section is a brief, focused summary of previous research that leads from the general problem (Move 1, Step 1) to the specific objectives of the present research study (Move 3). One of the most common mistakes writers make in Introductions is to develop a comprehensive review of the topic and include too much background information. Keep in mind that the purpose of this literature review is to establish the direct line of reasoning or specific context for the current investigation—not to provide an exhaustive summary of all the research on a given topic. Thus, the literature review in the Introduction section should not be confused with a review article, which is a comprehensive review of a topic. Typically the literature review in the Introduction section of a journal article can be accomplished in one or two short paragraphs (see Examples 5.1 and 5.2).

In an editorial on the problems of long Introductions, Peter Thrower (2008a), former editor in chief of the journal *Carbon* noted that long Introductions that thoroughly cover a topic "might . . . be appropriate for a popular science article in a newspaper but certainly not for [the scientific journal] *CARBON*" (184). As Thrower (2008a) stated,

> The depth of background history provided by some Introductions makes me wonder (cynically) why they don't start with the discovery of the electron, and then discuss chemical bonding, Bragg's work on crystal structures, etc. We would all find such an approach absurd, but many Introductions are nearly as bad.
>
> *(184)*

You will need to be judicious when reporting background research. As we discussed in Chapter 3, the general rule is to cite information that does not appear in the textbooks within the field. As you will see from reading the journal articles you have collected on the topic, the appropriate background information included in the Introduction section is fairly similar in recent articles on the topic. Often your Introduction will cover nearly all of the same background information with only a few additions of the findings from more recent papers (such as the findings from those recent articles themselves).

The content of the literature summary is organized as a funnel in which principal research claims from previous journal articles are presented with increasing specificity that leads the readers through a narrowing line of research that directs the focus to the current study (Cals and Kotz 2013a; Hofmann 2010; Swales 2004; Zeiger 2000). This organization is illustrated in the middle paragraph of Example 5.2, which is further analyzed in Example 5.6. In this example, note that statements of broader scope appear first and are organized by increasing specificity to a highly focused point.

Sometimes you may find that the present research depends on more than one lineage of background research. In such situations of a converging Introduction, you will establish multiple story lines but keep the flow from general to specific within each. For instance, in Example 5.2, the natural reservoir and reservoir sampling of *Salmonella* serotypes provided limited data, so the authors presented a converging line of research (i.e., *Salmonella* outbreak to specific food sources). Presenting converging lines of research often requires the author to select a primary storyline and briefly introduce the relevant pieces from a second line. The primary storyline should be the more prominent topic—which is also the focus of the problem statement in Move 1, Step 1. Introducing a secondary storyline disrupts the logical flow of information, so paragraph breaks are often used to indicate a new thread. In Example 5.2, the authors begin a new paragraph in the middle of the literature review (Move 1, Step 2) to transition between story lines.

EXAMPLE 5.6 THIS SUMMARY OF PREVIOUS RESEARCH FOCUSES FROM A GENERAL (OR BROAD) SCOPE TO A NARROW (OR SPECIFIC) SCOPE (FROM EXAMPLE 5.2)

Salmonella serotypes differ in their natural reservoirs and ability to cause human infections (*4–6*);

only a small proportion of >2,500 serotypes cause most human infections (*4,7*).

In 2009, only 20 serotypes comprised >82% of the ≈36,000 serotyped human-derived *Salmonella* isolates in the United States that were reported to the Centers for Disease Control and Prevention (*3*).

A few serotypes have been associated with specific animal reservoirs. For example, serotype Dublin, which caused 103 laboratory-confirmed human infections in 2009 (*3*), is found predominantly in cattle (*5*).

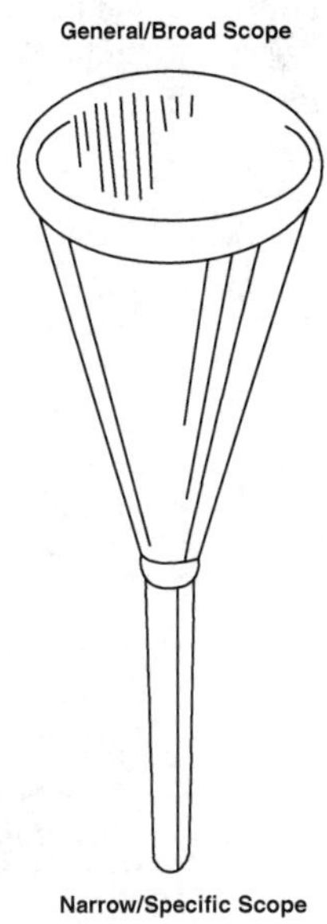

Move 2: Justifying the Present Study (to Establish a Research Niche)

Move 2 states the impetus and justification for the current research and establishes a RESEARCH NICHE, or unique position of the current investigation, in the existing scientific literature (Swales 1990, 2004). Move 2 serves as a transition from the summary of previous research presented in Move 1 to the current study that will be presented throughout the rest of the journal article. The research niche (Move 2) usually follows the background information (Move 1) and leads directly to Move 3, which introduces the present research (see Example 5.2).

In Move 2, authors typically take one of the following approaches: (1) identifying a gap in knowledge, an unaddressed issue, a new observation, or building on previous research findings; (2) identifying a problem with or limitation of previous research studies; (3) identifying an inconsistency in the scientific literature; (4) providing insight from new methods or advancing technologies; or (5) applying previous research to a clinical setting or the "real world" (Harmon and Gross 1996; Kanoksilapatham 2012; Lim 2012; Samraj 2002; Skelton 1994; Swales 1990, 2004). Although these approaches may overlap, authors usually select one of these approaches as the primary argument for their present research (see Table 5.1).

TABLE 5.1 Different approaches and statements for justifying the present study (Move 2).

Approach	Example	Source
Identify a gap in knowledge, an unaddressed issue, new observation or build on previous research findings	"To our knowledge, no systematic examination of *Salmonella* serotypes and food vehicles implicated in outbreaks has been reported."	Jackson, Brendan R., Patricia M. Griffin, Dana Cole, Kelly A. Walsh, and Shua J. Chai. 2013. "Outbreak-associated *Salmonella enterica* Serotypes and Food Commodities, United States, 1998–2008." *Emerging Infectious Diseases* 19: 1239–44.
Identify a problem or limitation of previous research	"Previous studies of the comparative virulence of various *Salmonella* serotypes have often been small [4–7], included only a limited number of serotypes [4–11], were performed in areas with substantially different distributions of predominant serotypes [4, 5], or used only the end point of isolation of *Salmonella* from blood [3–8]."	Jones, Timothy F., L. Amanda Ingram, Paul R. Cieslak, Duc J. Vugia, Melissa Tobin-D'Angelo, Sharon Hurd, Carlota Medus, Alicia Conquist, and Frederick J. Angulo. 2008. "Salmonellosis Outcomes Differ Substantially by Serotype." *The Journal of Infectious Diseases*, 198: 109–14.

Approach	Example	Source
Identify an inconsistency in the scientific literature	"Although eggs have been a frequent source of SE outbreaks in Denmark, a case-control study of sporadic SE infections in 1997–9 also failed to associate eating eggs with illness when exposure in the 5 days before illness onset was examined [9]. However, when exposure in the 1 day before illness onset was examined, eating eggs was strongly associated with illness."	Marcus, R., J.K. Varma, C. Medus, E.J. Boothe, B.J. Anderson, T. Crume, K.E. Fullerton, M.R. Moore, P.L. White, E. Lyszkowicz, A.C. Voetsch, and F.J. Angulo. 2007. "Re-Assessment of Risk Factors for Sporadic *Salmonella* Serotype Enteritidis Infections: A Case-Control Study in Five FoodNet Sites, 2002–3." *Epidemiology and Infection*, 135: 84–92.
Provide insight from new methods or advancing technologies	"In the 1990s, for the first time, sufficient national data became available on the serotypes of *Salmonella* species isolated from food animal carcasses to allow comparisons between *Salmonella* serotypes isolated from animals and those isolated from humans."	Sarwari, Arif R., Laurence S. Magder, Priscilla Levine, Ann Marie McNamara, Susan Knower, Gregory L. Armstrong, Ruth Etzel, Jill Hollingsworth, and J. Glenn Morris Jr. 2001. "Serotype Distribution of Salmonella Isolates from Food Animals after Slaughter Differs from That of Isolates Found in Humans." *The Journal of Infectious Diseases*, 183: 1295–9.
Apply previous research to a clinical setting or the "real world"	"Since most *Salmonella enterica* virulence genes are encoded on known pathogenicity islands, it is now possible to systematically attenuate the bacteria using genetic engineering tools in order to develop vaccines. These vaccines can be harnessed to carry foreign genes and can be used as recombinant vaccines."	Nyasha Chin'ombe, and Ruhanya, Vurayai. 2013. "Recombinant *Salmonella* Bacteria Vectoring HIV/AIDS Vaccines." *The Open Virology Journal*, 7: 121–6.

Move 3: Stating the Objective of the Current Research (to Fill the Research Niche)

The end of the Introduction is a key location where authors state the aim of the current journal article (Move 3), filling the niche stated in Move 2. Typically, authors begin Move 3 by announcing the purpose of the present research and by clearly identifying their role in the field by using the first-person "we" (Swales and Najjar 1987). Move 3 often begins with an explicit phrase such as "the purpose (or aim, goal, objective) of this study was to" (Kanoksilapatham 2012; Shehzad 2011; Skelton 1994; Swales 2004). As shown in Table 5.2, some variations of Move 3 include an overview of the research approach, or Methods, embedded in the purpose statement. In this construction, the experimental approach is often presented with the transitions "to," either beginning a sentence or following an overview of the Methods. Alternatively, the main purpose may be presented as a research question or hypothesis (see Table 5.2).

TABLE 5.2 Primary purpose statements (italicized) of examples of Move 3 are often made explicit, embedded with an overview of the Methods, or presented as a research question or hypothesis. Additionally, authors usually clearly identify their role by switching to the first person and using the word "we."

Approach	Example	Source
Explicit	"*The main objectives of this study were to* determine the risk factors and clinical features of sporadic SE infection."	Kimura, Akiko C., Vasudha Reddy, Ruthanne Marcus, Paul R. Cieslak, Janet C. Mohle-Boetani, Heidi D. Kassenborg, Suzanne D. Segler, Felicia P. Hardnett, Timothy Barrett, and David L. Swerdlow. 2004. "Chicken Consumption Is a Newly Identified Risk Factor for Sporadic *Salmonella enterica* Serotype Enteritidis Infections in the United States: A Case-Control Study in FoodNet Sites." *Clinical Infectious Diseases* 38 (Suppl 3): S244–52.
Embedded with methods	"We analyzed foodborne disease outbreak data *to determine associations between food commodities* and serotypes *to help inform future outbreak investigations, foodborne illness source attribution analyses, and control measures.*"	Jackson, Brendan R., Patricia M. Griffin, Dana Cole, Kelly A. Walsh, and Shua J. Chai. 2013. "Outbreak-Associated *Salmonella enterica* Serotypes and Food Commodities, United States, 1998–2008." *Emerging Infectious Diseases* 19: 1239–44.
Presented as a research question or hypothesis	"Thus, *we hypothesized that the presence of this system would also increase the ability of acid-sensitive strains of* S. typhi *to reach the intestinal tract in vivo.*"	Brenneman, Karen E., Crystal Willingham, Jacquelyn A. Kilbourne, Roy Curtiss 3rd, and Kenneth L. Roland. 2014. "A Low Gastric pH Mouse Model to Evaluate Live Attenuated Bacterial Vaccines." *PLoS ONE* 9: e87411.

The format that the authors use to state their aim, as a purpose statement or hypothesis, usually depends on the type of research investigation. The two broad types of research investigations are (1) EXPERIMENTAL (or HYPOTHESIS-) TESTING and (2) EXPLORATORY INVESTIGATION. A popular misconception is that all scientific research is experimental (or hypothesis-) testing. In this text we will use the descriptive term "experimental (or hypothesis-) testing" to refer to the conventional view of scientific research in which researchers examine the relationships between two variables. In contrast, we will use the term "exploratory investigation" to refer to research with more general (or exploratory) goals that cannot be converted into testable hypotheses. Exploratory investigations are commonly used in academic settings when researchers are exploring an unfamiliar phenomenon. Unlike experimental (or hypothesis-) testing research designs, exploratory investigations are not entirely predetermined, but the findings from one part of the investigation often determine the next phase of examination.

The two types of research investigations (experimental (or hypothesis-) testing and exploratory investigations) are also related to quantitative and qualitative research categorization, which is a common distinction in the social sciences. For example in psychology, experimental (or hypothesis-) testing research is also QUANTITATIVE RESEARCH because the data are quantitative or numerical measurements. In contrast, QUALITATIVE RESEARCH yields descriptive data and fits the exploratory investigation research design. But the distinction between quantitative and qualitative research is based on the type of data collected, and therefore quantitative research is not necessarily always experimental (or hypothesis-) testing research. For example, a researcher might collect survey results that are quantifiable but are exploratory in nature.

Stating a Hypothesis in Experimental (or Hypothesis-) Testing Research

Experimental (or hypothesis-) testing journal articles often state the objective of the research investigation (Move 3) as a hypothesis. A HYPOTHESIS is a formal statement that provides the researcher's prediction about the outcome of an experimental test, as was introduced in Chapter 2. The hypothesis may take slightly different forms depending on the conventions of the discourse community and journal. Typically, you will see hypotheses presented with various degrees of formality, which can be explicitly labeled as a hypothesis (see Example 5.7) or implied from the purpose statement (see Example 5.8). The defining features of hypotheses are the identification of participants (or subjects for nonhuman samples), variables, and predicted outcome prior to collecting (or presenting) data.

Not all experimental (or hypothesis-) testing journals articles included a formal, explicit hypothesis. But the purpose statement of experimental (or hypothesis-) testing research can easily be converted to a formal scientific hypothesis. For example, the experimental (or hypothesis-) testing research objective presented in Example 5.8 was not a formal hypothesis but rather a purpose statement. In

EXAMPLE 5.7 MOVE 3 COVERING THE OBJECTIVE (PURPOSE) OF THE PRESENT RESEARCH AND INCLUDING AN EXPLICIT HYPOTHESIS.

The Effectiveness of Breakfast Recommendations on Weight Loss: A Randomized Controlled Trial

In this study, we conducted a randomized controlled trial to determine whether advising good nutrition habits is more effective at producing weight loss if paired with advice to skip or eat breakfast. We tested the effect of breakfast recommendations on weight loss in free-living adults who were attempting to lose weight because this population is likely affected by public health breakfast recommendations. On the basis of previous findings (19), we stratified randomization by baseline breakfast eating habits and hypothesized that individuals who were advised to switch their usual breakfast eating habits due to their group assignment would lose more weight than the control group.

Aim: Indicates an experimental testing approach

Independent variable (manipulated by researchers)

Dependent variable (a quantitative measurement)

Population (Participants or Subjects)

Hypothesis (stated prediction of experimental outcome)

Adapted from Dhurandhar, Emily J., John Dawson, Amy Alcorn, Lesli H. Larsen, Elizabeth A. Thomas, Michelle Cardel, Ashley C. Bourland, Arne Astrup, Marie-Pierre St-Onge, James O. Hill, Caroline M. Apovian, James M. Shikany, and David B. Allison. 2014. "The Effectiveness of Breakfast Recommendations on Weight Loss: A Randomized Controlled Trial." *The American Journal of Clinical Nutrition* 100: 507–13.

EXAMPLE 5.8 MOVE 3 PRESENTS THE PURPOSE STATEMENT. IN THIS CASE, THE HYPOTHESIS IS IMPLIED.

Efficacy of High-Dose versus Standard-Dose Influenza Vaccine in Older Adults

The primary objective of this study was to show the efficacy of the high-dose, trivalent, inactivated influenza vaccine (IIV3-HD) as compared with a standard-dose vaccine for the prevention of laboratory-confirmed influenza illness in adults 65 years of age or older.

Prediction or outcome

Independent variables

Dependent variable

Subjects

Source: DiazGranados, Carlos A., Andrew J. Dunning, Murray Kimmel, Daniel Kirby, John Treanor, Avi Collins, Richard Pollak, Janet Christoff, John Earl, Victoria Landolfi, Earl Martin, Sanjay Gurunathan, Richard Nathan, David P. Greenberg, Nadia G. Tornieporth, Michael D. Decker, and H. Keipp Talbot. 2014. "Efficacy of High-Dose versus Standard-Dose Influenza Vaccine in Older Adults." *The New England Journal of Medicine* 371: 635–45.

Example 5.8, the high-dose influenza vaccine was compared to the standard-dose influenza vaccine, so the vaccines were the independent variables. The prevalence of disease (or conferred protection from the illness) was the measured outcome, so that was the dependent variable. The vaccinations were compared in adults 65 years of age or older, who were the subjects. The goal was to compare the two vaccines, with the implied hypothesis that the new vaccine (the "high dose, trivalent, inactivated influenza vaccine") was at least as effective in preventing disease as the old vaccine (the "standard-dose vaccine").

The philosopher Karl Popper (2002) noted that hypotheses clarify the focus of research investigations and must be both TESTABLE by experimental (or hypothesis-) testing research design and FALSIFIABLE, or able to be disproven by the outcome of research. Experimental (or hypothesis-) testing design requires the researcher to specify VARIABLES, which are events or conditions that can be observed and measured. Through experimental (or hypothesis-) testing investigation, researchers are to evaluate the impact of one variable on another. The two types of variables in experimental (or hypothesis-) testing research are:

1. The INDEPENDENT VARIABLE, or the treatment or condition that is adjusted by the researchers, and
2. The DEPENDENT VARIABLE, or the outcome or effect monitored for change.

For example, you could design experimental (or hypothesis-) testing research to evaluate the effect of time spent studying per week on student grade-point average (GPA). In this case, time spent studying is the independent variable because it may be adjusted by the researcher whereas GPA is the dependent variable because it is the variable that is monitored for change. Variables, as the name implies, must be variable (or able to change) and measurable. Therefore, time spent studying and GPA are variables because each can change and be measured. Experimental (or hypothesis-) testing research is always quantitative research because the data are in the form of measurements.

To complete the hypothesis, you need to identify the POPULATION, or the group being examined in the research. In our example, we can use college students as the population. Therefore the complete hypothesis might be this: College students who spend more time studying each week will have a higher GPA.

Occasionally a NULL HYPOTHESIS is also included, which states there is no relationship between the variables. In this example, the null hypothesis is that study time has no effect on GPA.

When you state a hypothesis, you must predict the outcome for the dependent variable. Often the predicted outcome is based on prior observations or experience. In this case I decided to construct the hypothesis to predict that increased time spent studying will result in a higher GPA. Even though the construction of the hypothesis has semantic meaning, the outcome of experimental (or hypothesis-) testing does not depend on how the hypothesis is constructed. Experiment testing will either SUPPORT or REFUTE the hypothesis. As we discussed in Chapter 2, scientific findings are always subject to future evidence, so authors use the term "support" a hypothesis rather than "prove." This leads us to one other type of hypothesis, the ALTERNATIVE HYPOTHESIS, which accounts for the opposite outcome of the hypothesis. In our example, the alternative hypothesis would be that increased time spent studying will result in a *lower* GPA. For simplicity, most authors just state the hypothesis and construct it so that the data support the statement because this format is clearer than refuting a negatively stated hypothesis.

Summarizing the Research

Summaries of the research in Move 3 may be classified as either indicative or informative, depending on if the results are included (Kanoksilapatham 2012; Swales and Najjar 1987). As shown in Example 5.9, INDICATIVE SUMMARIES include the (1) purpose and (2) methods and scope of the research. In contrast, INFORMATIVE SUMMARIES, as shown in Example 5.10, usually include information that corresponds to the each IMRAD section, the (1) purpose, (2) methods and scope, plus (3) the primary findings of the data, and (4) often the value or implications of the research.

You will find the type of summaries varies by field. Most style manuals recommend that the Introduction should conclude by announcing the primary findings (Swales and Najjar 1987) although some manuals and Instructions to Authors explicitly state that this information should not be included. For example, the *Uniform Requirements for Manuscripts Submitted to Biomedical Journals* states, "Do not include data or conclusions [in the Introduction] from the work being reported" (International Committee of Medical Journal Editors 2014, 14). Therefore, journals that follow those requirements would not have a statement of the primary findings at the end of the Introduction. But if not explicitly stated in the Instructions to Authors to do otherwise, you should include the major findings at the end of the Introduction and prepare an informative summary. Although revealing the primary findings at the beginning of a document seems unusual by literary standards, Ratnoff (1981) and Day and Gastel (2006) remind us that SMC are *not* mystery novels. As Ratnoff (1981) stated, "Reading a scientific article isn't the same as reading a detective story. We want to know from the start that the butler did it" (96).

EXAMPLE 5.9 INTRODUCTION SECTIONS CONCLUDE MOVE 3 WITH AN INDICATIVE SUMMARY.

In this study, we uncovered the general public's perceptions, knowledge, preventive behavior, and sources of information during a large, national *Salmonella* outbreak by a large-scale online survey.	Methods/Scope
As a result, we were able to answer our main research question: Which information should health organizations convey during a large-scale *Salmonella* outbreak, and by which channels, to maximize citizen compliance with preventive advice?	Purpose

Source: Van Velsen, Lex, Desiree JMA Beaujean, Julia EWC van Gemert-Pijnen, Jim E Van Steenbergen, and Aura Timen. 2014. "Public Knowledge and Preventive Behavior During a Large-Scale *Salmonella* Outbreak: Results from an Online Survey in the Netherlands." *BMJ Public Health* 14: 100.

EXAMPLE 5.10 INTRODUCTION SECTIONS CONCLUDE WITH MOVE 3 WITH AN INFORMATIVE SUMMARY THAT SUMMARIZES THE ENTIRE IMRAD JOURNAL ARTICLE.

The overall objective of this study was to investigate BARDOT's abilities to differentiate the top 20 *S. enterica* subsp. *enterica* serovars, which are responsible for 72.8% of all *Salmonella* isolates reported to the CDC (1), from other bacterial species and to detect and identify them in food.	Purpose
An initial study involved finding the most suitable selective agar medium for the measurement of elastic light scatter signatures and the generation of a signature library of *S. enterica* serovars. Next, the ability of BARDOT to detect *Salmonella* in spinach, in chicken breast with a high natural microbial background level, and in a low-moisture model food system (peanut butter) was examined.	Methods/Scope
The data show that this system can accurately detect most of the *Salmonella* serovars tested in the set of the 20 most prevalent serovars (positive predictive value [PPV] or classification precision level reaching 86%) and all serovars in the top 8 group (PPVs ranging from 68 to 93%),	Primary Findings
providing a valuable method for rapid screening of *Salmonella* for the benefit of the food-processing industry and regulatory agencies.	Value of the Research

Source: Singh, Atul K., Amanda M. Bettasso, Euiwon Bae, Bartek Rajwa, Murat M. Dundar, Mark D. Forster, Lixia Liu, Brent Barrett, Judith Lovchik, J. Paul Robinson, E. Daniel Hirleman, and Arun K. Bhunia. 2014. "Laser Optical Sensor, a Label-Free On-Plate *Salmonella enterica* Colony Detection Tool." *mBio* 5: e01019–23.

Writing Style in the Introduction Section

As we covered in Chapter 1, writers in SMC should strive for clear writing that serves as an information bridge or conduit between the writer and the readers (Miller 1979). Scientific style follows conventions of nuanced grammatical style that, notably for you as a writer, shifts within sections, paragraphs, and even sentences. In the Introduction, different styles are used for each of the three moves (establishing the research territory, identifying a gap in the existing literature, and stating the objective of the present research). Here we will cover scientific style conventions relevant for Introductions, which include (1) verb tense, (2) flow, (3) passive and active voice, (4) technical terms and jargon, and

TABLE 5.3 Common grammatical errors to avoid in scientific writing.

Name of Error	Description	Example of Error	Corrected Example
Anthropomorphism	Giving human qualities to nonhumans	Antibodies *want* to bind specifically to the matched antigens.	Antibodies bind specifically to the matched antigens.
Expletive at the beginning of a sentence (wordiness)	Starting with "there is" or "it is"	*There are* many studies focusing on autoantibodies in SLE.	Many studies focus on autoantibodies in SLE.
Incorrect word: affect/effect	Affect (most commonly used as a verb) vs. effect (most commonly used as a noun)	Competing antigen has a great *affect* on the binding.	Competing antigen has a great *effect* on the binding.
Incorrect clause: that/which	That (adds information that is essential to the meaning) vs. which (adds additional information and should be preceded by a comma)	The antibodies to SSA, *which* were stored at room temperature, failed to react with the sera.	*Intended meaning*: Only those antibodies to SSA *that* were stored at room temperature failed to react/with the sera.
Nonparallel construction (confusing construction)	Failure to use grammatically equal words, phrases, or clauses	The physician's goals include *managing* the pain and *prediction* of flares.	The physician's goals include *managing* the pain and *predicting* flares.
Nominalization (wordiness)	Unnecessarily using the noun form of a verb	Purified SSA *exhibited competition* with the sera for binding.	Purified SSA *competed* with the sera for binding.

Name of Error	Description	Example of Error	Corrected Example
Redundant categories (wordiness)	Attaching a category to a word that has its category implied	The sample was yellow *in color*.	The sample was yellow.
Redundant pairs (wordiness)	Using redundant words when one word is sufficient	The antibody was added to *each individual vial*.	The antibody was added to each vial.
Unnecessary clauses	Adding that, to, or which (wordiness)	Methods *that* are quantitative are preferred.	Quantitative methods are preferred.
Weak pronoun reference (confusing construction)	Use of "it," "this," "that," "these," "those," "they," and "which" without clear reference to a noun	Calf thymus is a rich source of SSA, SSB, and other antigens. *It* is homogenized in buffer to extract proteins.	Calf thymus is a rich source of SSA, SSB, and other antigens. Calf thymus is homogenized in buffer to extract proteins.

Adapted from The Cain Project in Engineering and Professional Communication. 2008. *Common Errors to Avoid in Scientific Writing.* OpenStax CNX. Accessed April 23, 2008, http://cnx.org/contents/4f8a53b4-8f24-491a-bef5-c16cc33f5b63@1@1. Download for free at http://cnx.org/contents/4f8a53b4-8f24-491a-bef5-c16cc33f5b63@1@1.

(5) scientific nomenclature and abbreviations. In addition, Table 5.3 summarizes the common grammatical errors in scientific writing.

Verb Tense

As a writer of SMC, you must carefully attend to verb tense because tense is used at two levels in formal scientific writing:

1. To indicate the status of a knowledge statement or claim, or
2. To indicate when an action occurred (past, present, or future).

Many contemporary sources provide the simple explanation that knowledge is current, so present tense is appropriate when referring to general claims or principles as shown in Table 5.4a (Cargill and O'Connor 2013; Cetin and Hackam 2005; Fahy 2008; *Nature* 2014). But an exception to this rule is to use past tense anytime that you reference actions that have occurred in the past. Typically this occurs when:

1. Attributing research findings to other researchers (see Table 5.4b) or
2. Referring to findings from your current research article (see Table 5.4c).

In practice, present tense is used in most of the Introduction section since authors are primarily presenting general research claims (current knowledge or understanding) from previously published journal articles (Iverson et al. 2007; Nwogu 1997; Swales 1990). However when following these guidelines, you will

TABLE 5.4 Summary of tense usage for presenting research findings.

Context	Tense	Example
(a) **Citing knowledge statements (or claims) that were reported in the primary literature (and considered the current knowledge or understanding)**	Present (or present perfect[1])	*Salmonella* organisms *infect* an estimated 1.4 million people annually in the United States, resulting in ~16,000 hospitalizations and > 500 deaths [1].
(b) **Attributing research findings to researchers who conducted specific studies**	Past (or past perfect)	*Mead et al. [1] estimated* that ~1.4 million nontyphoidal *Salmonella* cases occur in the United States each year.
(c) **Presenting findings (and methods/actions) from the current study being reported**	Past (or past perfect)	*We analyzed* foodborne disease outbreak data to determine associations between food commodities and serotypes to help inform future outbreak investigations, foodborne illness source attribution analyses, and control measures.

[1] Present perfect is used as a substitute for present tense when a previous action did not occur at a specific time point (such as "several researchers *have used* this method") or to denote an action that began in the past but continues to the present (American Psychological Association 2010).

TABLE 5.5 Typical verb tense in each section of journal articles.

Section	Subsection	Verb Tense	Examples
Introduction	Move 1	Mostly present	See Example 5.1
	Move 2	Often past	See Table 5.1
	Move 3	Past	See Table 5.2
Methods	–	Past	See Chapter 6
Results	–	Past	See Chapter 7
Discussion	–	Mostly present	See Chapter 8

see that tense may shift within paragraphs and even within sentences (Iverson et al. 2007). Additionally verb tense varies by section of IMRAD journal article (see Table 5.5).

Flow (or Continuity of Information)

In their popular article on scientific writing, Gopen and Swan (1990) recommend structuring scientific prose predictably and consistently. They recommend that authors place old information, or information that was already presented and links backwards to previous statements in the prose, at the beginning of each sentence (the topic position). Then authors should place new information at the end of the sentence, which is the stress position (see Example 5.11). The information at the end of one sentence should flow directly into the next sentence, using linking words or phrases as appropriate (see Table 5.6). Additionally, you should keep key terms consistent. By consistently following this structure, you will show the connection between information to readers so that they can focus on the content without having to decipher the information from a more complicated and inconsistent structure. Granted, this structure results in the passive voice, which we will address next.

EXAMPLE 5.11 RECOMMENDED FLOW OF SCIENTIFIC PROSE.

In this example, "old information" (single underline), which was previously discussed, appears at the beginning of the next sentence (the topic position). "New information" (double underline) appears at the end of the sentence (the "stress" position). (In this example, experts know that "IgE-mediated hypersensitivity reactions" are allergic reactions, so this sentence beginning connects the second sentence to the first.)

Typically, the onset of peanut allergy is in early childhood; 70% of reactions occur during the first known exposure.[5] These IgE-mediated hypersensitivity reactions require prior allergen exposure and sensitization, implying that prior exposure to peanut had already occurred in utero or through unknown exposures in the diet or environment, such as through skin or respiratory routes.[6]

Source: Frazier, A. Lindsay, Carlos A. Camargo Jr., Susan Malspeis, Walter C. Willett, and Michael C. Young. 2013. "Prospective Study of Peripregnancy Consumption of Peanuts or Tree Nuts by Mothers and the Risk of Peanut or Tree Nut Allergy in Their Offspring." *JAMA Pediatrics* 168: 156–62.

TABLE 5.6 Common linking words and phrases to connect sentences.

	Subordinators (Introduce a dependent clause that must be joined to a complete sentence)	**Sentence Connectors (Introduce a complete sentence)**	**Phrase Linkers (Introduce a noun phrase)**
Addition		Furthermore In addition Moreover	In addition to
Adversativity	Although Even though	However Nonetheless	Despite In spite of
Cause and Effect	Because Since	Therefore As a result Consequently Hence Thus*	Because of Due to As a result of
Clarification		In other words That is (i.e.)	
Contrast	While Whereas	In contrast However On the other hand Conversely	Unlike
Illustration		For example For instance	
Intensification		On the contrary As a matter of fact In fact	

* "Thus" may also be used in nonfinite clauses of result, as in this example: The accounting scandal deepened, thus leading to the firing of many high-ranking officials.

Source: Swales, John M., and Christine B. Feak. 2004. *Academic Writing for Graduate Students: Essential Tasks and Skills.* 2nd ed. Ann Arbor: University of Michigan Press.

Passive and Active Voice

One of the frequently debated conventions of scientific writing is the frequent use of passive voice. Despite some simplistic lists of grammatical "rules" provide sweeping recommendations to avoid passive voice, writers' use of passive voice persists and is the norm in many scientific discourse communities (Gross 1991; Riley 1991; Rodman 1981).

The "rule" to avoid passive voice and use active voice (see Table 5.7) arises, in part, from simplifying general advice on writing without analyzing the context in which those rules apply. For example, the popular book *The Elements of Style* by William Strunk ([1918] 1999) has a section entitled "Use the active voice." The section begins by stating, "The active voice is usually more direct and vigorous than the passive." These "rules" were reinforced by research from the 1960s to 1980s that compared active and passive voice. Yet, the examples and research were based on simple sentences that focused on people and actions, such as "John liked the small boy." Indeed, such sentences became complicated and longer when converted into the full passive voice construct, such as "The small boy was liked by John" (Rhodes 1997, 40).

Yet, Strunk ([1918] 1999) and others encourage the use of passive voice when the *object* is more important than the person doing the action (the *agent*). In such cases, as in some sections of scientific writing, the object should be placed in the subject position of the sentence. The use of passive voice enables writers to focus on the object of scientific communication rather than the person (Bazerman 1988; Ding 2002; Wilkinson 1992).

As you become more familiar with SMC, you will find that the use of passive voice varies by section as the type of information and presence of the author changes through shifts in person (see Table 5.8). For our current discussion of the Introduction section, passive voice is the predominant sentence construction

TABLE 5.7 Comparison of the active, full passive, and truncated passive voice. The full passive sentence includes the person who did the action (the agent in italics) added to the end of the sentence, but the truncated passive sentence does not. The number of words in the active voice sentence (five) is equal to the truncated passive voice sentence (five), but the full passive voice sentence has more words (seven) in order to add the agent at the end of the sentence.

		Topic Subject		End Focus
Active Voice		I	calculated	the standard deviation.
Passive Voice	**Full**	The standard deviation	was calculated	*by me.*
Passive Voice	**Truncated** (no agent)	The standard deviation	was	calculated.

Adapted from Rodman, Lilita. 1994. "The Active Voice in Scientific Articles: Frequency and Discourse Functions." *Journal of Technical Writing and Communication* 24 (3): 309–31.

in Move 1 but not for Move 2 or Move 3 (see Example 5.2). In the literature review section (Move 1), writers focus on the object being studied and generally avoid directly referring to the researchers, who are identified through citations (Rodman 1981). As illustrated in Example 5.12, use of the truncated passive voice construction makes the research findings, not the researchers, prominent and maintains the focus on the object of study (Cargill and O'Connor 2013). In SMC, the agents are often easily inferred from the context; that is, the authors of the journal article are the ones who performed the experimental methods and the authors of the cited paper are the ones who proposed a particular finding (Riley 1991; Rodman 1981).

As shown in Table 5.8, passive voice construction is most common in the Methods section (ranging from 43% to 80%) but varies significantly by discipline

EXAMPLE 5.12 ACTIVE VOICE INAPPROPRIATELY PLACES IRRELEVANT INFORMATION IN THE STRESS POSITION OF SENTENCES.

(a) Use of active voice may result in poor use of stress position when researcher name(s) are listed at the beginning of sentences:

Indeed, **Teunis et al.** have clearly demonstrated a dose-response relationship whereby persons challenged with a higher NoV dose have substantially greater illness risk (*16*).

(b) Rather, the recommended use of the beginning of sentences (the stress position) is with a topic that flows from the preceding sentence and enables continuity of information:

One of many concerns with all classic challenge studies is that the virus dose given to volunteers was several thousand-fold greater than the small amount of virus capable of causing human illness (estimated as 18–1,000 virus particles) (*16*). Thus, immunity to a lower challenge dose, similar to what might be encountered in the community, might be more robust and broadly protective than the protection against artificial doses encountered in these volunteer studies. Indeed, ~~Teunis et al. have clearly demonstrated~~ **a dose-response relationship** whereby persons challenged with a higher NoV dose have substantially greater illness risk (*16*).

Source: Simmons, Kirsten, Manoj Gambhir, Juan Leon, and Ben Lopman. 2013. "Duration of Immunity to Norovirus Gastroenteritis." *Emerging Infectious Diseases* 19: 1260–7.

TABLE 5.8 Use of passive voice in journal articles from different fields. The Methods section has the highest percentage of verbs in the passive construction.

Section	Percent of Passive Verb Constructs				
	Botany[1]	Chemistry/ Physics[1]	Medicine[2]	Psychology[1]	AVERAGE
Introduction	25	32	15	18	**23**
Methods	80	67	44	43	**59**
Results	28	42	11	27	**27**
Discussion	21	28	15	15	**20**

Sources:

[1] Rhodes, Susan. "The Active and Passive Voice Are Equally Comprehensible in Scientific Writing." PhD diss., University of Washington, 1997. https://digital.lib.washington.edu/researchworks/handle/1773/9033.

[2] Amdur, Robert J., Jessica Kirwan, and Christopher G. Morris. 2010. "Use of the Passive Voice in Medical Journal Articles." *American Medical Writers Association Journal* 25: 98–104.

(Amdur, Kirwan, and Morris 2010; Rhodes 1997). A lower percentage of passive construction is associated with the sections where a new argument is presented in the research article (Riley 1991). In the Discussion section, in particular, scientific writers use active voice and personal pronouns when presenting their argument (Bernhardt 1985). Most important, you should follow the style of the discourse community and try not to impose a different style based on how you think scientists should write or how nontechnical writing is composed (Ding 2002; Riley 1991; Wilkinson 1992). You must reorganize that the audience of journal articles is used to having information presented in a particular writing style. Thus, the use of passive voice, when used appropriately, does not result in lower reading comprehension, longer sentences, or longer reading times than the active voice (Rhodes 1997; Rodman 1994).

In review, passive voice should not necessarily be avoided *per se* although, you should be cautious of certain grammatical problems that commonly arise from using passive constructs. In particular, you should avoid nominalizations (using noun forms of verbs), dangling modifiers, anthropomorphisms (attributing human characteristics to objects or animals), and expletives (such as "it is" or "there is") at the beginning of sentences (see Table 5.3).

Technical Terms and Jargon

TECHNICAL TERMS—such as "lymphocytes", "IgM", "antigens", and "antibodies"—are specialized terms with specific meanings and are pervasive in SMC. Such technical terms are usually more specific and meaningful than nontechnical terms. Therefore use of technical terms is necessary for precise and effective communication among subject-matter experts within a discourse community (Wilkinson 1992). Yet, the use of technical terms is not appropriate for every context. Appropriate use of technical terms depends on the audience. Technical terms are considered JARGON, a pejorative label, when the terms are misused with an audience that is unfamiliar with the definitions of those terms (Jones 1998). For example, the technical terms I just listed would be appropriately used in discourse among immunologists, but some of the same terms may be considered jargon if used in the discourse of botanists.

For novice writers, technical terminology is a challenge. You should use technical terms that are accepted within the discourse community, which you can learn by reading sample journal articles of that discourse community. As a general guideline, you should practice looking up the definition of unfamiliar terms to learn the specialized vocabulary of your discourse community. Not only will you begin to understand the material, but doing so will enable you to recognize the different words that are nearly identical, such as "ilium" and "ileum." In this example, the "ilium" is the largest bone of your pelvis whereas the "ileum" is the last part of your small intestine. MedlinePlus (http://www.nlm.nih.gov/medlineplus), which is managed by the US National Library of Medicine and the US National Institutes of Health, is a particularly useful database of scientific and medical information that includes a link to *Merriam-Webster's Medical Dictionary*.

A related problem with technical terms arises when converting between singular and plural forms of Latin or Greek words, which follows different rules from that of English words. For example, although you may be writing about one bone or many bones, more than one vertebra are vertebrae. See Table 5.9 for rules on how to convert between singular and plural forms of scientific terms.

Scientific Nomenclature, Abbreviations, and Acronyms

The nomenclature and abbreviations in scientific prose can be overwhelming for those new to SMC. To illustrate the scope of this specialized nomenclature, the *AMA Manual of Style* (10th ed.) dedicates nearly 350 pages to scientific and medical terminology and abbreviations. But you do not need to learn to use all of those terms. Style manuals archive the terms necessary for communication within the field, which you can look up when you encounter one. Generally, you will become familiar with the common terms and relevant styles as you write within a specific discourse community.

Additionally scientific nomenclature usually has an underlying logic, which is explained in many style guides such as the *AMA Manual of Style* (10th ed.) and *Scientific Style and Format* (7th ed.). (As a reminder, the style guide for a particular discourse community is listed in each journal's Instructions to Authors.) For example, scientific names of organisms are italicized and are composed of two words (BINOMIAL) that use Latinized grammatical forms. The first word (the

TABLE 5.9 Basic rules for changing scientific terms from singular to plural form.

If the word ends in	Do this first	Then add	Example
-a		Add *-e*	*alga* becomes *algae*
-ax	Drop *-ax*	Add *-aces*	*thorax* becomes *thoraces*
-ex or -ix	Drop *-ex* or *-ix*	Add *-ices*	*cortex* becomes *cortices*
-ma		Add *-ta*	*stoma* becomes *stomata*
-is	Drop *-is*	Add *-es*	*diagnosis* becomes *diagnoses*
-nx	Change *-x* to *-g*	Add *-es*	*larynx* becomes *larynges*
-on	Drop *-on*	Add *-ia*	*mitochondrion* becomes *mitochondria*
-us	Drop *-us*	Add *-i*	*nucleus* becomes *nuclei*
-um	Drop *-um*	Add *-a*	*bacterium* becomes *bacteria*
-y	Drop *-y*	Add *-ies*	*biopsy* becomes *biopsies*

Source: Garrett and Penn (2008).

genus) is capitalized and the second word (the SPECIFIC EPITHET) is written in lowercase (e.g., *Streptococcus pneumoniae*). By convention, the complete name of a specific organism is written out the first time in the body text, but the genus is abbreviated for subsequent use (e.g., *S. pneumoniae*). Yet the use of scientific names is even more complicated. Multiple species in the same genus may be referred to by the genus name (e.g., *Salmonella* as in Example 5.2), or more specific subspecies of an organism may have something added to the name (e.g., "Dublin"). At the subspecies level, the nomenclature may be quite different and is rarely italicized. For example, a medically important subspecies of *E. coli* is designated as O157:H7 (but not in italics). The point is to know exactly what each term communicates and how to properly format the name. Finally, if a VERNACULAR (LAYMAN'S) TERM is also used interchangeably with the formal scientific name, you should select the most common term that is used within the discourse community and include parenthetical mention of the alternative term the first time that term is introduced in the body text (for example, *Ambystoma macrodactylum*, or long-toed salamanders).

Diseases, diagnostic tests, and scientific methods are often named after the person who made the discovery or the location where a discovery was realized. These terms, called EPONYMS, are often used as adjectives, and the current trend is to use them in a nonpossessive form to distinguish from actual possessives (Council of Science Editors 2006; Iverson et al. 2007). For example, the National Down Syndrome Society recommends using "Down syndrome" rather than "Down's syndrome" (American Psychological Association 2010). Generally the first letter of the term is capitalized, although derivative forms are not (e.g., "Gram stain" is capitalized but "gram-negative bacteria" is not). Occasionally you will encounter exceptions to these rules, so always refer to the Instructions to Authors for the corresponding style guide. You may also check usage of the term in recent journal articles published in the target journal. Recognize that formal usage in SMC frequently differs from less stringent use on organization websites and other popular media.

Authors of SMC must be meticulous when using and formatting scientific names, terms, eponyms, and abbreviations. Proper formatting requires precise use of capitalization, punctuation, typeface (italics and standard case), spacing, subscripts and superscripts, and may include the use of non-English characters (typically Greek or Latin letters). Subtle differences in style can have important differences in meaning. For example, the names of bacteria genes are generally written in italics with three lowercase letters and a capital letter (e.g., *rpoS*) whereas the translation of that gene into a protein, which is a completely different molecule, is written in standard case with the first letter capitalized (e.g., RpoS). The difference between writing *rpoS* and RpoS distinguishes a gene from a protein. Other terms may require careful formatting and the occasional use of the Greek alphabet (e.g., β_1 receptor). Refer to Appendix B for the Greek alphabet and instructions for inserting these letters (and other symbols) in Microsoft Word.

You should know how to find the correct formatting for nomenclature and terms. The first step should always be to read the relevant information about nomenclature and terms in the Instructions to Authors and the specified style guide. Often you do not need to read the entire content of the style guide but just the relevant passages. For example, if you were writing one of the articles on *Salmonella* from earlier in this chapter, you simply need to read the passage on the nomenclature of *Salmonella* in the *AMA Manual of Style* (10th ed.). You could find this information from the index under "nomenclature" or, more specifically, "*Salmonella* nomenclature." Additionally, you should check your use of nomenclature and terms against published articles from the target journal.

If you are going to be communicating scientific and medical terms, you need to make sure that you precisely understand any nuances between slightly different forms of the nomenclature. When you receive a new manuscript, you should begin by listing all of the unfamiliar terms (copying exactly how each term appears in the text). Note any slight deviation between similar nomenclature (such as capitalization, italics, special characters, and spacing) and note the relevant context (such as first usage or subsequent usage, beginning or middle of a sentence, or in a table or figure). Next, you should define each item so that you can read the manuscript and fully understand the material. This may mean that you have to research the meaning of specific terms, which is relatively easy with the internet. Finally, you should organize the unfamiliar terms into a hierarchy. Some terms may be general categories whereas others are specific names. Occasionally, scientific style may conflict with other writing conventions. In general, scientific style supersedes other conventions. For example, a sentence that begins with "*rpoS*" (the bacterial gene) would begin without a capital letter.

Finally, you should use abbreviations and acronyms cautiously. You can find lists of accepted abbreviations and acronyms in style guides and any journal-specific requirements for abbreviations in the Instructions to Authors. ACRONYMS are formed from the first letter of a longer term and pronounced as a word (such as NASA). When you use abbreviations, be sure to read your prose with the full term so that you use each abbreviation properly. For example, stating the full name for AIDS and HIV will help writers use each appropriately: AIDS (Acquired Immune Deficiency Syndrome) is the human disease caused by HIV (human immunodeficiency virus). Additionally, writing "HIV virus" is incorrect because the word "virus" is duplicated (which is basically saying, "human immunodeficiency virus virus").

Generally writers should limit abbreviations to those that are familiar to the discourse community, which you can identify by reading multiple recently published papers on the topic. You should resist the urge to introduce or create new abbreviations in your manuscripts because scientific and medical nomenclature is already sufficiently complex without having authors introduce abbreviations for their favorite phrases—consider how difficult that becomes for the reader (particularly nonnative English speakers). But Peter Thrower (2006), former editor in chief of the journal *Carbon*, noted that some authors might introduce abbreviations when trying to reduce the total word count of an article. Although I doubt that creating abbreviations is the most effective method to reduce word count since creating abbreviations does not save notable space in an article, I do think that the guidelines for abbreviation that Thrower presents are useful for writers who are inclined to create abbreviations. An adapted version of these guidelines for abbreviation are as follows:

1. When using an abbreviation, write the complete term in the initial use within the body text and then follow the term with the abbreviation in parentheses. For example: "X-ray diffraction (XRD) examination showed."
2. Only use an abbreviation in the text if you are going to use it again later. You must use the abbreviation at least one additional time in the manuscript. Again, however, an abbreviation should be used many times to make it worthwhile.
3. Never define an abbreviation in the manuscript Title.
4. Do not use abbreviations you created in Titles.

5. Be judicious when use abbreviations in Abstracts. If you use an abbreviation you created in an Abstract, you will need to follow step 1 in both the Abstract and again in the body text of the article. To make the abbreviation worthwhile in the Abstract, it should be used many times in that section.

Concluding Points

In this chapter, we have covered the content, organization, and style of the Introduction section. As an author, you may find the process of writing the Introduction section easier if you consider each of the three moves as individual sub sections. Often writers find it more efficient to write the Introduction in reverse order—beginning with Move 3 (stating the objective of the present research) and then going to Move 2 (stating the justification of) and finally to Move 1 (establishing a research territory). Writing the Introduction section in this order helps you to ensure that you include the relevant information in the literature summary and avoid the most common problems that arise when writing the first draft of an Introduction: excessive length, meandering prose, and an unclear research objective. Certainly you will need to check and possibly revise the content in Moves 2 and 3 for a perfect fit after you finish Move 1.

As you write the Introduction, ensure that you are properly citing the original source of information. I usually develop the list of references at the same time as I write Introductions since I have the articles readily available. Often the literature-search process (see Chapter 3) becomes intertwined with the writing process and you will find yourself researching specific topics as necessary. And you will find the overall process more efficient if you use the referencing style of the target journal when you first start writing. Alternatively, you may decide to use citation management software to manage the references and apply the appropriate citation style, which I also discussed in Chapter 3. Again, using the citation management software when you first begin writing is most efficient.

Finally, after writing the content and getting it properly organized, you should focus on the grammar and style—attending to tense, flow, voice, terminology, and nomenclature. Your first time writing an Introduction will be time-consuming and laborious, but the entire process becomes faster and easier as you become more familiar with the genre and specific discourse communities. The best thing to do is simply to get started with the easiest part—Move 3.

6

THE METHODS SECTION

In this chapter, you will learn to:

- Determine the appropriate content for the Methods section
- Organize the information presented in the Methods section
- Write the Methods section in appropriate scientific style and edit style problems that commonly arise in the Methods section

Overview

The content of the Methods section, as you might infer from its title, describes the experimental procedures used in the current research manuscript. This section is titled differently in various journals, such as "Methods," "Materials and Methods," or "Procedures." For consistency, we will use the term "Methods" to identify this section. As described by Peter Thrower (2008b), former editor in chief of the journal *Carbon*, the purpose of the Methods section is twofold:

1. To enable readers to repeat part, or all, of the procedure in order to either check dubious results or prepare materials for further investigation; and
2. To convince readers that the research has been designed systematically and performed thoroughly using an approach that is appropriate for addressing the research question or objectives.

In particular, peer reviewers, who are the gatekeepers of publishing, critically examine the research methods to determine if the approach is appropriate for the research question and if the data were collected systematically and analyzed

TABLE 6.1 Survey responses from journal editors, peer reviewers, and Nobel Laureates ($n = 29$) indicate that the Methods section contains the most flaws and is most often the section responsible for outright rejection of a manuscript.

Which section usually contains the most flaws?	
Methods	16 (55%)
Discussion	7 (24%)
Results	6 (21%)
Introduction	0 (0%)
Which section is most often responsible for outright rejection?	
Methods	15 (52%)
Results	8 (28%)
Discussion	6 (21%)
Introduction	0 (0%)

Adapted from Byrne (2000).

appropriately (Spigt and Arts 2010; Van Rooyen, Black, and Godlee 1999; Wolff 1970).

In a survey of journal editors, peer reviewers, and Nobel Laureates, researchers found that the Methods section of medical studies usually contains the most errors and is the section most often responsible for outright rejection of a manuscript (see Table 6.1; Byrne 2000). Thus the Methods section serves both functionally and rhetorically (Gross 1985; Huth 1982; Nair 2005). Due to this dual nature of the section, you need to include both precise information summarizing the experimental approach used to collect and analyze the data and metastatements explaining the rationale for the general approach of the investigation.

What to Expect

The content of the Methods section is not necessarily presented chronologically in regard to when the procedures were actually conducted but rather is organized and presented to establish a rational narrative in which the outcome (data) from one experimental subsection leads logically to the next (Chiswick 2002). Although you may write the Methods section before writing the Results (and, in practice, many researchers begin with this section because they find it easiest to summarize the procedures that they just completed), you should consider the organization of the Methods tentative until you finalize the organization of the

Results section (which we will cover in Chapter 7). Ultimately you will want the procedures in the Methods section to be presented in the same order that the data are presented in the Results section to establish a consistent and parallel organization throughout the journal article. Below, we will examine the content of each subsection of the Methods section.

A complete Methods section contains sufficient information to enable the replication of an entire research experiment. Yet in contrast to a set of instructions, the Methods section is usually not a one-source document for the entire procedure. Rather, the Methods section in journal articles is a brief summary of the experimental methods used in the research. Novice writers often create Methods sections that are far too detailed for some steps, yet they omit other, highly important details. These initial drafts are far too long and need to be carefully edited to meet the information needs of the audience. The typical published Methods section ranges in length from 500 to 1,500 words depending on the novelty of the research approach because new methods must be described in detail whereas standard methods simply need to be mentioned (Brand and Huiskes 2001). This chapter examines the organization, content, and style of the Methods section so that you are able to write this section effectively.

How to Prepare

Generally speaking, an experienced scientist who is fairly familiar with the scientific literature within the discourse community tends to write the Methods section properly. Determining the appropriate amount of information can be difficult for novice writers and professional communicators who are not as familiar with particular experimental procedures. But you can use published journal articles to determine the appropriate level of detail for each experimental procedure (even if you do not know what the particular details mean). As you will see in this chapter, using published examples from a target journal is a useful tool for novice writers.

To write the Methods section, you need to know (1) how the participants (or subjects) were selected, (2) the research procedures that were performed (along with the relevant conditions or parameters of each procedure), and (3) how the data were analyzed. If you did not conduct the research yourself, then the investigator performing the experiment should have compiled and summarized this information in a LABORATORY NOTEBOOK, which is the detailed record of the exact actions performed. You should also collect recent journal articles from your target journal to use as models for writing each subsection of your Methods section. In this chapter we will examine the Methods section issues conceptually, but to write the Methods at an appropriate level of detail for the audience of your target journal, you should refer to recently published articles using similar methods.

Organization and Content of the Methods Section

A general overview of the research approach typically precedes the Methods section—either at the end of the Introduction (see Chapter 5) or at the beginning of the Methods section in some discourse communities. Refer to Appendix C for a glossary of terms used to describe studies (particularly in the medical field). Following this overview, the Methods section is usually subdivided into the following parts: (1) selection of participants or subjects, (2) investigation procedure, and (3) data analysis (International Committee of Medical Journal Editors 2014; Kanoksilapatham 2005; Nwogu 1997; Pique-Angordans and Posteguillo 2006; Skelton 1994). We will examine each of these parts below.

Part 1: Selection of Participants or Subjects

By convention, the first subsection of the Methods usually begins by describing the selection of participants or subjects. The term "PARTICIPANTS" usually refers to humans who have autonomy and freely agreed to be involved in the research whereas the term "SUBJECTS" is reserved for animals, plants, bacteria, viruses, and inanimate objects that were exposed to the experimental procedure by the researchers. The subsections for human participants are different from those for nonhuman subjects. Each will be explained below.

Studies with Human Participants

In studies with human participants, the first major subsection is the criteria for selection (or exclusion) of participants (International Committee of Medical Journal Editors 2014; Masic 2011). As show in Example 6.1, the Participants subsection should include the specific details unique to the study, such as the exact source and number of participants. These details are in contrast to the relative lack of description of common procedures. For example, the process of random selection, as a common research method, was not detailed since trained researchers should be able to replicate that part of the procedure.

Example 6.2 provides a more detailed Participants subsection in a medical study. In this example, you will note that the authors provide the precise parameters of individuals who met the study selection criteria as well as detailed criteria for exclusion. Whereas this particular medical study required extensive selection and exclusion criteria, the survey research in Example 6.1 provides less selection criteria.

After providing the criteria for participation, you should provide the details of the recruitment procedure. Often you will identify the time frame, process of selection, number of individuals who were contacted, and the number of individuals who participated. Finally you should identify the demographic variables relevant to the research study. The demographic variables included in the Methods

EXAMPLE 6.1 CONTENT OF THE PARTICIPANTS SUBSECTION FOR A SURVEY.

Specific details unique to the study include the exact source and number of participants. This detail is in contrast to the relative lack of description of common procedures (such as the random selection of participants).

Prescribing "Placebo Treatments": Results of National Survey of US Internists and Rheumatologists

To further inform ethical discussions about the appropriateness of recommending placebo treatments, we examined the attitudes and behaviours regarding placebo treatments among a national sample of clinically active internists and rheumatologists in the US.

End of the Introduction section typically provides an overview of the research approach

METHODS

Study Population

First subsection of Methods explains the selection of participants or subjects of the research

Using the 2006 American Medical Association masterfile, we randomly selected 1200 physicians listed with the primary specialties of internal medicine (600) or rheumatology (600): a group of physicians who commonly treat patients with debilitating chronic clinical conditions that are notoriously difficult to manage. In June 2007, an independent survey research firm posted a confidential, self administered survey, a $20 (£11, ∊15) incentive, and a letter outlining the voluntary nature of participation. Participants were assured that their identities would not be disclosed to investigators. Those who did not respond to the first survey were sent a second six weeks later. Of the 1200 physicians who were sent questionnaires, 679 responded (overall response rate 57%), of whom 334 specialised in internal medicine (56% response rate) and 345 in rheumatology (58% response rate).

Context and selection criteria of participants

Common research methods do not require citation nor description

Describe unique steps of the procedure

Total participants and relevant demographic variables

Source: Tilburt, Jon C., Ezekiel J. Emanuel, Ted J. Kaptchuk, Farr A. Curlin, and Franklin G. Miller. 2008. "Prescribing 'Placebo Treatments': Results of National Survey of US Internists and Rheumatologists." *BMJ* 337: 1097–100.

EXAMPLE 6.2 DETAILED CONTENT OF THE PARTICIPANTS SUBSECTION FOR A MEDICAL STUDY.

Low-Dose Hydrochlorothiazide and Preservation of Bone Mineral Density in Older Adults: A Randomized, Double-Blind, Placebo-Controlled Trial

METHODS

Participants — Subheading

Our study was conducted at Group Health Cooperative of Puget Sound, a large staff-model health maintenance organization in Seattle, Washington, serving 370 000 people. [Context of study] The target population for the trial was healthy men and women 60 to 79 years of age who were normotensive, were free of serious heart disease, were not taking hormone replacement therapy or bisphosphonates, and had baseline bone mineral density at the total hip that was within two standard deviations of the normal value for their age (Z-score ± 2). [Criteria for selection] We excluded potential participants if they 1) had contraindications to thiazide therapy (allergy, proteinuria, abnormal serum creatinine concentration, hyponatremia, hypokalemia, gout, or low-density lipoprotein cholesterol level > 4.91 nmol/L [190 mg/dL]), 2) had conditions that would complicate thiazide treatment (alcohol abuse, serious coronary artery disease, electrocardiographic evidence of moderate ischemia or arrhythmia, use of any diuretics or antihypertensive medications, hypotension, or congestive heart failure), 3) had conditions and diseases known to influence bone loss (immobility, weight > 135 kg, hepatic disease, malabsorptive conditions, hypercalcemia, metabolic bone disease, or use of corticosreroids or antineoplastic drugs), or 4) had conditions and diseases that made completion of the trial unlikely (malignant cancer or other life-threatening disease, dementia, or anticipated change in residence). [Criteria for exclusion]

Age-eligible participants were recruited through direct mailings and a local media campaign from February 1993 to September 1994. Those who expressed interest were prescreened by telephone (n = 3520), and potentially eligible participants were invited to a screening visit (n = 865). [Recruitment procedure and time frame] Of these, 320 (205 women, 115 men) were enrolled, including 304 white persons (95%), 9 Asian-American persons (2.8%), 3 Hispanic-American persons (0.9%), 2 African-American persons (0.6%), and 2 persons whose ethnicity was nor specified (0.6%). [Total participants enrolled in the study and relevant demographic variables] The Human Subjects Review Committees at Group Health Cooperative and the University of Washington, Seattle, Washington, approved the study, and each participant provided written informed consent. [Mention approval of research methods for human subjects]

Source: LaCroix, Andrea Z., Susan M. Ott, Laura Ichikawa, Delia Scholes, and William E. Barlow. 2000. "Low-Dose Hydrochlorothiazide and Preservation of Bone Mineral Density in Older Adults: A Randomized, Double-Blind, Placebo-Controlled Trial." *Annals of Intern Medicine* 133: 516–26.

section are often those highly relevant to the research question, such as the medical specialization of participants in Example 6.1. A more complete summary of participant demographics is often provided at the beginning of the Results section (see Chapter 7).

Typically the Participants subsection concludes by explicitly stating that the research protocol was reviewed and approved by the human-subjects committee at the research institution. As in Example 6.1, you should mention any incentive that was provided to participants.

Studies with Nonhuman Subjects

The Subjects subsection is less structured than the Participants subsection and covers a much broader range of research. Each discourse community has different conventions for summarizing the animals, plants, bacteria, viruses, or inanimate objects being studied. This subsection is often labeled as the particular category of subjects (such as "Animals" or "Bacterial Strains"). When describing nonhuman subjects, you should include the following information:

- Identify the exact species, subspecies, genetic strains, or materials that were the subjects studied (for species, use the scientific name);
- State the relevant features or characteristics of each subject along with any specific genetic manipulation or selection criteria;
- Include a description of the growth, maintenance, and care conditions of subjects; and
- Identify the origin or source of each subject that was obtained from another researcher or organization.

Finally, if animals are used in the research, you should indicate that the institution's animal care committee approved the research (see Example 6.3).

Part 2: Investigation Procedure

Part 2 of the Methods section, which covers the investigation procedure, is often divided into multiple subsections detailing each stage of the research investigation. Unfortunately, we do not have a single, universal structure within this subsection because each research investigation is, by definition, unique. Further subdivision is common in exploratory research studies because such research

EXAMPLE 6.3 CONTENT OF AN ANIMAL SUBJECT SUBSECTION.

Influenza A (H7N9) Virus Transmission Between Finches and Poultry

METHODS

Animals

Study birds were of both sexes and included 3- to 6-month-old society finches (*L. striata domestica*) (Birds Express, South El Monte, CA, USA); 5-week-old white leghorn hens (*Gallus gallus domesticus*) (McMurray Hatchery, Webster City, IA, USA); and 16-week-old bobwhite quail (*Colinus virginianus*) (B&D Game Farm, Harrah, OK, USA).	Summary of animal subjects and relevant characteristics (including source in parentheses)
The birds were quarantined for 1 week, and prechallenge swab samples were confirmed influenza virus-negative by egg isolation. Food and water were provided ad libitum.	Relevant study conditions and care of subjects
Animal experiments were approved by the St. Jude Children's Research Hospital Animal Care and Use Committee and complied with all applicable US regulations.	Mention approval of research methods for animal subjects

Source: Jones, Jeremy C., Stephanie Sonnberg, Richard J. Webby, and Robert G. Webster. 2015. "Influenza A (H7N9) Virus Transmission Between Finches and Poultry." *Emerging Infectious Diseases* 21 (4): 619–28.

employs several different research procedures that are pieced together to investigate a phenomenon. In contrast, experimental (or hypothesis-) testing research typically employs a single procedure or targeted approach to answer the research question, so such investigations may not necessarily subdivide the study design section but rather label it with a descriptive title.

The stages of this subsection are rarely presented chronologically, but rather they are presented so that the order of each data set forms a logical narrative in the Results section. In the Methods section, the narrative tells the reason for each different investigation and how the data set from each investigation appropriately address the research purpose. In my own experience, constructing this narrative meant that the order of experiments was reorganized to show progress from the research question to the proposed conclusion. I had to group related experiments together—regardless of when they were actually performed—and even exclude presenting tangential lines of investigation that I attempted in the laboratory but that did not provide any useful insight into the research question. Actually, experiments are also commonly done after the first draft of the scientific paper is written because additional data are required to justify an important point. (In fact, scientists not only do this on their own, but reviewers and journal editors may ask that certain additional experiments be performed and integrated into the research.) In any case, these additional experiments are not necessarily added to the end of the narrative but where the information best fits into the "story."

The procedural content of the Methods section is more similar to a cooking recipe than an instructions manual (Day and Gastel 2006; Harmon and Gross 1996; Katz 2009; Thrower 2008b; Zeiger 2000), in which you should identify the ingredients and the processes or conditions and expect that other researchers in the same field that are familiar with the standard techniques of the discipline could reconstruct or replicate each part of the experiment (Day and Gastel 2006; Zeiger 2000).

As summarized in Table 6.2, the amount of technical detail that you include for a procedure depends on the novelty of the approach within the discourse community (Kanoksilapatham 2005; Nair 2005; Penrose and Katz 2010). For example, standard

"I think you should be more explicit here in step two."

procedures only need the relevant settings or details to replicate the exact conditions (see Example 6.4). Similarly, procedures detailed previously or kits purchased from companies and used according to the instructions simply need to be referenced (see Example 6.5 and 6.6). In contrast, a substantially modified or new procedure should describe the relevant materials and conditions (see Example 6.7). The American Psychological Association (2010) provides the following instructions:

> Include in these subsections only the information essential to comprehend and replicate the study. Insufficient detail leaves the reader with questions; too much detail burdens the reader with irrelevant information. Consider using appendices and/or a supplemental website for more detailed information. *(29)*

TABLE 6.2 Categories of procedures and the technical content required in the Methods section.

Category[1]	Content Required
Standard procedure	Identify procedure by conventional name and include relevant settings or details of the exact conditions. (See Example 6.4)
Procedure detailed in previous study or using a commercial product	Identify procedure by name and reference the journal article or source that details the method. If the procedure is not obvious from the name, you might include a brief overview so that the readers will understand the general approach without having to access another source. You should include any specific populations or conditions that were unique to your application of the procedure. (See Examples 6.5 and 6.6)
Procedures detailed in previous study with modifications	Include all of the information of the previous category "procedure detailed in previous study" but clearly indicate the changes made to the previous procedure. Often authors will reference the previous study and then follow with a statement of the modification. (See Example 6.7)
Substantially modified or new procedures	Describe the procedure in detail. (See Example 6.7)

[1] Categories adapted from Penrose and Katz (2010).

EXAMPLE 6.4 STANDARD PROCEDURE.

Molecular Cloning and Characterization of a Novel Carboxylesterase-Like Protein That is Physiologically Present at High Concentrations in the Urine of Domestic Cats (*Felis catus*)

Analyses of Urinary Proteins — Descriptive Subheading

A sample of urine was taken by direct manual pressure of the cat bladder. The urine samples were sedimented by centrifugation for 5min at 500*g*, to discard occasional cell debris, and were stored at −20°C for later analysis. The urinary protein concentrations were measured by the Bradford method using BSA as the standard. Aliquots of urine (20µl) were resolved by SDS/PAGE under non-reducing conditions and stained with Coomassie Brilliant Blue R-250.

Standard procedure

Specific conditions of the procedure

Rationale for procedure

Adapted from Miyazaki, Masao, Katsuyoshi Kamiie, Satoshi Soeta, Hideharu Taira, and Tetsuro Yamashita. 2003. "Molecular Cloning and Characterization of a Novel Carboxylesterase-Like Protein That is Physiologically Present at High Concentrations in the Urine of Domestic Cats (*Felis catus*)." *Biochemical Journal* 370: 101–10.

EXAMPLE 6.5 PROCEDURE DETAILED IN PREVIOUS STUDY.

dksA Is Required for Intercellular Spread of *Shigella flexneri* via an RpoS-Independent Mechanism

Construction of Bacterial Mutants — Descriptive Subheading

The *dksA* mutation from SA2287 was moved into wild-type SA100 by P1 transduction (28) to generate SA5287.

Procedure described previously (requires citation)

Source: Mogull, Scott A., Laura J. Runyen-Janecky, Mei Hong, and Shelley M. Payne. 2001. "*dksA* Is Required for Intercellular Spread of *Shigella flexneri* via an RpoS-Independent Mechanism." *Infection and immunity* 69: 5742–51.

EXAMPLE 6.6 PROCEDURE USING A COMMERCIAL PRODUCT. THE COMPLETE NAME OF THE PRODUCT (WITHOUT TRADEMARK SYMBOLS) AND THE NAME AND LOCATION OF THE COMPANY ARE INCLUDED.

Sequence Variability and Geographic Distribution of Lassa Virus, Sierra Leone

Nucleic Acid Extraction — Descriptive Subheading

RNA was extracted from 10 mg of the spleen from each rodent using TRIzol Reagent (Life Technologies, Carlsbad, CA, USA) following the manufacturer's recommendations. RNA samples were stored at -80°C.

- RNA was extracted from 10 mg of the spleen from each rodent — Summary of the method
- using TRIzol Reagent (Life Technologies, Carlsbad, CA, USA) — Commercial procedure
- following the manufacturer's recommendations. — Indicate that the protocol provided by the manufacturer was followed (or note any variation)

Adapted from Leski, Tomasz A., Michael G. Stockelman, Lina M. Moses, Matthew Park, David A. Stenger, Rashid Ansumana, Daniel G. Bausch, and Baochuan Lin. 2015. "Sequence Variability and Geographic Distribution of Lassa Virus, Sierra Leone." *Emerging Infectious Diseases* 21 (4): 609–18.

EXAMPLE 6.7 SUBSTANTIALLY MODIFIED OR NEW PROCEDURE.

A Low Gastric pH Mouse Model to Evaluate Live Attenuated Bacterial Vaccines

In Vitro Acid Resistance Assay — Descriptive Subheading

Acid resistance was determined essentially as described previously [39, 40] with the following modifications. Wild-type strains were propagated either under normal inoculation conditions (grown to an optical density (OD_{600}) of 0.9 in LB broth with 0.1% glucose with aeration) or under acid resistance-inducing conditions (in a static anaerobic TSB culture with 0.4% glucose for 17 h). Cultures were normalized to the same OD_{600} value, then pelleted and washed once in EG medium, pH 7.0 containing no growth supplements [41]. Cells were pelleted a second time and resuspended at a density of 1×10^9 CFU/ml in EG medium at pH 3.0 containing 0.1% casamino acids. Because the

- Acid resistance was determined essentially as described previously [39, 40] — Procedure described previously (requires citation)
- with the following modifications. — Specify that the referenced procedure was modified
- Wild-type strains were propagated … — Precise conditions of a novel procedure

wild-type strains use a variety of decarboxylase enzymes to resist low pH, casamino acids were supplied in order to provide all strains with the same combination of amino acids. Low pH challenge was conducted at 37°C and samples were collected immediately after resuspension (t = 0) and hourly for 4 h. Samples were serially diluted and plated onto LB agar to assess viability during challenge.

Source: Brenneman, Karen E., Crystal Willingham, Jacquelyn A. Kilbourne, Roy Curtiss 3rd, and Kenneth L. Roland. 2014. "A Low Gastric pH Mouse Model to Evaluate Live Attenuated Bacterial Vaccines." *PLoS One* 9: e87411.

Part 3: Data Analysis

The Methods section usually concludes with a summary of how the data were analyzed (see Example 6.8). You should also explain why certain data analysis methods were selected if not obvious to the audience. As in the previous subsections, the amount of detail provided depends on the amount of information necessary for other researchers to replicate the work. In most cases, you will not need to explain the calculation of common statistical measures, but you may need to provide the threshold values, such as degree of confidence. Again, the information reported depends on the discourse community and conventions of the target journal.

EXAMPLE 6.8 DATA ANALYSIS SUBSECTION.

Influenza A (H7N9) Virus Transmission Between Finches and Poultry

Statistical Analyses — Descriptive Subheading

Mean infectious titers were compared by using the 1-tailed Student t-test in Excel (Microsoft, Redmond, WA, USA) or GraphPad Prism v5 (La Jolla, CA, USA). Area under the curve (AUC) analysis for cumulative shedding was performed by using GraphPad Prism v5.

Summary of the analysis

Data analysis procedures and software used for the analysis

Include company (name and location) that sold the software

Source: Jones, Jeremy C., Stephanie Sonnberg, Richard J. Webby, and Robert G. Webster. 2015. "Influenza A (H7N9) Virus Transmission Between Finches and Poultry." *Emerging Infectious Diseases* 21 (4): 619–28.

Writing Style in the Methods Section

In this section, we will examine the writing style of the Methods section. The primary issues for writing the Methods section are (1) verb tense, (2) passive versus active voice, and (3) use of personal pronouns.

Verb Tense

Since the Methods section reports on actions that occurred previously, you should write this section in the past tense. Moreover, unlike the Introduction section, the Methods section is generally written in past tense. For more information on tense, refer to the discussion in Chapter 5.

Passive and Active Voice Revisited

In practice, the Methods section is neither exclusively active nor exclusively passive voice but, as in Example 6.1, writers use a combination of the two styles to avoid monotony. As shown in Table 5.8 (see Chapter 5), the Methods section contains the highest percentage of passive construction of all sections of the scientific article (Amdur, Kirwan, and Morris 2010; Nwogu 1997; Rhodes 1997; Riley 1991; Rodman 1994). However, the frequency of active voice versus passive voice construction varies by discourse community. As illustrated in Example 6.9, passive voice emphasizes the object of research by placing the subject at the beginning of the sentence followed by the verb of action (Knight 2003). Although some communication scholars propose that the passive voice obscures the researchers from being the ones who perform the action (Bazerman 1988; Ding 2002; Gross 1985; Wilkinson 1992), others have noted that taken in the context of the complete journal article, the authors of the article obviously conducted the research stated in the Methods section (Rodman 1981).

EXAMPLE 6.9 COMPARISON OF THE USE OF PASSIVE AND ACTIVE VOICE IN THE METHODS SECTION.

Both examples have the same number of words but shift the emphasis, the subject of the sentence, between the object of study (passive voice) to the researchers (active voice).

Passive Voice

Cells *were placed* in a thin chamber (10 x 20 x 1.5 mm) consisting of two glass coverslips separated by a U-shaped brass spacer and held together with vacuum grease. The coverslips *were siliconized* to reduce adhesion of the cells to the glass.

Active Voice

We placed cells in a thin chamber (10 x 20 x 1.5 mm) consisting of two glass coverslips separated by a U-shaped brass spacer and held together with vacuum grease. *We siliconized* the coverslips to reduce adhesion of the cells to the glass.

Adapted from Wilkinson (1992).

Guidelines for voice is often inconsistent between the recommendations in the Instructions to Authors or style manuals and the common practice in a target journal. Since no universal rule exists for voice (Nair 2005), you should examine several recently published articles from your target journal to determine the currently accepted use of voice within the discourse community.

Personal Pronouns

Along with our discussion of active voice, we must include an analysis of the use of personal pronouns that authors use to refer to themselves or others in journal articles (Bernhardt 1985; Van Way 2007). The use of first-person pronouns is very common in most scientific disciplines. As summarized in Table 6.3, authors of journal articles nearly exclusively use plural first-person pronouns (i.e., "we," "us," and "our"; Kuo 1998). The first-person plural is so common in scientific communication that the word "we" is often referred to as the "scientific we" (Van Way 2007, 638) and compared to the "royal we" used by the Queen of England.

Almost always you will use plural, not singular, first-person pronouns (such as "we") because most research is conducted as a team of researchers (Van Way 2007). But use of singular first-person pronouns (such as "I") is acceptable and somewhat common in single-authored research published in the soft sciences (Van Way 2007). While professional societies and journals embrace self-reference, the American Psychological Association (2010) recommends that writers avoid referring to themselves in third person (such as, "the authors of this paper").

Use of first-person pronouns also varies by discipline (Hyland 2001). As summarized in Table 6.4, use of first-personal pronouns is approximately three times more frequent in the soft sciences than in the hard sciences.

TABLE 6.3 Use of personal pronouns in scientific writing.

Category of Personal Pronouns	Occurrences[1]	Percent
First-person singular (I, me, my)	0	0.0%
First-person plural (we, us, our)	1750	87.9%
Second person (you, your)	8	0.4%
Third-person singular (he, him, his, she, her)	38	1.9%
Third-person plural (they, them, their)	59	3.0%
Indefinite (one, one's, ones)	135	6.8%
Total	**1990**	**100.0%**

[1] From a corpus of 36 journal articles consisting of 224,328 total words. Table adapted from Kuo (1998).

TABLE 6.4 Frequency of first-person pronouns by discipline.

Discipline	Frequency of First-person Pronouns (Cases per journal article)
Hard Sciences	
Physics	17.7
Microbiology	15.5
Electrical engineering	11.6
Mechanical engineering	2.6
Average (hard sciences)	11.9
Soft Sciences	
Marketing	38.2
Philosophy	34.5
Applied linguistics	32.3
Sociology	29.4
Average (soft sciences)	33.6

Table adapted from Hyland (2001).

Concluding Points

Peer reviewers and journal editors report that the Methods section is usually the most flawed section of manuscripts submitted for publication, and it is often the section responsible for a manuscript's outright rejection (Byrne 2000). The irony is that this section is also the least prominent section in the published article because most readers rely on peer reviewers' assessment of the research methods (Berkenkotter and Huckin 1995) and reducing space in the printed journal enables editors to fit more articles into the same page count. So even though the Methods section traditionally follows the Introduction section in the IMRAD organization, journals often print the section in smaller typeface, move the section to the end of the article (following the discussion), or place the section entirely online at the publisher's website. As science rhetoric scholars Joseph Harmon and Alan Gross (1996) explain, the Methods section is

> usually segregated from the rest of the text in a way that it can be skipped without losing the author's overall narrative thread. Indeed, it will be skipped or skimmed by most readers unless a strong disbelief in the results prompts thought of replication, or else a strong belief in them inspires any new converts to adapt the original to their own research purpose. . . . But replication solely for the sake of replication is rare in contemporary science because it is often a complex, time-consuming, and expensive process, with little reward whether the repeater succeeds or fails.
>
> *(66–7)*

7

THE RESULTS SECTION AND EFFECTIVE PRESENTATION OF DATA

In this chapter, you will learn to:

- Organize and present data in the Results section
- Write in a scientific style and edit style problems that commonly arise in the Results section
- Present data effectively in tables and graphs, and through statistics and figures

Overview

The Results section is the core of a scientific article because this is the section that provides evidence to answer the research question (Cetin and Hackam 2005; Thrower 2010). As explained by Michael Katz (2009), from Case Western Reserve University School of Medicine, the Results section is the "enduring substance" (97). Whereas theories and generalities that are presented in the Discussion section may be revised in future articles with additional, more revealing data, the Results section documents the relevant data under the conditions presented in the Methods section. Updated theories must account for such data and observations reported in the Results sections of previous articles.

To illustrate the concept of enduring evidence versus transient theory, Katz (2009) described the "central dogma of molecular biology" proposed by Francis Crick in 1958. The central dogma theory proposed a model for the transfer of genetic information in cells (from DNA to RNA and then to protein) and clearly stated that information cannot be transferred from protein to either DNA or protein. At the time, the scientific community lacked experimental evidence for transfer of information by proteins. Decades later, new data and observations made by Alain Bussard (2005), Honorary Professor at the Pasteur Institute, showed that a unique class of proteins (called "prions") function as genetic elements to store and transmit information. Examples of prion-induced disease include scrapie in sheep bovine spongiform encephalopathy (BSE) in cattle and Creutzfeldt-Jakob disease in humans. As Thomas Kuhn (2000) explained, the latest theories modify previous ones and provide a more comprehensive account of *all* observed data. Therefore, as you can see, the data (not theories) are the "enduring substance" of science. In this chapter, we will examine how researchers report and archive this evidence in the Results section.

What to Expect

The Results section provides the data, or the evidence from the current research investigation, that directly address or answer the research question presented in the Introduction section (Fahy 2008; Kurmis 2003). The Results section extends the line of evidence summarized in the Introduction section by adding new data about the phenomenon. Specifically the content of the Results section builds on the background research findings presented in the Introduction (Move 1, Step 2, see Chapter 5) and occupies the research niche specified in Move 2 of the Introduction.

Journal articles following the IMRAD organization separate the evidence or data in the Results section from the meaning or theory in the Discussion section. The reason for this organization is to clearly distinguish between observations and human interpretation. Yet this distinction is somewhat misleading because the data presented in the Results do undergo selection and presentation—which requires some analysis and interpretation. In practice, researchers purposely select data that are the most meaningful and relevant evidence to answer the research question, and, furthermore, researchers transform raw data into clear and concise prose or visuals to effectively communicate the data to the reader (Harmon and Gross 1996). Therefore the Results section, as any communication, is subject to human bias. Yet ethically researchers and writers have the responsibility to share and disclose the complete breath and depth of the data relevant to the investigation so that the readers will have as thorough of an understanding of the issue as permitted by the evidence. The spirit of the Results section is to present the summarized data completely and objectively, and reserve any curtailing or critique until the Discussion section.

The research design, which differs between fields, influences the content and organization of the Results section. The two major types of research design are both EMPIRICAL, or rely on direct observation and data to answer a question, but each design has different objectives and therefore a unique way for data to be presented. For example, the data in experimental (or hypothesis-) testing journal articles (see Chapter 5), which are often found in medical journals, all support a single, focused, predetermined research question (or hypothesis). In contrast, data from exploratory investigations in basic academic research fields, such as biology and biochemistry, typically have a broad research goal rather than a specific, testable hypothesis. Journal articles of exploratory investigations usually consist of a series of separate investigations and the data or outcome from each of these "mini-investigations" provides the justification or rationale for the next step. Therefore, the Results sections of exploratory investigations typically include a short review and reasoning for each procedure, followed by the data from that particular procedure, and then a brief interpretation of the data to explain the rationale for the next procedure.

The two types of research designs will help you to plan the overall organization of your Results section. But no single organization strategy can accommodate every possible context as research is, by definition, exploratory and original (Thrower 2010). Therefore, as a writer, you need to consciously and continually evaluate the best organization of the data based on your audience's expectations and needs. Specifically, you should consider the following questions:

1. How does the audience expect the information in the Results section to be presented? (By examining recent articles on similar topics within the discourse community, you will find certain patterns of data presentation. We will examine the common patterns in this chapter.)
2. What will the audience consider the most important, interesting, or convincing data? (You should present this information first, much like journalists do in news reporting.)
3. Does the audience need the outcome from one data set to understand or fully appreciate another? (Then you will need to present that data set first and then follow with the dependent data.)
4. Does all of the information provided directly relate to the purpose of the research stated in the Introduction? (If not, you should exclude that information.)

When you begin writing the Results section, you should follow the examples in this chapter—either experimental (or hypothesis-) testing or exploratory investigation—that have the greatest similarities to your research data. But you should remain flexible in order to adapt the guidelines presented in this chapter for your particular situation. In some situations, you might even find that a hybrid of the approaches described here will work best for a particular journal article.

Despite popular misconceptions, the Results section should be short and concise. Typically, the relevant data are briefly summarized in the text and presented in detail in figures or tables. The length of the Results section will vary depending on the length of the entire journal article and norms of the discourse community. For example, the Results sections of the experimental research journal articles from *The New England Journal of Medicine* that are used as examples in this chapter range in length from 430 to 785 words. Since the Results section of exploratory research journal articles includes more detail and takes a narrative approach, the section may reach up to 2,000 words. So you can see, the Results section may vary significantly in length.

Data in the Results section are rarely presented in the same order that they were collected. Rather, they are organized logically—and distracting or unnecessary data are omitted—to help focus readers' attention on the relevant information (Cals and Kotz 2013a; Cetin and Hackman 2005; Harmon and Gross 1996; Huth 1982). According to Selma Cetin and David Hackam of the Children's Hospital of Pittsburgh and University of Pittsburgh School of Medicine, "It is perfectly appropriate to describe the results in a manner that makes sense, as opposed to describing the experiments in the temporal order in which they were performed" (166). You must maintain the focus of the content in the Results section because this section determines the entire "story," or narrative, of the journal article. Every section of the IMRAD journal article is constructed around the organization of data in the Results section. For example, the experimental procedures presented in the Methods section are organized in the same order that the data are presented in the Results. Also, the Discussion extrapolates on the data in the order it was presented in the Results. A challenge when writing the Results section is to keep the content focused on the purpose of the article and avoid reporting tangential findings (International Committee of Medical Journal Editors 2014).

How to Prepare

When writing a Results section, you should first identify the design of the research since that will often influence the organization of the data. In the majority of original research journal articles, the Results section provides data for either:

- Experimental (or hypothesis-) testing of a research question, hypothesis, or supposition that is often explicitly stated in the Introduction, or
- Exploratory investigations of a research problem for which a formal experimental hypothesis was not articulated (or easily constructed from the purpose statement).

After you identify the research design of the overall study, which was discussed in Chapters 5 and 6, then you can organize your data as follows:

- For experimental testing data, you should present the strongest (or most conclusive and convincing) and broadest evidence first. Organize subsequent data from general to specific findings.
- For exploratory investigations, which are common in basic research, the Results are often constructed into a logical narrative in which the results

from each mini-investigation provides the rationale for the next. To organize your data, you will construct the logical narrative—which is not necessarily the order in which the experiments were performed or data were collected. In this narrative, you should provide the broadest evidence first. Then as you progress through the Results section, the data address finer details and more specific points. In a sense, the Results section of exploratory investigations take a somewhat abbreviated, IMRAD-structure for each mini-investigation with a brief introduction based in the findings from the previous mini-investigation, a short review on the methods, the detailed results, and a simple conclusion, a brief summary leading to the next mini-investigation. This pattern then repeats for each mini-investigation presented in the Results.

We will talk more about the structure for each type of design later in this chapter, but the main thing to do before you begin writing is to outline the data you plan to present in order to follow one of these organizational strategies.

To analyze the logical flow of information in the Results section, you should initially focus on the overall meaning or relevance of the data from each part of the research or mini-investigation. For each part, you should evaluate the *relevance* of the data rather than focus on the specific details of the data (McMillan 2012). The relevance of the data is designated as the RESULTS whereas DATA are the raw measurements that are obtained directly from experiments and observations (Hofmann 2014; Zeiger 2000). Results summarize the main point of the data (see Example 7.1). We will use this distinction throughout this chapter.

EXAMPLE 7.1 HIGHLIGHTING "RESULTS" AND SUBJUGATING DATA.

Results Explicitly Stated	☑ **Recommended**
The mean resting blood pressure	Test variable
was higher in the 30 tennis players than in the 20 control	Results or key summary of data
participants [94 ± 3 (SD) vs. 85 ± 5 (SD) mmHg, $P > 0.02$].	Data (facts that support the results summary statement subjugated to support the statement)

Data Lacking Results Statement	☒ **Not Recommended**
In the 20 control participants, the mean resting blood pressure	Test variable
was 85 ± 5 (SD) mmHg.	Data
In comparison, in the 30 tennis players, the mean resting blood pressure	Test variable
was 94 ± 3 (SD).	Data

Sample sentences adapted from Zeiger (2000).

A useful method for organizing the Results section, described by Malcolm Chiswick (2002), professor of Child Health and Paediatrics at the University of Manchester, is to take a blank sheet of paper and write the aims of the study and any hypothesis at the top of the page. Underneath the aims, list all the results that you have collected. Next, you should cluster the results into natural groupings—placing related information together regardless of the order in which it was collected. Develop an order to or flow of the data that follows the logic of strongest to weakest (for experimental testing research) or a logical and sequential narrative that describes a series of observations (for exploratory investigations). After you have completed this outline, you should ensure that all of the results answer or address the aims or hypothesis written at the top of the page. Remember to exclude any results that do not directly contribute to the aims or hypothesis of the present investigation. As Chiswick explained, "Using this approach you are more likely to produce a concise and logically ordered results section that answers the aims of the study without burdening the readers with extraneous material that distracts from key findings" (416). Having used the brainstorming method Chiswick described, I will add that taking an hour or two to organize your results in this way will save significant time later during the writing and editing process.

As many of the top scholars emphasize, you should not try to report everything that has been done in the laboratory in a single journal article. Rather, you should report only the relevant information to answer the research question. You should also be cautious of any "pet" data as writers of SMC often become attached to data that have a nice illustration or were particularly challenging to obtain—even if the results do not contribute to the goal of the journal article. Such "pet" content will ultimately be removed from the manuscript (by collaborators, peer reviewers, or editors) because the information is irrelevant to the purpose. Therefore, you can save time and effort by remaining flexible as you evaluate the content for the Results section.

Once you have established a logical organization for the results, you are ready to begin writing the Results section. As always, several recent articles on similar topics and with similar research designs—experimental (or hypothesis-) testing or exploratory investigations—from the target journal are useful models that will help you ensure that you are using an appropriate organization and style for the discourse community. Published examples of Results sections from your target journal should be used to determine if the experimental or exploratory research organization is appropriate and provides examples of table and figure design, which are also often well described in the Instructions to Authors and style guides.

Content, Organization, and Writing Style of the Results Section

In this section, we will examine the content and style for each of the two broad categories of research design articles: experimental (or hypothesis-) testing and exploratory investigations. For either research design, you should approach the Results section as a composite of multiple subsections with a prototypical structure that should be followed consistently to facilitate clear and effective communication.

Within this section, I have incorporated a discussion of scientific writing style. The style issues presented in this chapter build on the content introduced in Chapters 5 and 6. As you will see in the examples included in this chapter, passive voice is frequently used in the Results section to present data. The active voice may also be used in exploratory investigation narratives to provide the rationalization for each mini-investigation. Additionally, as covered in Chapter 5, the majority of the Results are written in past tense because the actions (data collection and observation) occurred in the past. Yet, the tense may shift if you make a general knowledge statement or cite a claim because knowing a "fact" occurs in the present. The only exception is if you directly attribute the claim to a researcher or experiment because the claim occurred in the past.

Experimental (or Hypothesis-) Testing Research

In experimental testing research, the focus of the entire article is to answer the research question or test a hypothesis, whether explicit or implied, that was presented in the Introduction. All data in the Results section provides the evidence. Since this type of journal article usually includes only one research question, writers adheres closely to the conventional IMRAD structure and only present data in the Results section.

The Results section of experimental (or hypothesis-) testing journal articles has standardized content and organization (Moher et al. 2010, Nwogu 1997, Skelton 1994). The Results section of these articles are organized into the following five subsections:

1. Participants,
2. Experimental validation (if necessary),
3. Major findings,
4. Minor findings, and
5. Harms to participants and unintended effects.

Typically the Results in subsections 2, 3, and 4 use descriptive subheadings that label each mini-investigation rather than the categories I have listed above.

Subsection 1: Participants

The first subsection of experimental design Results provides a summary of the relevant demographics and characteristics or features of the study participants. The Participants subsection (see Example 7.2) should include the following:

1. Total number of participants enrolled in the study and the time frame or duration,
2. Number of participants randomly assigned and included in the test (or treatment) group,
3. Number of participants randomly assigned and included in the control (or nontreatment) group,
4. Number of participants completing the experimental protocol and included in follow-up observations and data analysis (consider adding a FLOW DIAGRAM as in Example 7.2 which is described below), and
5. A table summarizing the relevant baseline demographics and clinical characteristics of each group.

We will examine some differences in the discussion below.

As you write the Participants subsection, you may notice that many discourse communities use the term "PARTICIPANTS" to designate humans that, as a

EXAMPLE 7.2 PARTICIPANTS SUBSECTION FROM THE RESULTS SECTION OF AN EXPERIMENTAL TESTING RESEARCH ARTICLE.

Efficacy of High-Dose Versus Standard-Dose Influenza Vaccine in Older Adults

RESULTS

Participants	Subheading
A total of 31,989 participants enrolled in the study over 2 years.	Total participants enrolled in study and time duration
Of the total participants, 15,991 were randomly assigned to IIV3-HD	Number of participants in test group
and 15,998 were randomly assigned to IIV3-SD	Number of participants in control group
(Figure 1). Of the participants who underwent randomization, 31,983 (>99.9%) received study vaccine. All 31,983 participants were included in the full analysis set, and 31,803 (99.4%) were included in the per-protocol analysis set.	Summary of participants completing experimental protocol (and reference to a participant flow diagram)
Baseline characteristics of the recipients of IIV3-HD and the recipients of IIV3-SD were similar (Table 1).	Baseline data (typically summarized in the first table)

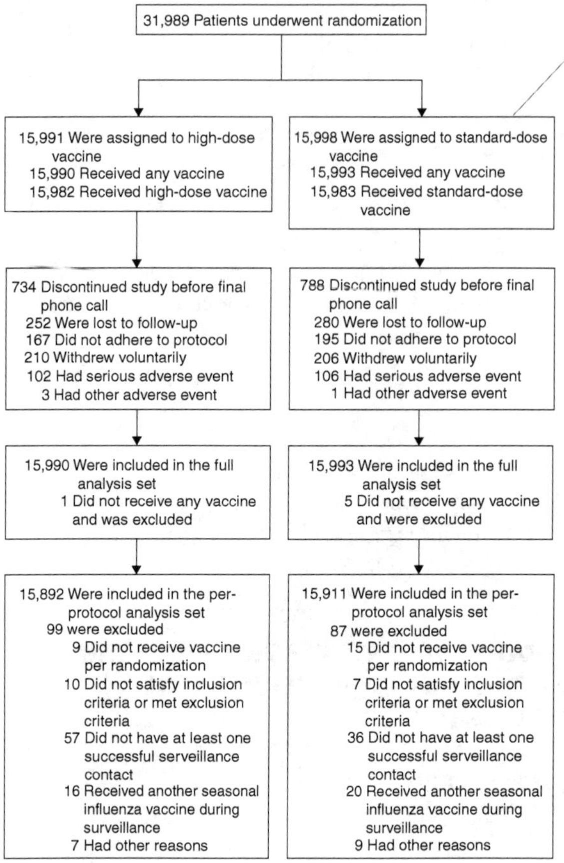

Participant flow diagram showing the participants undergoing randomization in test (left) and control (right) groups and their progress through the experimental protocol

FIGURE 1. Enrollment and Follow-up of Study participants.

TABLE 1 Baseline Demographic and Clinical Characteristics of the High-Dose and Standard-Dose Vaccine Groups.*

Summary of relevant demographics and clinical characteristics

Characteristic	**IIV3-HD (N = 15,990)**	**IIV3-SD (N = 15,993)**
Female sex—no. (%)	9,131 (57.1)	8,963 (56.0)
Mean age—yr	73.3 ± 5.8	73.3 ± 5.8

Characteristic	IIV3-HD (N = 15,990)	IIV3-SD (N = 15,993)
Racial background—no. (%)†		
White	15,103 (94.4)	15,167 (94.8)
Asian	118 (0.7)	105 (0.7)
Black	670 (4.2)	612 (3.8)
Other	97 (0.6)	106 (0.7)
Hispanic ethnic group—no. (%)†	958 (6.0)	982 (6.1)
At least one prespecified chronic coexisting condition—no. (%)‡	10,750 (67.2)	10,752 (67.2)
At least two prespecified chronic coexisting conditions—no. (%)	5,385 (33.7)	5,403 (33.8)
Cardiac and respiratory disorders—no. (%)		
Coronary artery disease	2,735 (17.1)	2,732 (17.1)
Atrial fibrillation	1,103 (6.9)	1,112 (7.0)
Valvular heart disease	744 (4.6)	741 (4.6)
Congestive heart failure	451 (2.8)	446 (2.8)
Chronic obstructive lung disease	1,500 (9.4)	1,495 (9.4)
Asthma	1,415 (8.8)	1,408 (8.8)
Received influenza vaccine the previous season—no. (%)	11,758 (73.5)	11,773 (73.6)

In some discourse communities, only females are presented (to save space since males can be easily calculated by the audience)

Provide mean and standard deviation for continuous variables (such as age)

Provide both raw numbers and percentages (in parentheses) for categorical variables

* There were no significant differences between the treatment groups. IIV3-HD denotes high-dose, trivalent, inactivated influenza vaccine, and IIV3-SD standard-dose, trivalent, inactivated influenza vaccine. Plus-minus values are means ±SD.

† Racial background and ethnic group were self-reported. The category "Other" includes American Indian or Alaska Native, Native Hawaiian or other Pacific Islander, and mixed origin. Information on racial background was missing for two recipients of the high-dose vaccine and three recipients of the standard-dose vaccine. Percentages may not total 100.0% because of rounding.

‡ Prespecified chronic coexisting conditions include the listed cardiac and respiratory disorders, as well as sickle cell disease, diabetes mellitus, hypothyroidism, epilepsy, stroke, spinal cord injury, Parkinson's disease, chronic kidney disease, chronic hepatitis, cirrhosis, human immunodeficiency virus-acquired immunodeficiency syndrome, cancer, longterm systemic glucocorticoid therapy, and other potentially immunosuppressive therapies (per-group frequencies of these conditions are provided in the Supplementary Appendix).

Adapted from DiazGranados, Carlos A., Andrew J. Dunning, Murray Kimmel, Daniel Kirby, John Treanor, Avi Collins, Richard Pollak, Janet Christoff, John Earl, Victoria Landolfi, Earl Martin, Sanjay Gurunathan, Richard Nathan, David P. Greenberg, Nadia G. Tornieporth, Michael D. Decker, and H. Keipp Talbot. 2014. "Efficacy of High-Dose versus Standard-Dose Influenza Vaccine in Older Adults." *The New England Journal of Medicine* 371: 635–45.

result of proper research protocols, have autonomously agreed or volunteered to be included in the research investigation. By comparison, the term "SUBJECTS" connotes individuals who are subjected to an experiment without their consent and is generally reserved for nonhumans lacking autonomy (such as animals). Furthermore, the American Psychological Association (2010) recommends paying careful attention to labels when referring to participants and using active voice to demonstrate the decisive roles of participants, or agency, in research.

The CONSORT guidelines, for reporting randomized clinical trials, encourages the use of a participant flow diagram (see Figure 1 in Example 7.2). A flow diagram illustrates the number of participants enrolled in a study and those that proceeded to each stage of the experimental protocol (Moher et al. 2010). The guidelines recommend that writers report the number of participants progressing to each major phase of the experiment, which includes (1) initial enrollment, (2) allocation (or assigning individuals to test and control groups), (3) follow-up observations or tests, and (4) analysis of data (Moher et al. 2010). Additionally, authors are asked to document participant attrition by specifying the reasons and number of individuals that failed to proceed to each phase of the experiment (see Figure 7.1). You should include the number of participants and reasons that they were excluded, dropped out, or were lost from follow-up investigation (Iverson et al. 2007). Finally, if applicable, you should note if any part of the experimental protocol was prematurely halted or abandoned for a group and the reason for the change.

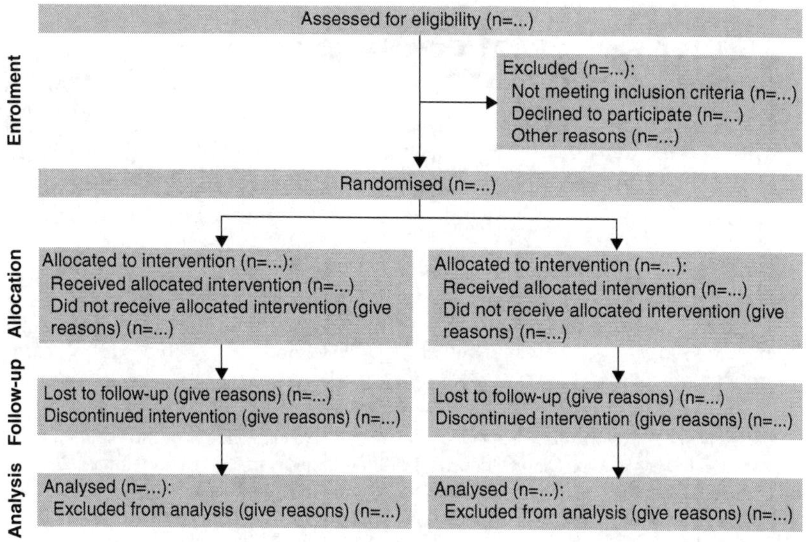

FIGURE 7.1 Participant flow diagram recommended by the CONSORT guidelines.
Source: Moher et al. (2010).

The CONSORT guidelines (Moher et al. 2010) and *Publication Manual of the American Psychological Association* (American Psychological Association 2010) both recommend that authors include the dates of recruitment and follow-up in the Results section (see Example 7.3). In practice, however, the time frame is often placed in the Methods section since it details the experimental protocol. Further complicating this issue is that the CONSORT guidelines' example of a time frame statement to include in the Results section (see Example 7.3) is a heavily edited passage that was taken from the Methods section of an article published in the *Annals of Internal Medicine*. Additionally, the example from *The New England Journal of Medicine* (see Example 7.2) similarly includes the time durations but details the time frame information in the Methods section. Since the location of time duration and time frame information lacks consistency between style manuals and practice, you will need to determine the proper location for time duration and time frame information by referring to recently published articles in your target journal as well as the Instruction to Authors.

The relevant characteristics of each group are usually summarized in the first table of a journal article (see Example 7.2). BASELINE DATA, which summarize the relevant demographics and preexperimental characteristics of each participant group (such as test and control), are necessary to provide the reader with an understanding of the makeup of the participants in the study and to show equivalency (or lack thereof) between each group (Furler et al. 2012; Tse, Williams,

EXAMPLE 7.3 A PARTICIPANT RECRUITMENT AND FOLLOW-UP TIME FRAME.

Style guides often recommend placing the time frame in the Results section, although in practice such statements are often located in the Methods section.

Age-eligible participants were recruited from February 1993 to September 1994.	Dates that the study took place
Participants attended clinical visits at the time of randomization (baseline) and at 6-month intervals for 3 years.	Follow-up

Source: Moher, David, Sally Hopewell, Kenneth F. Schulz, Victor Montori, Peter C. Gotzsche, P. J. Devereaux, Diana Elbourne, Matthias Egger, and Douglas G. Altman. 2010. "CONSORT 2010 Explanation and Elaboration: Updated Guidelines for Reporting Parallel Group Randomized Trials." *BMJ* 340: c869.

and Zarin 2009). Furthermore, baseline data enable readers to evaluate the generalization of the findings to other groups (Furler et al. 2012; Vandenbroucke et al. 2007).

As a writer, you should only report baseline data for characteristics directly relevant to the study topic and hypothesis. The minimum requirement for reporting baseline data during registration of clinical trials is the age and sex of the participants (Tse, Williams, and Zarin 2009). Yet medical articles often include other characteristics that likely affect the outcome of the results (see Table 7.1). Baseline characteristics should not be reported unless they are relevant to the research topic—especially for sensitive demographics such as race/ethnicity, income or socioeconomic status, education, employment, marital status, and religion (Ma et al. 2007; Rivara and Finberg 2001). Since the appropriate terminology to describe groups changes over time, you should refer to the latest style manuals (particularly the *Publication Manual of the American Psychological Association*) for the currently accepted terms (American Psychological Association 2010; Kaplan and Bennett 2003; McKenzie and Crowcroft 1996; Winker 2004).

Reporting of baseline data varies by category. For categories, such as sex and race, both raw numbers and percentages are included (see Example 7.1). Continuous measures, such as age, are reported with the mean and standard deviation (see Example 7.1) unless the distribution is skewed, in which cases the median and percentile range are reported (Vandenbroucke et al. 2007). In cases when

TABLE 7.1 Prevalence of demographic variables reported in medical journal articles.

Demographic Variable	Percentage of Papers Reporting that Variable
Age	99
Sex	99
Weight/BMI	39
Ethnicity	37
Smoking	24
Education attainment	23
Marital status	13
Any area-based measure (e.g., index of deprivation or disadvantage, rurality distance from health centers, etc.)	10
Employment status	10
Alcohol [consumption]	4
Occupational status	2
Income	2
Language spoken	0

Source: Furler et al. (2012).

ordered categories are used to describe a continuum, such as stages of cancer (I to IV), you should report the stage number and percentage of participants in each category rather than descriptive statistics such as a mean (Vandenbroucke et al. 2007).

Subsection 2: Experimental Validation (If Necessary)

If necessary, experimental research journal articles provide data to validate an aspect of the research design. Such experimental validation may establish participant qualifications or the efficacy of the experimental protocol (as shown in Example 7.4). When included, the experimental validation information is typically placed before the primary results in order to circumvent possible uncertainty or questions that readers may have regarding the validity of the research tools or approach (Huth 1982).

EXAMPLE 7.4 EXPERIMENTAL VALIDATION PRECEDING THE PRESENTATION OF MAJOR FINDINGS.

In this experimental research article, the authors establish participant qualifications and the efficacy of the experimental protocol before reporting the major findings regarding the effectiveness of standard-dose (IIV3-SD) and high-dose (IIV3-HD) vaccines. (This subsection followed the Participants subsection presented in Example 7.2.)

Text	Annotation
Illness Surveillance and Collection of Nasopharyngeal Swabs	Descriptive subheading
In the full analysis set (according to treatment assignment), 3745 participants in the IIV3-HD group (23.4%) had at least one protocol-defined influenza-like illness, 758 (4.7%) had at least one modified CDC-defined influenza-like illness, and 8168 (51.1%) had at least one respiratory illness.	Confirming respiratory and influenza illnesses in high-dose (HD) vaccine group
In the IIV3-SD group, 3827 participants (23.9%) had at least one protocol-defined influenza-like illness, 838 (5.2%) had at least one modified CDC-defined influenza-like illness, and 8270 (51.7%) had at least one respiratory illness.	Confirming respiratory and influenza illnesses in standard-dose (SD) vaccine group
In the IIV3-HD group, nasopharyngeal swabs were collected within the protocol-specified time frame for 80.0% of protocol-defined influenza-like illnesses, 73.6% of modified CDC-defined influenza-like illnesses, and 67.2% of respiratory illnesses.	Confirming protocol adherence in HD vaccine group
In the IIV3-SD group, nasopharyngeal swabs were collected within the protocol-specified time frame for 79.3% of protocol-defined influenza-like illnesses, 73.7% of modified CDC-defined influenza-like illnesses, and 66.8% of respiratory illnesses.	Confirming protocol adherence in SD vaccine group

Source: DiazGranados, Carlos A., Andrew J. Dunning, Murray Kimmel, Daniel Kirby, John Treanor, Avi Collins, Richard Pollak, Janet Christoff, John Earl, Victoria Landolfi, Earl Martin, Sanjay Gurunathan, Richard Nathan, David P. Greenberg, Nadia G. Tornieporth, Michael D. Decker, and H. Keipp Talbot. 2014. "Efficacy of High-Dose Versus Standard-Dose Influenza Vaccine in Older Adults." *The New England Journal of Medicine* 371: 635–45.

Subsection 3: Major Findings

Major findings are the primary or most convincing evidence addressing the main research question. As illustrated in Example 7.5, the major findings subsection is a concise summary of the data necessary to confirm the point. Such data often corresponds to a single table or graph, as in Example 7.5.

EXAMPLE 7.5 MAJOR FINDINGS FROM AN EXPERIMENTAL RESEARCH ARTICLE ESTIMATING THE EFFECTIVENESS OF A TWO-DOSE VACCINE AGAINST CHOLERA.

(This subsection follows the Participants subsection.)

Text	Annotation
Use of *Vibrio cholerae* Vaccine in an Outbreak in Guinea	Journal article title
Vaccine Effectiveness	Descriptive subheading
Protection against cholera was greater in the participants receiving full vaccination (two doses) than in participants who did not receive vaccination	Results or summary statement of key findings
(adjusted effectiveness, 86.6%;	Report key statistic summarizing data
95% confidence interval [CI], 56.7 to 95.8; *P* = 0.001)	Report the precision of statistics as the confidence interval and the range of data
(Table 2).	Refer to visual in parentheses to point reader to full report of the data
The effectiveness of incomplete vaccination (a single dose) was inconclusive (42.8%; 95% CI, -83.6 to 82.2; *P* = 0.35).	Maintain consistent order and units of data with previous reports (see first sentence)

TABLE 2 Effectiveness of Complete and Incomplete Vaccination

Vaccination Status	Controls (N = 160)	Case Patients (N = 40)	Crude Estimate of Vaccine Effectiveness*	*P* Value	Adjusted Estimate of Vaccine Effectiveness†	*P* Value
	no. of patients (%)		*% (95% CI)*		*% (95% CI)*	
No vaccination	23 (14)	15 (38)	Reference		Reference	
Incomplete vaccination ‡	36 (23)	14 (35)	38.9 (-55.2 to 76.0)	0.30	42.8 (-83.6 to 82.2)	0.35
Full vaccination	101 (63)	11 (28)	84.0 (59.7 to 93.6)	<0.001	86.6 (56.7 to 95.8)	0.001

* Vaccine effectiveness was based on the odds of vaccination among case patients and controls; crude vaccine effectiveness was calculated as (1-odds ratio) x 100.

† Estimates were adjusted according to the number of people living in the household, whether drinking water was treated, and whether a latrine was shared with a person who had cholera.

‡ Vaccination was considered incomplete if only one dose was taken or if one or both doses were spat out or vomited.

Adapted from Luquero, Francisco J., Lise Grout, Iza Ciglenecki, Keita Sakoba, Bala Traore, Melat Heile, Alpha Amadou Diallo, Christian Itama, Anne-Laure Page, Marie-Laure Quilici, Martin A. Mengel, Jose Maria Eiros, Micaela Serafini, Dominique Legros, and Rebecca F. Grais. 2014. "Use of *Vibrio cholerae* Vaccine in an Outbreak in Guinea." *The New England Journal of Medicine* 370: 2111–20.

Each subsection of a data set should highlight the importance of the data—telling readers the interesting pattern or phenomenon that the data shows—and then make reference to a visual summary of the data (Penrose and Katz 2010). For each data set, you should organize information so that the most important concepts are presented first in the primary emphasis positions and details are presented in secondary support positions. You can effectively present information for each data set as follows:

1. Begin with a descriptive subheading summarizing what the data describe (see Example 7.5).
2. Lead the paragraph with a clear results summary statement. Readers should get all of the important information in the first sentence of each data-reporting paragraph (Brand and Huiskes 2001). Here is the recommended structure (see Example 7.6):

   a. Begin with the feature being studied (typically a noun).
   b. Follow with a comparing term (adverb or adjective) and indicate the direction and magnitude of a difference between variables or groups (see Example 7.7). (Also see Table 7.2 for descriptive terms that are often improperly used in scientific writing and that should be used only in particular contexts.)
   c. Clearly identify the test group(s) and the control group.

   In this arrangement, you will describe the data trends, which should not be confused with interpretation of the data that is reserved for the Discussion section (see Chapter 8).

3. Provide key statistics that summarize the data. However, note that statistics do not replace data and should not be presented in place of data (Hofmann 2014). When reporting data, include the range of values and precision as the confidence interval[1] (see Example 7.5). Although you may select the experiments to report in an article, you are obligated to report complete data sets (including outlying or inconsistent data) once you have decided which experiments to report (Katz 2009).
4. Highlight key data values in the text (International Committee of Medical Journal Editors 2014) but subordinate or downplay the quantitative measurements by presenting them in parentheses (Brand and Huiskes 2001; Hofmann 2014; Penrose and Katz 2010), as illustrated in Example 7.5.
5. Reference a visual summary (figure or table) of the data set, ideally in parentheses at the end of the first sentence reporting the data (see Example 7.5). Do not try to report all of the raw data in the text because it will be

confusing to readers and can obscure the key point or results (Zeiger 2000), as in Example 7.8. Generally multiple data measures and data comparison between groups are more effectively reported in visuals (figures or tables) than in text (see Example 7.9). You should not "parrot" or repeat the information from the visual in the text (Matthews, Bowen, and Matthews 2000). Additionally, you should not repeat the data set in both figures and tables (International Committee of Medical Journal Editors 2014). As an author,

EXAMPLE 7.6 GRAMMATICAL STRUCTURE OF SENTENCES HIGHLIGHTING RESULTS AND SUBJUGATING DATA.

The mean resting blood pressure was higher in the 30 tennis players than in the 20 control participants [94 ± 3 (SD) vs. 85 ± 5 (SD) mmHg, $P > 0.02$].

- Feature being studied (noun)
- Adjective summarizing the affect and degree shown by the data
- Clearly identify test and control groups
- Key statistics summarize quantitative data (in the same order that the groups were mentioned)
- Provide a measure of the effect size (see footnote at the end of this chapter)

Sample sentence adapted from Zeiger (2000).

EXAMPLE 7.7 COMMUNICATING THE DIRECTION AND MAGNITUDE OF AN ASSOCIATION BETWEEN VARIABLES.

	Example Statement	Description
Poor	Mortality and age are <u>correlated</u>	This sentence does not tell the audience the direction of the association (whether age and mortality are positively or negatively related) nor the degree
Better	As age increases, mortality <u>increases</u>	This version provides the audience with the direction of the association but not the degree
Best	Among the elderly, mortality <u>roughly doubles</u> for each successive five-year age group	This version provides both the direction and the magnitude of the age–mortality association

Adapted from Jane Miller (2004).

TABLE 7.2 Precise language use in the Results section.

Use "fewer" for numbers and "less" for quantity:

- Fewer than 10 participants
- Less stress

Restrict use of terms with statistical meaning (reserve for use with statistics only):

- Correlations
- Normal
- Random (This implies a randomization procedure.)
- Sample
- Significant (Never use "significant" but use "statistically significant" when appropriate.)

Avoid terms that imply a cause-and-effect relationship (reserve any interpretation for the Discussion):

- Causes
- Effects

Avoid terms that imply a subjective evaluation or interpretation (reserve interpretation for the Discussion):

- Interestingly
- Remarkably
- Strikingly

Sources: International Committee of Medical Journal Editors (2014); Kotz and Cals (2013a); O'Connor and Holmquist (2009).

EXAMPLE 7.8 PRESENTING RAW DATA IN TEXT IS OFTEN CONFUSING FOR READERS AND OBSCURES THE MAIN POINT.

Among breakfast skippers, baseline mean (± SD) weight-, age-, sex-, site-, and race-adjusted weight changes were 20.71 ± 1.16, 20.76 ± 1.26, and 20.61 ± 1.18 kg for the control, breakfast, and NB groups, respectively. Among breakfast eaters, baseline mean (± SD) weight-, age-, sex-, site-, and race-adjusted weight changes were 20.53 ± 1.16, 20.59 ± 1.06, and 20.71 ± 1.17 kg for the control, breakfast, and NB groups, respectively.

Although these data are presented in parallel order, presenting raw data in a table format is recommended

Source: Dhurandhar, Emily J., John Dawson, Amy Alcorn, Lesli H. Larsen, Elizabeth A. Thomas, Michelle Cardel, Ashley C. Bourland, Arne Astrup, Marie-Pierre St-Onge, James O. Hill, Caroline M. Apovian, James M. Shikany, and David B. Allison. 2014. "The Effectiveness of Breakfast Recommendations on Weight Loss: A Randomized Controlled Trial." *American Journal of Clinical Nutrition* 100: 507–13.

EXAMPLE 7.9 COMPARISON OF DATA PRESENTED (A) IN THE TEXT AND (B) IN A VISUAL.

In this example, the authors reported the results of the QFT-IT test for tuberculosis (TB) in children living with adults with smear-positive results for the bacterium causing TB, which were reported as negative or positive (scanty to +++). This example illustrates the complexity of multiple raw data values and comparisons between data presented in text. In such cases, raw data should be presented visually in figures or tables.

(a) Data set presented in the text

Forty-five percent (4/11), 80% (16/20), 68% (15/22), and 90% (17/19) of children in contact with adults with scanty, +, ++, and +++ AFB, respectively, had positive results from the QFT-IT test ($p = 0.03$), as shown in Figure 2.

☒ Not Recommended

Presenting raw data in text makes comparison between groups difficult

(b) Data set presented in a visual

☑ Recommended

Visual representation organizes raw data and facilitate comparison between groups

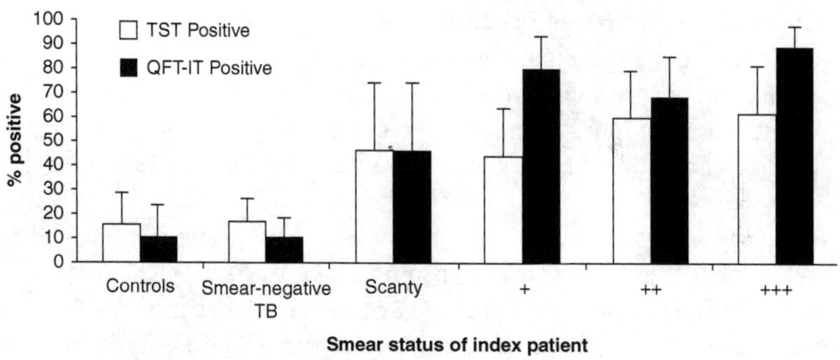

Figure 2. Proportion of children with positive tuberculin skin test (TST) (>10 mm) and QuantiFERON Gold in Tube (QFT-IT) test results, by adult smear positivity. Error bars show 95% confidence intervals.

Source: Nakaoka, Hiroshi, Lovett Lawson, S. Bertel Squire, Brian Coulter, Pernille Ravn, Inger Brock, C. Anthony Hart, and Luis E. Cuevas. 2006. "Risk for Tuberculosis among Children." *Emerging Infectious Diseases* 12: 1383–8.

you should select the single best format to present the data (which we will discuss at the end of this chapter).

6. Conclude the subsection by highlighting any inconsistent observations or outlying data (if observed). Although some novice writers may feel uncomfortable reporting inconsistent or outlying observations or data, Joseph Harmon of Argonne National Laboratory and Alan Gross of the University of Minnesota tried to quell such concerns. As they stated, "Ironically, presenting and discussing negative results or outlining the limitations of positive results further bolsters an article's credibility" (Harmon and Gross 1996, 68). In science and medical research, the absence of negative or outlying data values may lead readers to question the unusual consistency or perfection of the data. This may also lead readers to question the integrity of the authors and thereby their confidence in the rest of the journal article—particularly because such inappropriate suppression of data is not uncommon in science (Fanelli 2009; Kirby and Houle 2004; Martinson, Anderson, and de Vries 2005; Swazey, Anderson, and Louis 1993; Wenger et al. 1997) and medicine (Altman and Moher 2014; Chalmers 1990; Coultas 2007; Friedman 1990). Furthermore, as Iain Chalmers (1990) from the Radcliffe Infirmary stated, "Failure to publish 'disappointing' or 'uninteresting' research results, or failure to report results in sufficient detail, may either lead patients to receive ineffective or dangerous forms of care or result in a delay in recognizing that other forms of care are beneficial" (1405–6). According to Drummond Rennie (2014) of the University of California, San Francisco, and Deputy Editor (WEST) of *JAMA*, "Bias and poor reporting threatens to overwhelm the credibility of research and to ensure that our treatments are based on fiction, not fact" (xiii). Finally, inaccurate reporting "does an injustice" to other researchers and those who have provided funds and other resources for research (Chalmers 1990, 1406).

By following these six steps, you can provide a comprehensive summary of each data set in the text that is remarkably compact and concise.

The scope and length of the complete Results section depend on the number of experiments conducted and data sets collected that are necessary to answer the research question. The Results section is composed of stacking outcomes from each data set—following this same six-step organization pattern. Using a parallel and predictable structure for each data set subsection—as well as for the data themselves, as in Example 7.5—effective communicators enable readers to devote their full attention to the information being communicated without having to untangle or decipher the relevant information from an inconsistent presentation format. Unlike writers of creative writing who focus more on ornate language, you should not vary the sentence or paragraph structure merely for the sake of variation because doing so creates unnecessary distraction and complicates the connection and comparison between parallel information. Rather, as a writer of SMC, you should use consistent organization and grammatical structure to reinforce equivalent information and enable the reader to easily compare data between tests or groups.

Subsection 4: Minor Findings

The label "minor" as used here, depends on the relevance of the results to the research question or hypothesis. The content, organization, and style of minor findings are the same as for major findings. The main point is that each data set is presented in order of decreasing importance or relevance to the primary research question and minor findings are often somewhat ancillary to the primary goal.

In the Results section, writers do not usually use the term "minor" to distinguish the results of less relevant data sets from those of more relevant sets. But identifying the minor findings from major findings will help you organize your Results section. After identifying the minor findings, you should evaluate your reasons for including those findings. Specifically you should ask the following questions:

- Why are the findings considered minor rather than major?
- Were the findings planned as part of the original (intended) research design?
- Would the findings be more appropriate in an article with a different purpose or hypothesis?
- Am I reporting the findings because I (or another person on the research team) have a passionate attachment to the research method, data set, figure, or table?

Furthermore, you should consider any unintended consequence of reporting the findings, which includes asking the following:

- Does reporting the findings cause the Results section to deviate from the flow of reporting primary evidence?
- Do the findings inappropriately bias the article?
- Would reporting the findings potentially confuse a casual reader?

If findings are tangential to the primary purpose of the research, then including these findings is discouraged.

Subsection 5: Harms to Participants and Unintended Effects

After reporting all the results from the data sets, you should conclude the Results section by addressing any harm to participants or unintended effects (called "ADVERSE EVENTS") that were observed (see Example 7.10). Adverse events of medical therapies, in particular, are often inadequately reported thus inappropriately biasing the benefits of a treatment (Ioannidis et al. 2004; Moher et al. 2010). Clear and unbiased reporting of adverse events—particularly from randomized clinical trials and other research with therapeutic potential—is the ethical responsibility of writers and researchers. To appropriately report adverse event, you should:

- Identify all adverse effects observed and provide the exact number of participants affected in each group and the severity of each event;
- Distinguish participants with multiple adverse effects those who exhibit only one adverse effect;
- Provide the frequency and duration of events for recurring adverse effects or those with periodic outbreaks; and
- Finally, identify participants who withdraw from the experiment due to harms or other unintended treatment effects, as well as any deaths that occurred during the study.

EXAMPLE 7.10 ADVERSE EVENTS DESCRIBED AT THE END OF A RESULTS SECTION.

Safety

During the safety surveillance period (approximately 6 to 8 months after vaccination), 1323 participants (8.3%) in the IIV3-HD group and 1442 participants (9.0%) in the IIV3-SD group had at least one serious adverse event. The relative risk for having at least one serious adverse event with IIV3-HD, as compared with IIV3-SD, was 0.92 (95% CI, 0.85 to 0.99).	Summarize adverse effects in all groups
During the surveillance period, 83 (0.5%) of the participants in the IIV3-HD group died, as did 84 (0.5%) of the participants in the IIV3-SD group. Six recipients of IIV3-HD died within 30 days after vaccination. Two deaths were deemed accidental (smoke inhalation and traumatic head injury) and the other four were caused by heart failure, cerebral bleeding, pneumonia, and myocardial infarction and occurred in participants who had established risk factors for those conditions. Site investigators classified these six events as unrelated to the study vaccine. No deaths occurred within 30 days after vaccination in the IIV3-SD group.	Account for participants in each group that died during study protocol and provide reasons for death, (do not use euphemisms to mask death or other adverse events)

Three IIV3-HD recipients had serious adverse events categorized by their site investigators as related to vaccination: cranial-nerve VI palsy starting 1 day after vaccination; hypovolemic shock associated with diarrhea starting 1 day after vaccination; and acute disseminated encephalomyelitis starting 117 days after vaccination. All three events resolved before study completion; none resulted in discontinuation from the study. No serious adverse events occurring in IIV3-SD recipients were considered to be related to vaccination by the investigators.	Identify and account for all adverse events—noting time of onset
A total of 99 participants (0.6%) in the IIV3-HD group and 103 participants (0.6%) in the IIV3-SD group discontinued the study owing to serious adverse events, none considered to be related to vaccination. Cardiac disorders and infections were the most frequent types of serious adverse events in both groups (see the Supplementary Appendix for serious adverse events according to organ systems).	Note participant withdrawal for adverse events—even if deemed unrelated to intervention

Source: Luquero, Francisco J., Lise Grout, Iza Ciglenecki, Keita Sakoba, Bala Traore, Melat Heile, Alpha Amadou Diallo, Christian Itama, Anne-Laure Page, Marie-Laure Quilici, Martin A. Mengel, Jose Maria Eiros, Micaela Serafini, Dominique Legros, and Rebecca F. Grais. 2014. "Use of *Vibrio cholerae* Vaccine in an Outbreak in Guinea." *The New England Journal of Medicine* 370: 2111–20.

You may include a figure or table to summarize adverse events (in particular, a Kaplan-Meier plot, which we will cover at the end of this chapter, to illustrate survival between groups if appropriate for the study).

Exploratory Investigations

In exploratory investigations, each journal article covers a broader research question and reports the findings of several mini-investigations that cumulatively advance the understanding of the research topic in a meaningful step. The Results section of exploratory investigations is more flexible than the Results section of experimental (or hypothesis-) testing research because the entire research protocol (or all the mini-investigations) of exploratory research is rarely predetermined. Rather, this research usually followed multiple lines of investigation, often simultaneously, with some paths providing fruitful data and others being dead ends. The Results section, however, is where you and the research team construct a logical, post hoc narrative (story) around the fruitful data from the relevant mini-investigations (Cetin and Hackam 2005; Harmon and Gross 1996; Huth 1982; Kotz and Cals 2013a). To construct "readable" scientific articles, Hertzel Gerstein (2014) of McMaster University and Hamilton Health Services gives this advice:

> Tell a story. Storytelling is perhaps the oldest of the social activities practiced by humans. The closer your document is to a logical and easily remembered narrative the more effective it will be. . . . Facts that are woven into the concept have a greater impact than isolated facts.
>
> *(4)*

From a communications perspective, Richard Johnson-Sheehan and Scott Rode (1999) from the University of New Mexico examined the similarities between scientific discourse and narratives. They found that "across scientific eras . . . scientists have used narratives to invent rational interpretations of their experiences with nature" (337). Mary Coney (1992) of the University of Washington adds that the goal is to "present that [scientific] knowledge in a credible form to the audience" (154).

Writers of exploratory research often reorder some results of the mini-investigations from the actual order in which the data were obtained to create a logical organization of the results, which becomes the narrative. Additionally, writers of exploratory investigations often include justifications for each mini-investigation from the scientific literature to tie together each mini-investigation and logically lead from one to the next. For brevity, however, writers rarely report irrelevant data from mini-investigations reaching a dead end unless those findings provide important clarification or insight (Matthews, Bowen, and Matthews 2000; Nair 2005). For example, in my research of *Shigella flexneri* (a bacterium related to *E. coli*), I organized the Results section to present the data of each mini-investigation in an order that was based on the infection process (see Example 7.11). This linear storyline provides a structure that audiences can follow to understand the findings.

Each subsection corresponds to the results from a single data set and includes (1) a short review stating the reason for each procedure, (2) an overview of the methods, (3) the data from the mini-investigation, and (4) a summary or a brief interpretation of the data that leads to the rationale for the next mini-investigation. This pattern then repeats for each mini-investigation until cumulatively making a sufficient contribution that addresses the purpose of the research stated in the Introduction section.

You should compose each subsection for the Results section of exploratory research as follows (see Example 7.12):

1. Begin with a descriptive subheading.
2. Provide a brief background summary that explains why the mini-experiment was performed (Kanoksilapatham 2005). Typically authors include a

EXAMPLE 7.11 EXCERPT FROM A RESULTS SECTION SHOWING A STORYLINE NARRATIVE THAT ORGANIZES RESULTS INTO A LOGICAL ORDER.

The results of two subsections are organized into an overall organization structure that reflects the order of the environmental conditions that the bacteria are exposed to during infection. The order and the rationalization for each mini-investigation are often developed post hoc.

dksA Is Required for Intercellular Spread of *Shigella flexneri* via an RpoS-Independent Mechanism

RESULTS

Acid Sensitivity of the dksA and rpoS Mutants — Descriptive subheading

DksA regulation of RpoS . . . may be important for survival of *S. flexneri* during passage through the host stomach. — Establish the storyline for the mini-investigation

S. flexneri is highly resistant to acidic conditions and is capable of surviving at pH 2.5 for several hours (12, 38). This ability has been attributed to expression of RpoS-dependent genes, and an *rpoS* mutant is extremely acid sensitive (46). — Establish the scientific foundation

Since *S. flexneri* encounters acidic conditions during transit through the stomach and possibly in other locations within the host, — Storyline (continued)

we compared the in vitro acid resistances of *dksA* and *rpoS* mutants by examining the survival of stationary-phase cultures in LB at pH 2.5. — Overview of methods

Both the *rpoS* and *dksA* mutants were significantly more sensitive to acid than the parental strain. The *dksA* mutant exhibited a sharp reduction in acid survival during the first hour of incubation in LB at pH 2.5 (Fig.1) and was approximately 100 times more sensitive than the wild type. The *rpoS* mutant was even more sensitive to acid than the *dksA* mutant, exhibiting sensitivity below the detectable level of 0.001% survival. — Results

Since the *dksA* mutant was more resistant to acidic conditions than the *rpoS* mutant, we postulate that DksA regulation of RpoS may only partially contribute to reduced virulence of *S. flexneri* during passage through the host. — Summary of the conclusion of the mini-investigation in the context of the storyline

This regulation may also contribute to survival of other environmental stresses encountered during the infection process. — Establish a transition to the next mini-investigation

Excerpt	Annotation
Oxidative Sensitivity of the dksA *and* rpoS *Mutants*	Descriptive subheading
Following the acidic conditions in the stomach, *Shigella* bacteria are exposed to another Rpo-S dependent environmental stress,	Continue the storyline to provide the rationale for the next mini-investigation
oxidative stress (8), during intercellular infection in the intestine.	Establish the scientific foundation
To test the sensitivity of the *dksA* mutant to oxidative stress, we exposed undiluted, stationary-phase cultures to oxidizing agent cumene hydroperoxide.	Overview of methods
All strains tested exhibited sensitivity to oxidative stress, but the *dksA* and *rpoS* mutants were significantly more sensitive than the wild type (Fig. 3).	Results

Adapted from Mogull, Scott A., Laura J. Runyen-Janecky, Mei Hong, and Shelley M. Payne. 2001. "*dksA* is Required for Intercellular Spread of *Shigella flexneri* via an RpoS-Independent Mechanism." *Infection and Immunity* 69: 5742–51.

brief summary of the relevant research findings directly related to the mini-investigation.

3. Briefly summarize the goal or objective of the mini-investigation.

    a. Provide an overview of the methods for the mini-investigation (Kanoksilapatham 2005). This overview does not provide the depth of detail covered in the Methods section but rather provides sufficient context for the data that will be reported. As Plaxco (2010) from the Interdepartmental Program in Biomolecular Science and Engineering at the University of California Santa Barbara advises authors, you should explain the investigation in "broad terms, as most of your readers will initially skip over the methods section and jump right to the 'meat'" (2264).
    b. Justify or support the approach by referencing previous research that used the approach in a similar context and influenced the choice of procedures (Kanoksilapatham 2005).

4. State the results and highlight key data; results and key data are the substance of each Results subsection. To write this section, follow the information provided in the "Experimental (or Hypothesis-) Testing Research" section under "Subsection 3: Major Findings." Begin with step 2, in which you do the following:

    a. Lead with a clear results summary statement,
    b. Provide key statistics that summarize the data,

c. Highlight key data (either quantitative or qualitative) in the text,
d. Reference a visual summary (figure or table) of the data set, and
e. Highlight any inconsistent observations or outlying data (if observed).

5. Interpret the results by briefly summarizing, explaining, or making generalizations of the findings in the subsection in order to establish the foundation for the next Results subsection (or to transition the Discussion section, if it is the final Results subsection).

This relatively consistent organizational structure for subsections of exploratory research follows an abbreviated IMRAD structure within the Results section. When you reach the Discussion section, the cumulative impact of these findings will be described in greater detail.

When writing the Results section, you should critically evaluate each data set for relevance to the primary narrative. Writers of exploratory research, in particular, are of greater risk for creating an unfocused Results section (and subsequently problematic Discussion section) because the connections between each data set must be established for the reader and logically lead to the next. In this

EXAMPLE 7.12 RESULTS SUBSECTION FROM AN EXPERIMENTAL INVESTIGATION RESEARCH ARTICLE.

Sources of Dengue Viruses Imported into Queensland, Australia, 2002–2010

RESULTS — Section heading

Increasing Incidence of Dengue Outbreaks and Serotype Diversity — Descriptive subheading

Previous reports (3,11) and anecdotal evidence indicated that there has been an increase in the number of dengue outbreaks occurring in Queensland. To investigate this apparent trend, we combined recent and historical outbreak data (3) over a 20-year period. A 5-year moving average does indeed show trends of increasing dengue outbreak incidence and increasing diversity of DENV [Dengue virus] serotypes that cause such outbreaks (Figure 1, panels A, B). All 4 DENV serotypes caused outbreaks; DENV-2 was the most common cause (50.0%), followed by DENV-1 and DENV-3 (19.4% each) and DENV-4 (11.1%) (Figure 1, panel C). The increase in outbreak incidence reflects changes in international travel over this period, which has increased 3.5-fold

Brief background

Goal or objective statement

Overview of methods (use active voice)

Overview of results

Data to support results statement are presented in figures (or tables)

Highlight most important data

since the early 1990s (12). This increase is consistent with increased importations of virus carried by viremic travelers and the recognized increase in DENV infections throughout the world (13).	Brief interpretation (further analysis is presented in the Discussion section)

Source: Warrilow, David, Judith A. Northill, and Alyssa T. Pyke. 2012. "Sources of Dengue Viruses Imported into Queensland, Australia, 2002–2010." *Emerging Infectious Diseases* 18: 1850–7.

regard, you should review some of the key questions introduced previously for evaluating each data set. In particular, you should ask yourself:

- Does reporting the findings cause the Results section to deviate the flow of the narrative from focusing on the primary evidence?
- Would the findings be more appropriate in another journal article with a different purpose or hypothesis?
- Are you reporting the findings because you (or another person on your research team) have a passionate attachment to the research method, data set, figure, or table?
- Would reporting the findings potentially confuse a casual reader?

Depending on your answers to these questions, you may decide (appropriately) to exclude a data set from a particular journal article or you might expand the scope presented in the Introduction. Alternatively, you may find (as researchers often do) that the narrative requires further research (in the laboratory, literature, or both) to complete and connect the story. As you can see, reporting the results in exploratory investigations is often more complicated than in experimental testing. As with any communication, if you do not clearly understand (or cannot articulate) the connection between the data you are providing to the reader, then your readers will likely be confused. Journal editors and reviewers expect *you* to articulate the connection between data in *your* article.

Summarizing and Presenting Data

Data in the Results section provide the evidence for the claims that you make in the Discussion section of journal articles. Each piece of evidence in the Results section contributes to developing the case for the Discussion and Conclusion. Often this evidence is provided as data summaries or visuals documenting a phenomenon or process (Lynch and Woolgar 1990). The format of this evidence depends on the information to be communicated and, despite popular misconceptions, is rarely interchangeable with another format (Amann and Knorr-Cetina 1990; Desnoyers 2011; Doumont and Vandenbroeck 2002; Durbin 2004;

Gross 1983, 1990, 2007; Kostelnick 1998, 2007; MacGregor 1982; Szoka 1982). These are the categories of data summaries:

- TABLES—usually summarize quantitative or categorical data in numbers and statistics, words, or symbols. Data are presented in tables when the exact measurements or values are important.
- GRAPHS (or CHARTS)—summarize quantitative data in which the trends or relationships between the data are important.
- STATISTICS—summarize a data set or are used to extrapolate from a sample to a larger population.

These are categories of visuals documenting a phenomenon or process:

- PHOTOGRAPHS and INSTRUMENT OUTPUT—provide direct evidence of a phenomenon and serve as documentation for future claims.
- DIAGRAMS—substitute for photographs to show the relevant features of a visual phenomenon.

In most cases, you should present the data in one of the preceding categories. But in rare situations, if you have an extremely simple data set, you might consider reporting the data in the text. Yet even with relatively simple data sets, text is often a poor format for presenting data (see Examples 7.8 and 7.9). Rather, use the text to highlight the results (see Example 7.1) and report key or outlying data.

Before we begin examining visuals in more detail, we need to examine their role in journal articles. Unlike visuals in popular media that are used to attract a general audience, visuals in SMC provide evidence for the purpose of convincing an expert audience of the validity of the conclusions (T. Miller 1998). Effective visual communication requires careful attention to the message being communicated through the data. You should have a single, clear point for each visual that you can articulate in a sentence and use as a guide through the selection and design of the visual (Szoka 1982). Furthermore, each visual must be a stand-alone communication in which a reader can clearly decipher the message from the visual without reading the article (Kotz and Cals 2013b). Finally, to effectively communicate visually, you must adhere to the standards for the graphical format. As Kosslyn (1989) of Harvard University stated, "[Visual] displays necessarily use symbols, which are marks that are interpreted in accordance with convention." Yet few scientific authors are properly trained to communicate visually (Bayer 1991; Frankel and DePace 2012). As a result, approximately 30% to 50% of graphs published in scientific journals contain errors in the design and presentation of information (Cleveland 1984b; Cooper, Schriger, and Close 2002; Krebs et al. 2001). Therefore, you should follow the design guidelines presented in this section and in the Instructions to Authors and style guides rather than blindly try to copy examples from published articles.

As you are planning your visuals, you should note that many journals (such as *Science* and *Nature*) and grant-funding organizations (such as Howard Hughes Medical Institute, the National Institutes of Health, and the Wellcome Trust) require authors to share complete data sets used as evidence in journal articles with other researchers (Nature Neuroscience 2007). The Committee on Responsibilities of Authorship in the Biological Sciences (2003) stated that:

> an author's obligation is not only to release data and materials to enable others to verify or replicate published findings (as journals already implicitly or explicitly require) but also to provide them in a form on which other scientists can build with further research.
>
> *(34)*

Although data sharing has been a long-standing expectation of publication, the practice is evolving with multiple online repositories for data (Campbell et al. 2002; Harris 2011; Vines et al. 2014). Each field and data type has a different database resource that is the primary repository, such as GenBank for DNA sequences or ClinicalTirals.gov for clinical trial data. Usually the Instructions to Authors specifies the appropriate database for different data sets and provides guidelines for referencing the data set in the journal article. The journal *Scientific Data* provides a current list of recommended data repositories for each data format (http://www.nature.com/sdata/data-policies/repositories), which also includes repositories for data from figures. To save time later in the publication process, you should identify the appropriate database and prepare the data set for submission as you prepare the visuals.

The following section focuses on the information planning and design phases of visuals. The primary objectives of this coverage are for you to be able to identify the correct reporting format for the data and to present data clearly. The ultimate formatting of visuals, which includes lines and line weights, data-point symbols, color and shading, and acceptable file formats, varies by journal and discourse community. Therefore, you must also refer to recent examples of visuals from your target journal, the Instructions to Authors, and discipline-specific style guides to provide you with those formatting requirements.

Tables

Tables display precise measurements or characteristics in a systematic matrix for easy identification and comparison of items (Buehler 1977; Horton 1991; Wright and Fox 1970). Although they often contain numerical data, tables may also contain words and symbols and therefore can communicate nonnumeric information. The three general categories of tables include:

- Quantitative data tables—contain numerical data and are used when the specific, precise values are important and the amount of data presented would be confusing to present in the text (see Example 7.13). Because data tables show only numerical values, you must explain and interpret the data trends in the text. If only the trend of the data is important, you should use a graph rather than a table.

- QUALITATIVE DATA TABLES—contain words or symbols to identify and compare items (see Example 7.14).
- REFERENCE TABLES—contain a list of similar items and serve as a reference for easy comparison (see Example 7.15).

Very simple data sets of a single variable may be efficiently listed in the text (Kozak and Hartley 2012). But tables are used to effectively communicate more complex data sets, particularly of multiple variables and groups, and to allow comparison of specific data.

EXAMPLE 7.13 QUANTITATIVE DATA TABLE.

Number and percentage of norovirus gastroenteritis outbreaks, by genotype, strain, setting, and mode of transmission in Minnesota, Ohio, Oregon, Tennessee, and Wisconsin (August 2012–April 2013*).

	No. (%) Outbreaks with Sequence Data			No. (%) Outbreaks with No Sequence Data, n = 279	Total No. (%), N = 637
Outbreak Mode of Transmission	**GII.4 Sydney, n = 226**	**Non-GII.4 Sydney, n = 132**	**RR (95% CI)**		
Person to person	172 (76.1)	93 (70.5)	1.08 (0.95–1.23)	216 (77.4)	481 (75.5)
Foodborne	35 (15.5)	28 (21.2)	0.73 (0.47–1.14)	28 (10.0)	91 (14.3)
Water	0	1 (0.8)	NA	0	1 (0.2)
Environmental	0	2 (1.5)	NA	0	2 (0.3)
Other/ unknown	19 (8.4)	8 (6.1)	1.39 (0.62–3.08)	35 (12.5)	62 (9.7)

* RR, rate ratio; NA, not available.

Adapted from Leshem, Eyal, Mary Wikswo, Leslie Barclay, Eric Brandt, William Storm, Ellen Salehi, Traci DeSalvo, Tim Davis, Amy Saupe, Ginette Dobbins, Hillary A. Booth, Christianne Biggs, Katie Garman, Amy M. Woron, Umesh D. Parashar, Jan Vinjé, and Aron J. Hall. 2013. "Effects and Clinical Significance of GII.4 Sydney Norovirus, United States, 2012–2013." *Emerging Infectious Diseases* 19: 1231–8.

EXAMPLE 7.14 QUALITATIVE DATA TABLE USED TO COMMUNICATE NONNUMERIC INFORMATION.

Comparison of wild-type and mutant *S. flexneri* in Henle cell monolayers

Strain	Phenotype	Result of assay for:	
		Invasion[a]	Plaque formation[b]
SA100	Wild type	+	+
SA2287	DksA⁻	+	–
SA5287	DksA⁻	+	–
SA5287/pSAM1	DksA⁺	+	+
SA710	RpoS⁻	+	+

[a] +, at least 30% of the Henle cells contained three or more intracellular bacteria in the invasion assay.

[b] +, formation of wild-type-size plaques in Henle cell monolayers; –, pinpoint or no plaques.

Source: Mogull, Scott A., Laura J. Runyen-Janecky, Mei Hong, and Shelley M. Payne. 2001. "*dksA* Is Required for Intercellular Spread of *Shigella flexneri* via an RpoS-Independent Mechanism." *Infection and Immunity* 69: 5742–51.

EXAMPLE 7.15 REFERENCE TABLE USED TO ORGANIZE AND COMPARE SIMILAR ITEMS.

Primers used to map the putative recombination event upstream of the *pbp2x* gene of *Streptococcus pneumoniae*, Spain, 2004–12.*

Primer	Sequence, 5′ → 3′	Position†
F1	AAATCCGAAGAAGGCCAAAT	278234–278253
R1	GTACTTGAGATTGGCGTGTTTG	278834–278813
F2	TCAATGACTGTGATGCCTGTT	290699–290719
R2	TGTCAGACAAATAGGACAAGGAGA	291315–291292
F3	GTCAATGACACCAACCTCTTG	282168–282188
R3	GCTATGAGCCATTCTAGCAAAGA	283037–283015
F4	TGAATGTAAAGACACACGAGGAA	273278–273300
R4	CAGTGATAACGAATACCATACAGAA	274128–274104
F5	CAGCTCTATGAACACCGGACT	289177–289197
R5	TTCCTAGTCGTAACCATCATTTCA	289927–289904
F6	CCTTGGATACGGGTATTCGTT	287148–287168
R6	GCAGTCGCTTGACCTTTTCT	287756–287737
F7	GTGGACAGGAAGCAAAGCTC	275192–275211
R7	GGCAGTCAGATTTGCAGACA	276056–276037

* The penicillin-binding protein 2x gene (*pbp2x*) (SPG_0305) of *S. pneumoniae* G54 strain (serotype 19F-ST63; GenBank accession no. CP001015) is located between positions 292638 and 294890 of its genome. F, forward; R, reverse.

† Corresponding to the genome of the G54 strain.

Adapted from Ardanuy, Carmen, Adela G. de la Campa, Ernesto García, Asunción Fenoll, Laura Calatayud, Emilia Cercenado, Emilio Pérez-Trallero, Emilio Bouza, and Josefina Liñares. 2014. "Spread of *Streptococcus pneumoniae* Serotype 8-ST63 Multidrug-Resistant Recombinant Clone, Spain." *Emerging Infectious Diseases* 20: 1848–56.

How to Prepare

Every table should have a single purpose that you can articulate in a sentence and that you use to guide the organization of data. First, you should identify how the audience will need to use the data or information that you provide in the table (Wainer 1997). (If you are trying to communicate multiple ideas in one table, then consider dividing the information into multiple tables.) Then, the content should be properly focused and limited to the information necessary to accomplish the goal.

What to Expect

Some readers prefer to look at tables and not read the text, so tables must be a complete, stand-alone unit of communication. Self-contained tables provide readers with all of the information needed to understand the point of the table. That means the table should have a descriptive title, descriptive column and row labels, precise numerical values in every cell, clearly identified units of measurement in columns or rows, and, if necessary, footnotes for elaboration. Additionally, data table design should be easy to comprehend, only requiring the audience to search for the information (Wright and Fox 1970). Occasionally, IMPLICIT TABLES are used to save space but require the audience to perform some calculations or interpretations from the information provided in the table in order to gather complete information. For example, the table in Example 7.2 provides the number and percentage of only female participants. In this implicit table, the number or percentage of male participants cannot be referenced but must be calculated by the readers. Implicit tables, therefore, include all the information needed to calculate the necessary information but require the readers to deduce some items. For clarity, you should use explicit tables unless an implicit table is standard (as in the medical journal article example).

This section guides you through the organization and design process, explaining how to (1) plan a table, (2) present data, (3) format publication-quality tables, and (4) integrate tables into the text.

Planning a Table

In published SMC, tables are used to present meaningful summary metrics or information and should not be used as a database of raw data or to show stepwise calculations from raw data to meaningful metrics (Jane Miller 2004). Often data tables created during the data collection phase of research need to be redesigned and reorganized to clearly communicate the information in publication (Ng and Peh 2009c; to compose a table in Microsoft Word, refer to Appendix D).

A basic understanding of table structure will help you to organize and present content in a table. Tables are an intermediate communication format between text and graphs, and the conventions of both text and graphs influence table organization (Buehler 1977; Gross 1983). The elements at the top of the table, the BOXHEAD, identifies the content of each vertical COLUMN (see Figure 7.2). The left-hand column, or STUB, identifies the content of each horizontal ROW. Similar to a graph, the intersection of the vertical (column) and horizontal (row) is a single DATA CELL, which provides a value that meets both criteria listed in the column head and row heading (see Figure 7.3). As you can see from this matrix structure, tables are particularly effective at presenting an organized summary of repetitious information when several similar items are evaluated under one or more similar conditions.

The first major decision for designing a table is determining which features to list in the boxhead (top) and stub (left side). Sometimes writers have difficulty deciding which categories to list in the columns versus the rows, but adhering to certain conventions enhances the readability of tables. To help you begin the process, I have summarized the most important criteria, which should be considered in the following order:

1. **Enable comparison of data vertically between columns.** Often, dependent variables are listed in columns (boxhead) and independent variables are listed in rows (stub) (Iverson et al. 2007). Then, all similar measurements are listed in individual rows within each column. You can confirm that you properly have followed these conventions by ensuring that the data or other content in each column are the same type and format and, if numerical, have the same unit of measurement. As a general rule, put any redundant information in the column head and not in the data field. Therefore, you

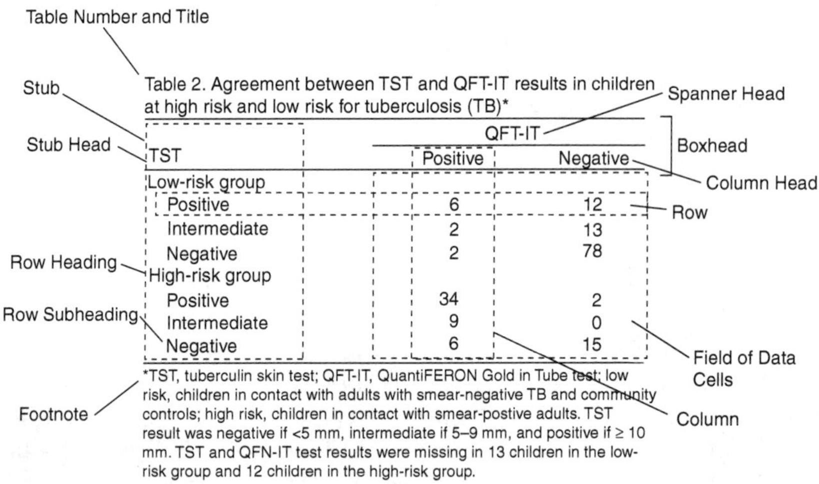

FIGURE 7.2 Annotated anatomy of a table. (Table from: Nakaoka, Hiroshi, Lovett Lawson, S. Bertel Squire, Brian Coulter, Pernille Ravn, Inger Brock, C. Anthony Hart, and Luis E. Cuevas. 2006. "Risk for Tuberculosis among Children." *Emerging Infectious Diseases* 12 (9): 1383–8.)

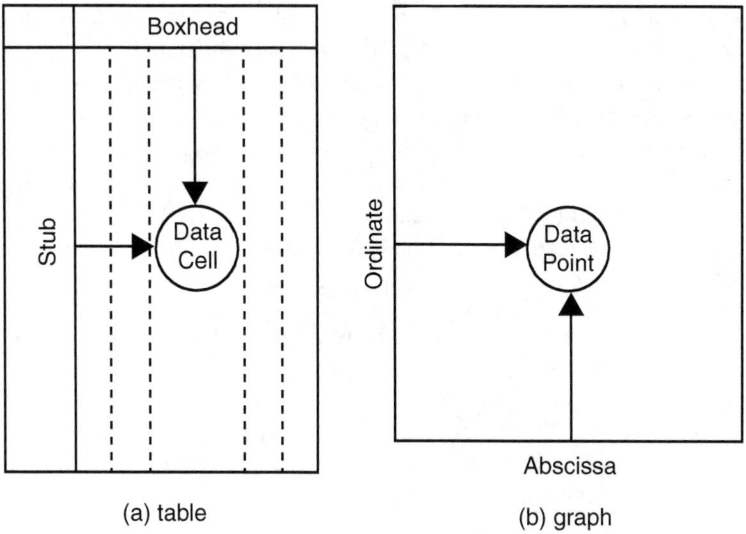

FIGURE 7.3 Matrices of (a) tables are similar to (b) graphs in that the intersection of the horizontal and vertical scales references a specific coordinate with a unique relationship to both dimensions.

Adapted from Buehler (1977).

should include the unit of measurement in the column head found in the boxhead and not in the data cells.

2. **Divide groups in parallel subgroups and nest under headings.** The division of information should be parallel (or at equivalent levels) within the columns and within the rows. Both columns and rows can be nested (as in Figure 7.2). When nesting information, all of the lower level subgroups must be part of the higher level group—much like constructing the subgroupings in an outline. Generally you should limit the nesting of groups to one level below the primary heading because tables can quickly become overly complex and therefore difficult to read. Limited use of horizontal rule lines helps to delineate subgroups between columns, as illustrated in Figure 7.2.
3. **For column order, place initial measurements or standard information to the left of newer or derived information.** Since English is organized from left to right, readers will expect newer information located to the right of older information (Gopen and Swan 1990; Wright and Fox 1970). One application of this principle is that any calculations or derived statistics should be placed to the right (or below) the original data values. As illustrated in the first data column in Example 7.13, the data values appear above the percentages, the derived values, which are placed in parentheses. Also in Example 7.13, the derived statistics from the first two data columns (the "RR" or "rate ratio")

is placed in the third data column to the right of the raw data, and the data totals are summarized in the column furthest to the right.

4. **For row order, use a logical and predictable organizational scheme that is appropriate for the categories.** Organizing the order of rows requires you to consider the best organization scheme for the information (Jane Miller 2004; Wainer 1997; Wright and Fox 1970). You might be able to organize the same content in different orders. But once you select an organization scheme, you should use it consistently. Some typical organization schemes are listed below:

   a. If an inherent organization exists for the information, such as time, then you should use that.
   b. You can highlight the most important information by placing it above less relevant information. You can do this by organizing categories from highest to lowest value in the key data measure, or arrange the categories in the stud from greatest to least in importance or relevance.
   c. Alphabetical order is useful for long reference tables, such as listing organisms by scientific name. This organization scheme reflects the names rather than a physical quality of the subject so it should be reserved for tables designed to look up information rather than directly compare between the data.

   If necessary, use indentation for subgroups below the row head (see Figure 7.2). Finally, reserve the bottom row for generic categories such as "unknown" or "other" or to report totals.

When organizing tables, you need to be judicious and actively evaluate the placement of information in the columns and rows to determine the most effective presentation. Some journals even limit the number of columns to 10 because that is the approximate maximum number of columns that can be displayed on the page (Hamilton 2009; Iverson et al. 2007). In situations with a large number of columns, you may need to design separate tables or switch the content with rows since printed tables can accommodate more rows than columns.

Presenting Data

After you have organized the categories into columns and rows, you are ready to add data to each data cell. The goal is to report data equivalently and consistently to facilitate comparisons between values. Data throughout the table should be reported in standard measurements, often using base units such as meters for length or kilograms for weight (see Appendix E). But the appropriate measurement to report depends on the phenomenon, measurement instrument, and audience expectations for the type of data. For example, you may report time in hours, days, or months and temperature as degrees Celsius or Fahrenheit. Common and accepted measures are often listed in the Instructions to Authors and style guide and may also be found in tables of similar data published in the target journal.

For numbers that are integers, use decimals rather than fractions. Furthermore, you should adjust the scale of the measurements to include the least number of zeros (such as millimeters for very small objects or kilometers for long distances; see Appendix E). Once you have selected the measurement and scale that are appropriate for the majority of the data in a table, convert all data measurements to that scale and report values to a consistent number of decimal places. But, do not add additional digits after the decimal place if the value would exceed the proper SIGNIFICANT FIGURES, a reflection of accuracy in measurement, because doing so would imply greater precision than what the measurement obtained. In general, significant digits are maintained when adding and subtracting numbers. But the number of significant digits is limited to the number of digits in the denominator when dividing (as in determining the percentage; Iverson et al. 2007).

After you have converted the data into a uniform measurement and scale, the values should be consistent down each column. Therefore, list the unit of measurement in the column heading rather than in each data cell (as in Example 7.5). If reporting numbers less than one, be sure to include a zero before the decimal point so that the readers do not miss seeing the decimal and therefore greatly misread the data value. Then, align numbers along the units column or decimal point and separate thousands by a comma (as in 1,000) or by another method (such as a space as done in some journals). The proper style for the journal will be specified in the Instructions to Authors.

Generally you should not leave data cells empty because doing so can lead to misinterpretation by the reader (Council of Science Editors 2006). As illustrated in Example 7.13, a "0" reports the value from a measurement whereas "NA" identifies when data was not collected, available, or reported. Additionally, you should define all abbreviations in footnotes at the bottom of the table (even common abbreviations such as NA, which may mean "not available," "not applicable," or "not analyzed").

Formatting Publication-Quality Tables

Journals have specific formatting requirements for tables. Tables intended for publication should not look like the spreadsheet output from a software program nor should they be designed using one of the decorative table-design options in word processing programs. When designing tables, you will need to refer to the Instructions to Authors and to tables recently published in the target journal. Journals often require a basic design as illustrated in Examples 7.13, 7.14, and 7.15 (Iverson et al. 2007). Here is a list of key table-design guidelines:

1. **Use the table editor in word processing programs to create tables.** Many journal editors prefer tables for publication to be submitted in Microsoft Word format and some do not accept Microsoft Excel format. When

creating a table, you need to specify the appropriate number of columns and rows, merging headings that span multiple columns, and reporting data in individual data cells. Using the table editor in Microsoft Word (see Appendix D) enables proper alignment of content and facilitates the publication process (Frels, Onwuegbuzie, and Slate 2010). You should never use tabs, spaces, or carriage returns to construct a table (Iverson et al. 2007) because the alignment may be imprecise and these nonprinting characters will complicate the typesetting process.

2. **Limit use of rule lines.** The basic format for tables includes only three horizontal rule lines (as in Example 7.15): One above the boxhead, one below the boxhead, and one at the bottom of the table field but above the footnotes (Hamilton 2009). Additional, but limited, use of horizontal rule lines may be needed to nest columns under a common heading or to separate different sections of the data within the body of the table (see Example 7.14). In most journals and fields, vertical rule lines are uncommon.
3. **Use text alignment (left, center, right) to align content in data cells.** Do not use spaces or tabs to align text. Content in the stub is aligned along the left margin and indented for subgroups (see Figure 7.2). Text in the boxhead and data cells in the field is typically centered. Numerical data are typically aligned along the decimal place—either centered or aligned to the right margin of the column.
4. **Include a descriptive title for each table.** The title should explain the content of the table but not interpret the results (Ng and Peh 2009c).
5. **Order footnotes sequentially.** Footnotes may be short phrases or complete sentences, but they typically always end with a period. Footnotes should be ordered by beginning in the boxhead from left to right and then continuing to each subsequent row following the same pattern (from left to right). Journals may use superscript letters or symbols for footnotes (see the Instructions to Authors). Within each table, footnotes are organized alphabetically or, for symbols, the journal will specify the proper order (Hamilton 2009). Numbers are less commonly used for footnotes to avoid possible confusion with data values (Council of Science Editors 2006). Footnotes are listed at the bottom of each table below the bottom horizontal rule line. Since footnotes appear at the bottom of the table rather than the bottom of the page, you will need to manually add footnotes in superscript text (rather than using an automated footnoting feature in a word processing program).

Typically, journals require authors to place each table on a separate page at the end of the manuscript (or in a separate file for independent uploading to online submission systems). Most journals editors prefer that you include all of the content for a table (including the title and footnotes) on a separate page or document. Large tables are rare but occasionally necessary in SMC (Hartley 1991). In general, limit tables to a single page and keep the table in the vertical, or portrait orientation. If the number of columns necessitates a wider space, then you may reorient a table horizontally in landscape orientation. If the amount of information in a table flows to another page, write "continued" at the bottom of each page from which it continues. On each following page, repeat the boxhead containing all the column headings and then continue with the new rows.

Integrating Tables into the Text

As we discussed previously, the purpose of a table should be self-explanatory or evident from the information that you include in the table itself, title, and footnotes. In other words, a well-designed table does not require the readers to interpret, speculate, or refer to the text to understand the meaning of the content (American Psychological Association 2010; Kotz and Cals 2013b; Ng and Peh 2009c). But you should not simply insert a table into a document and expect readers to infer your reason for including the table. Rather, you should state the main point of a table in the text.

To effectively integrate a table in prose, (1) explicitly state the main point and incorporate a parenthetical reference (such as, "see Table 1") to tell the reader when to look at the table, (2) identify and explain the important data or information from the table to support your point, and (3) acknowledge any outlying data values.

The text of the manuscript will determine the order of tables. You should begin the numbering of tables in the order that each is first mentioned in the text. For clarity, use consistent terms, measurements, and units between the text and table. Some readers will not interrupt their reading of the text to view a table, so you must thoroughly communicate the important concept in the text.

Below are several of the most common errors found in tables in SMC (Ng and Peh 2009c):

- Tables in the manuscript are not cited in the text,
- Tabular data do not agree with the data given in the text,
- Inaccurate arithmetic for columns (e.g., numbers do not add up),
- Information in the tables is parroted in the text, and
- Information in the tables is duplicated in graphs.

After you have created a table, you should use this checklist to ensure consistency, accuracy, and proper use of each table with the rest of your manuscript.

Graphs (or Charts)

Graphs are the most prevalent visual in scientific communication, with greater than 90% of the articles published in *Science* containing one or more graph (Mogull and Stanfield 2015). Graphs provide a visual representation of quantitative data and are particularly effective for consolidating and representing an entire data set into a single visual communication. Unlike text and tables, graphs do not have a defined, linear viewing order and permit readers to visualize the overall trend of the data (Doumont and Vandenbroeck 2002). Graphs employ established conventions and symbols, which underlie their effectiveness in communicating quantitative information (Kosslyn 1989). When used appropriately, graphs facilitate the communication process (Kostelnick 2007; Meyer, Shamo, and Gopher 1999).

Scholars have argued that compared to text and tables, graphs are a more advanced form of communication and that the widespread use of graphs in the sciences partially contributes to the power and authority of scientific discourse (Amann and Knorr-Cetina 1990; Bastide 1990; Bazerman 1988; Cleveland 1984b; Krohn 1991; Latour 1990; Lynch 1990; Smith et al. 2002; Wickman 2013). Yet many studies report that graph selection is often inappropriate and design is inadequate in published scientific literature (Connor 2009; Cooper, Schriger, and Close 2002; Cooper, Schriger, and Tashman 2001; Krebs et al. 2001; Levine, El-Nahas, and Asa 2010; Schriger et al. 2006). Therefore, to enable you to communicate accurately and effectively through graphs, this section addresses (1) selecting the proper graph; (2) organizing and plotting data; (3) designing clear, data-rich graphs; and (4) integrating graphs into the text.

Selecting the Proper Graph

Although many different types and variations of graphs exist, a few standard types of graphs are used primarily in SMC (Cleveland 1984b; Mogull and Stanfield 2015; Ng and Peh 2009a; Schriger et al. 2006). Generally speaking, the most common graphs are bar, line, and scatter graphs (Arsenault, Smith, and Beauchamp 2006; Cleveland 1984b; Cooper, Schriger, and Close 2002; Howarth and Turner 1987; Mogull and Stanfield 2015; Schriger et al. 2006). Pie graphs are uncommon in expert-to-expert communication because text or a table more precisely communicates percent composition, so the printing of a pie graph would be less precise and occupy more space (Cooper, Schriger, and Tashman 2001; Schriger and Cooper 2001; Schriger et al. 2006; Stengel, Calori, and Giannoudis 2008). Some discourse communities have standard variations of these common graph types. For example, the Kaplan-Meier plot is a variation of a line graph showing the survival from different medical treatments (see Figure 7.4). Because the selection and construction of graph variants are based on the three standard graph types, we will limit our discussion to these graphs (see Table 7.2).

Different types of data are typically represented in different graphs, and graphs are rarely, if ever, interchangeable. Selecting the proper graph facilitates communication because readers are accustomed to reading and interpreting the data differently in specific graphs (Doumont and Vandenbroeck 2002; Kostelnick 2007). The proper choice of graph depends on the data set, the important feature

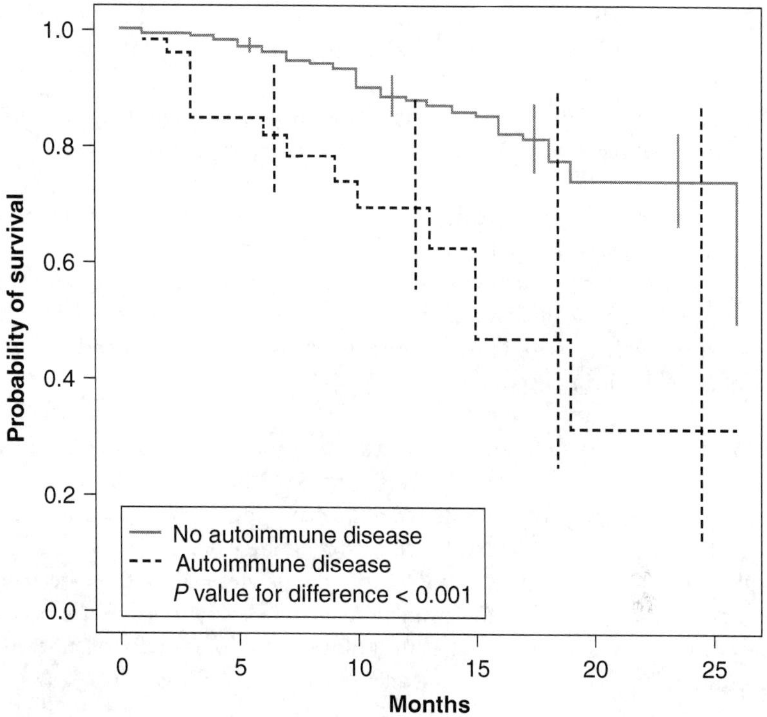

FIGURE 7.4 A Kaplan-Meier plot is a variation of a line graph commonly used in medical discourse communities to show survival of different groups or conditions. This graph shows survival of one or more groups over time, just like a line graph. The main difference is that a smooth line is replaced by a series of steps that represent each measured interval.

Source: Connor (2009).

reported, and conventions of presentation within the discourse community (Doumont and Vandenbroeck 2002; Kostelnick 1998; Spriestersbach et al. 2009; Tufte 2001). These criteria direct you toward a single graph type (see Table 7.2). Examining both the research question and variables will help keep your graph selection appropriate.

To determine the appropriate graph for data, first evaluate the research question (Doumont and Vandenbroeck 2002). In this case, the research question is defined as the object of each individual mini-experiment yielding a single data set (that you are graphing) rather than the entire research paper. Typically you will show one of the following, in which the type of data is either:

- Comparisons between groups or conditions (use a bar graph),
- Changes over time (use a line graph), or

- Correlations between variables, which are typically limited to two variables because of the nature of experimental design (use a scatter graph).

After you have determined the research question, you can identify the appropriate graph type (see Table 7.3). The type of data is either:

- CONTINUOUS on a measurement scale in which the data can be of any incremental value (such as temperature or weight); or
- CATEGORICAL (or DISCRETE), in which the observations fit into a group. These groups may be NOMINAL, or unable to be ordered (such as male or female), or ORDINAL, which has an inherent order (such as low, medium, and high dose of a medication).

In some cases, as you examine the data, you may notice that continuous data are collected and reported as ordinal values—that is, although the dose of medication may be theoretically administered at any concentration, certain dose levels were selected in the experiment. When reporting these data, you should select a graph that most accurately reflects the experimental design—in this case, selecting a graph that shows discrete groups rather than a continuous scale because the experiment was comparing the effect of different doses (a discrete measure) rather than determining the optimal dose (a continuous measure).

Bar, Line, and Scatter Graphs

In BAR GRAPHS, data are represented as the length of bars along the vertical *y*-axis. Since the exact position of the bar top is the informative feature of a bar graph, you should represent the data in a two-dimensional graph (not three-dimensional, as shown in Figure 7.5) to enable precise estimation of the bar length (Bryan 1995; Connor 2009; Stengel, Calori, and Giannoudis 2008). For proper design of bar graphs, see Example 7.9.

LINE GRAPHS, by convention, measure the change in a variable (plotted along the vertical *y*-axis) over time (plotted along the horizontal *x*-axis; Tufte 2001). Line graphs are, in essence, a sequential scatter graph with the points connected (Doumont and Vandenbroeck 2002; Franzblau and Chung 2012). In line graphs, multiple groups or conditions are represented by different lines. The overlap (or lack thereof) of the lines in these graphs shows the relationships between groups over time (see Figure 7.6). To further represent the similarities or differences between the groups, line graphs should include ERROR BARS (see Figure 7.7), which are the lines extending in the positive and negative directions above and below data points to show the margin of error (Allen, Erhardt, and Calhoun 2012; Boers 2004; Lane and Sandor 2009; Paniello et al. 2011). Error bars may represent one of several statistical measures: standard deviation (SD), standard error of the mean (SEM), or confidence interval (CI). Typically the SD is used in SMC because it

TABLE 7.3 Standard graph selection matrix based on number of variables and objective of the research question.

<table>
<tr><th rowspan="2"></th><th colspan="2">Number of Continuous Variables</th></tr>
<tr><th>One</th><th>Two</th></tr>
<tr><td>Comparison between discrete groups or conditions</td><td>Bar graph
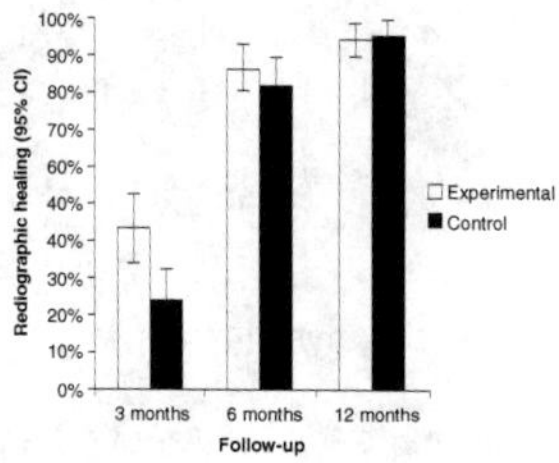
</td><td>NA</td></tr>
<tr><td>Change in one or more groups over time</td><td>Line graph
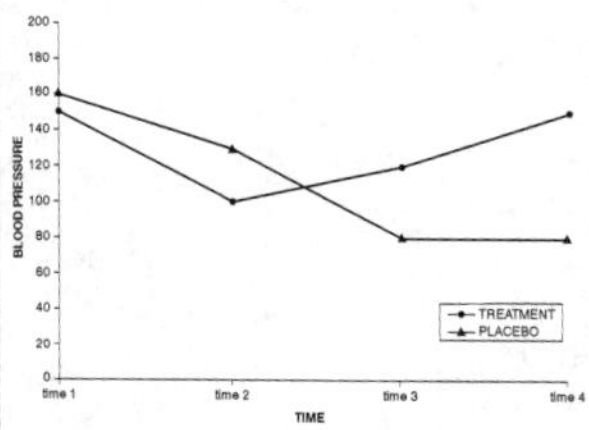
</td><td>NA</td></tr>
<tr><td>Correlation between two variables
(excluding time)</td><td>NA</td><td>Scatter graph
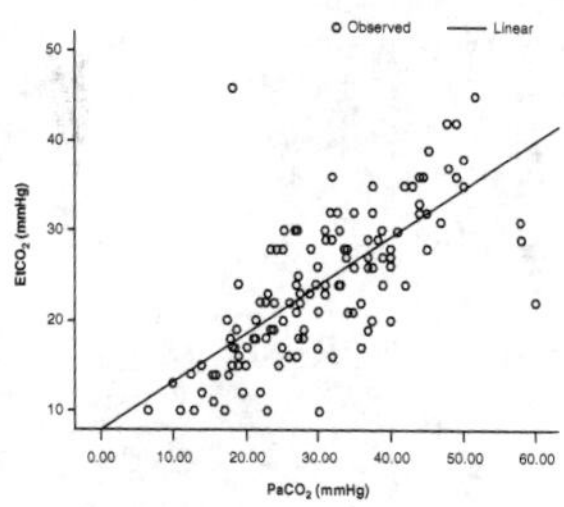
</td></tr>
</table>

NA = not applicable.

Table adapted from Doumont and Vandenbroeck (2002). Bar graph adapted from Stengel, Calori, and Giannoudis (2008); line graph from De Amici, Klersy, and Tinelli (1997); and scatter graph from Ng and Peh (2009a).

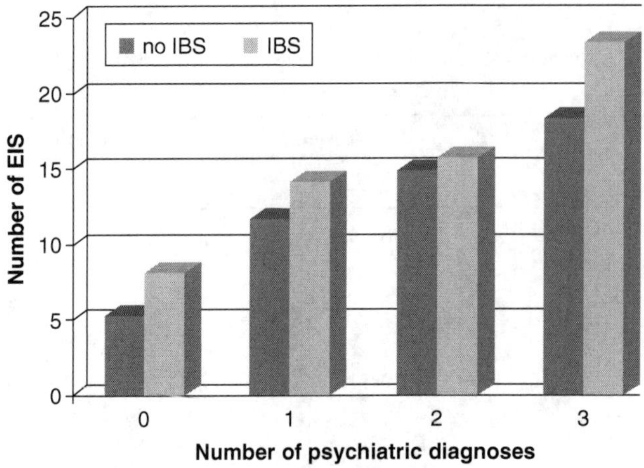

The relationship between number of extra-intestinal symptoms (EIS) and number of psychiatric diagnoses. IBS, irritable bowel syndrome.

FIGURE 7.5 Inappropriate use of three dimensions in a bar graph makes the exact *y*-axis value difficult to read.

Source: Connor (2009).

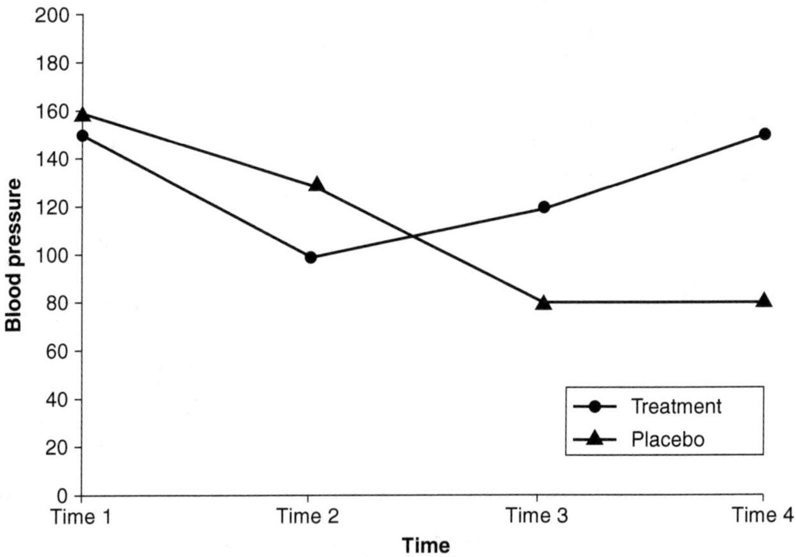

FIGURE 7.6 Sample line graph showing the change between two groups (treatment and placebo) over time. In this basic example, the measurements along the *y*-axis are similar at time intervals 1 and 2 and begin to diverge at intervals 3 and 4. For clarity, error bars should be added to the line graph.

Source: De Amici, Klersy, and Tinelli (1997).

shows the variability (the spread or dispersion) of the data values in the group (Paniello et al. 2011). Both the CI and the SEM show the stability of the central tendency (often the mean). The CI (often set at 95%) is the more informative of the two although the SEM is typically a smaller value and therefore may be clearer when graphed (Boers 2004; Paniello et al. 2011). Because error bars may be one of several measures, you must specify in the caption which measure is being represented. Another common problem of line graphs with error bars is that the error bars may overlap, and therefore readers will be unable to determine which error bar corresponds to which data point (see Graph C in Figure 7.7). You can clarify each error bar by labeling the error bars with the exact values, providing a separate table of values, or, for journals published in color, using a different color for each line (and corresponding error bar).

In SCATTER GRAPHS measurements are plotted along two different continuous scales (the horizontal x-axis and vertical y-axis), which will show the direction and strength of the relationship between the two variables (Cleveland, Diaconis, and McGill 1982; Doumont and Vandenbroeck 2002). Scatter graphs only show correlation between two variables; these graphs do not indicate causation or that

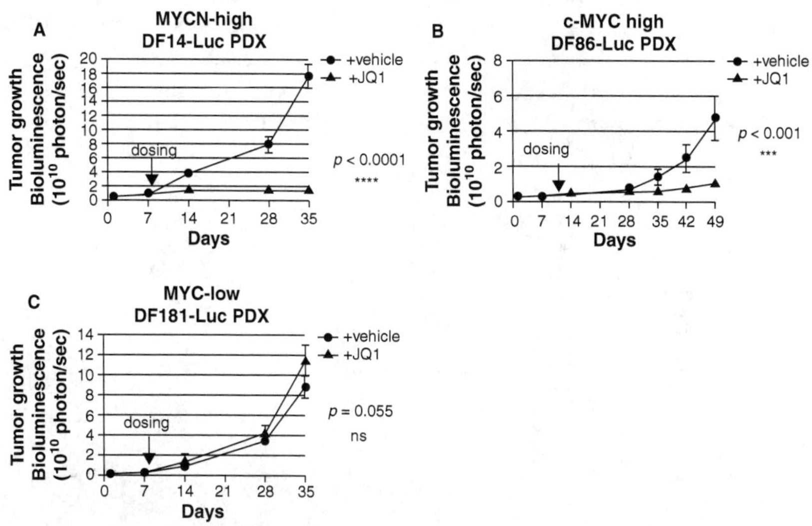

FIGURE 7.7 Line graphs with error bars showing the variability of measurements. The nonoverlapping regions showing differences between groups in graph A begins at day 14 and in graph B begins at day 35. The overlapping error bars in graph C show that no significant difference exists between the two groups.

Source: Baratta, Maria Giuseppinam, Anna C. Schinzel, Yaara Zwang, Pratiti Bandopadhayay, Christian Bowman-Colin, Jennifer Kutt, Jennifer Curtis, Huiying Piao, Laura C. Wong, Andrew L. Kung, Rameen Beroukhim, James E. Bradner, Ronny Drapkin, William C. Hahn, Joyce F. Liu, and David M. Livingston. 2015. "An In-Tumor Genetic Screen Reveals that the BET Bromodomain Protein, BRD4, Is a Potential Therapeutic Target in Ovarian Carcinoma." *PNAS* 112: 232–7.

one variable directly influenced the other (Velickovic 2015). If two variables are related, the data points will form a distinctive pattern such as a straight line or smooth curve (Anscombe 1973). For example, in a positive linear correlation (see Figure 7.8a), an increase in the values for the independent variable (plotted along the x-axis) correlates with an increase in the dependent variable (plotted along the y-axis). If the dependent variables are randomly distributed on the graph without any visible pattern or relationship to the independent variable, no correlation exists between the variables (see Figure 7.8c).

The closeness of the clustering of data points shows the strength of the correlation. The CORRELATION COEFFICIENT (represented as r) is a statistical measure that summarizes the degree of the association between two variables (Creswell 2012; Overholser and Sowinski 2008). Correlation coefficients range from −1 to +1, with values closest to −1 or +1 representing the strongest correlations between the two variables. A value of 0 indicates no correlation between the two variables (see Figure 7.8c). The negative or positive value indicates the direction of the relationship. For positive r values, the values on the y-axis increase as the values on the x-axis increase (see Figure 7.8a). For negative r values, the values on the y-axis decrease as the values on the x-axis increase (see Figure 7.8b). The correlation coefficient should be presented with a scatter graph (when appropriate). For example, in Figure 7.9, all four graphs have the same correlation coefficient ($r = 0.816$). But the interpretation of a strong linear correlation is only appropriate in the top-left graph because curved patterns and extreme outliers distort the correlation coefficient (Anscombe 1973; Velickovic 2015). Often scatter graphs

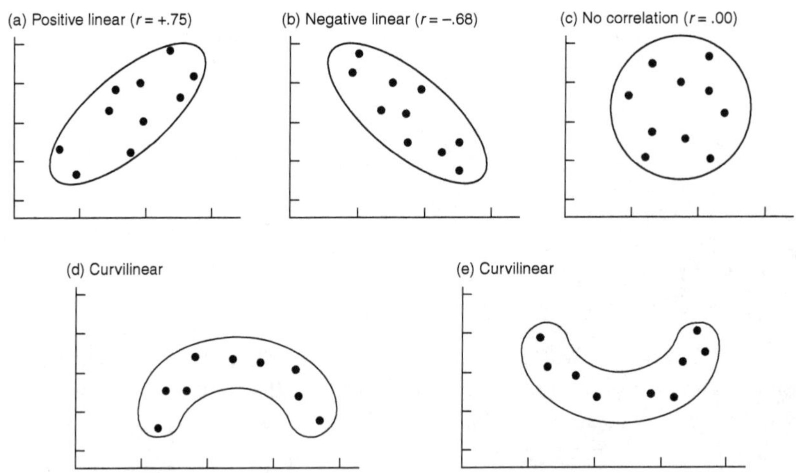

FIGURE 7.8 Patterns of data points in scatter graphs. Correlation coefficients (r) can be calculated for linear to random distributions (a, b, and c) but not curves (d and e).

Source: Creswell (2012).

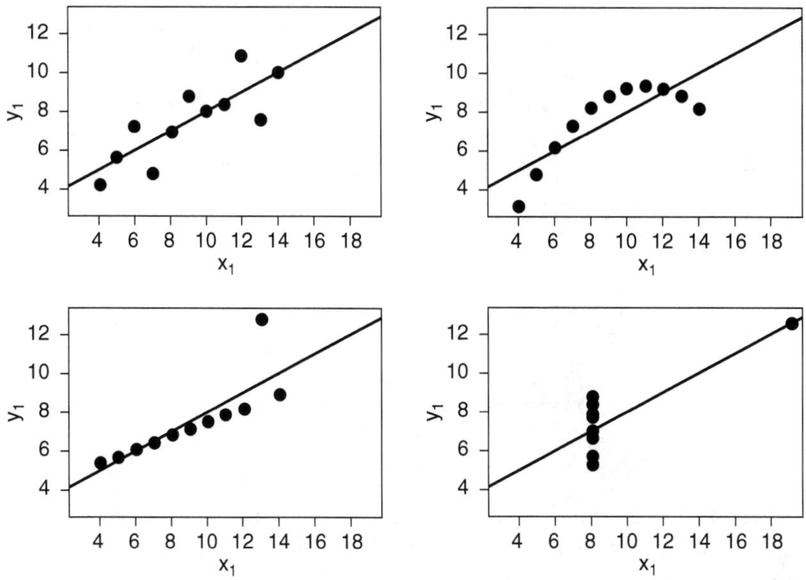

FIGURE 7.9 Each graph shown has the same correlation coefficient ($r = 0.816$). But the correlation coefficient is only appropriate in the top-left graph of a linear plot lacking extreme outliers.

Sources: Anscombe (1973) and Velickovic (2015).

include a computer-drawn LINEAR REGRESSION LINE to help the readers visualize the correlation (see Figure 7.10). The linear regression line should only be drawn within the plotted data values because no data exist for the pattern beyond those values (Cleveland and McGill 1985; Kuo 2002).

Organizing Data for Publication-Quality Graphs

In SMC, graphs communicate data to an expert audience to support an argument (Kelleher and Wagner 2011; Kostelnick 1998). Unlike graphs for general audiences, the goal in designing graphs for experts is clarity and less emphasis is placed on using graphs to attract readers to the article (Kostelnick 2007; Tufte 2001). Aesthetics of graphs in SMC is focused on clear, formal, and standardized presentation of data. Researchers and practitioners need clear, unambiguous graphs so that they can use the findings to further research or practice. The importance of data in SMC is typically on the direction, magnitude, and certainty of relationships between groups or variables—all of which are effectively illustrated in graphs (Connor 2009; Tufte 2001).

Although graph design partially depends on software tools, you cannot necessarily rely on the standard output or default settings to create clear, effective, and

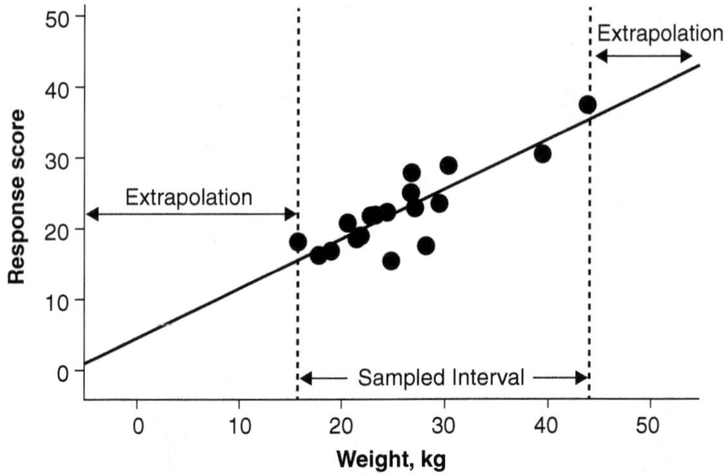

FIGURE 7.10 Linear regression line in a scatter graph should only extend within the sampled interval (not be extrapolated beyond the data values).

Source: Kuo (2002).

publication-ready graphs. In fact, Charles Kostelnick (2007) of Iowa State University noted, "The effects of these technological tools . . . have been mixed, . . . partly because the defaults, templates, and wizards can actually undermine, rather than foster, rhetorical adaptation" (284). Jason Connor (2009), a statistical reviewer for *The American Journal of Gastroenterology*, explained that

> with today's software it is very easy to produce graphics. But much like good writing, good graphics are not produced merely by the click of a button. They require substantial forethought, meticulous craftsmanship, and repeated revision until the final product is clear, concise, and informative. *(1629)*

The specific goals of creating effective graphs are to (1) show unbiased representation of the data; (2) enable comparison between the values represented by the data points, lines, or bars; and (3) explain the full context of the data set by including descriptive labels, keys (or legends), titles, and captions (Cleveland 1984b; Howarth and Turner 1987; Jane Miller 2004; Puhan et al. 2006).

Effective design of graphs depends on proper organization and presentation of the data set. By convention the dependent (or measured) variable is plotted along the vertical Y-AXIS (see Figure 7.11). The data plotted along the horizontal X-AXIS varies based on the type of graph. In a bar graph (or box plot), the *x*-axis is the names of the groups or categories. In a line graph, the *x*-axis is time (a continuous variable). And in scatter graphs, the *x*-axis is any other continuous variable other than time and each data point represents a unique individual measurement.

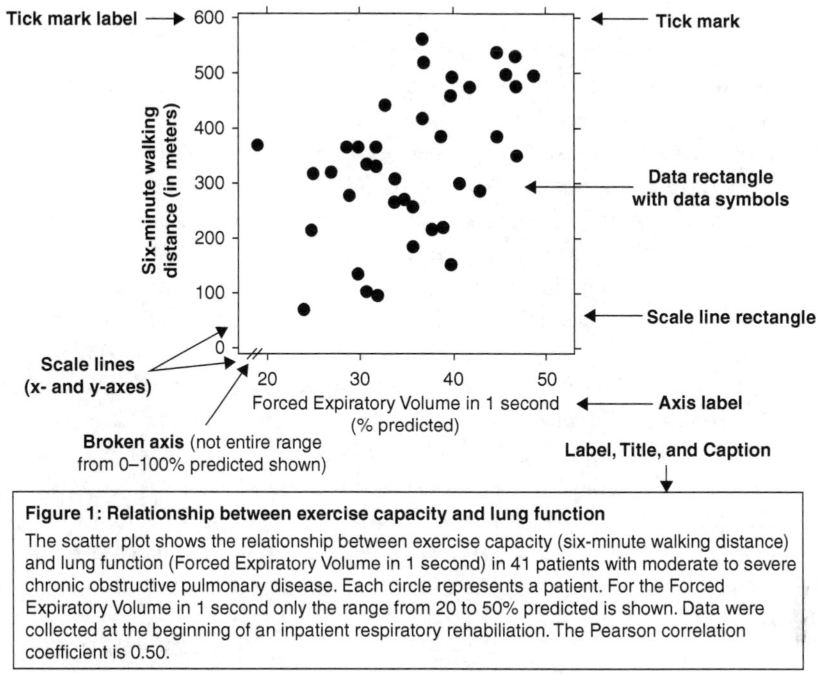

FIGURE 7.11 Anatomy of a graph.

Adapted from Puhan et al. (2006).

After you have determined the correct variables for each axis, you are ready to plot the data and set the scale for each axis. When plotting the data, you should keep the graphical display directly proportional to the actual data by ensuring that the dimensions of the graph area do not substantially exceed the data values (Allen, Erhardt, and Calhoun 2012; Paniello et al. 2011; Tufte 2001). To do so, follow these guidelines:

1. **Set the scale of the axis so that the maximum value slightly exceeds the greatest data value** (Ng and Peh 2009a; Tufte 2001). The maximum value of a continuous scale is determined by selecting units that yield 5 to 10 standard values on each axis. For example, if the data points on an axis range from 0 to 75, use 80 as the maximum value rather than 100. This way, you would have 8 tick mark labels at consistent intervals of 10 units each. Using this method, the audience will be able to most accurately estimate the exact data values represented in the graph (Jane Miller 2004), and the data will occupy the entire graph area and not result in a large white space at the top of the graph (Ng and Peh 2009a).
2. **For data sets of only positive values, set the base of the scale at zero.** If the range of data is significantly greater than zero, you should include zero as

the base of the axis but add hatch marks between the zero and the numerical values to show a broken axis and suppressed zero (see *x*-axis in Figure 7.11).

3. **For data sets with negative values, follow guideline 1 to determine the minimum, negative value.**
4. **For data sets with data that span an extremely large range of values, use a logarithmic (log) scale.** A log scale uses a multiple of a number, called the BASE NUMBER, to transform an extremely large range of values into a manageable representation. In SMC, the standard log scale is a multiple of 10 (or base 10 represented in subscript, as in "$\log_{10}$") although any base number multiple may be used (Cleveland 1984a). In a $\log_{10}$ scale, each measure and tick mark label would increase 10-fold, such as 1, 10, 100, 1,000, 10,000, and 100,000. For example, the growth rate of bacteria (which double in number by dividing from one cell into two each time interval) is a measure that increases exponentially and is appropriately graphed on a log scale. Alternatively, radioactive decay or half-life (in which 50% of the material degrades during each time interval) is another measure appropriately graphed on a log scale. When using a log scale, be sure to label the axis as a logarithmic scale or "log," and include the appropriate base number as a subscript.

Although each graph should be graphed individually, you should try to maintain consistency between all graphs in the same manuscript and especially in multipanel graphs in which multiple graphs are presented in the same figure (see Figure 7.12). To maintain consistency, use the same scales for all graphs to enable comparisons between the variables and groups (Kelleher and Wagener 2011; Jane Miller 2004). Additionally you should present variables or groups in the same order and use symbols consistently for the same groups or variables. For example, Figure 7.7 is a multipanel graph that might have been graphed with consistent *y*-axes to enable comparison between graphs.

Designing Clear, Data-Rich Graphs

Graphs should represent data clearly by prominently displaying the points, lines, or bars that correspond to the data values (Allen, Erhardt, and Calhoun 2012; Lane and Sandor 2009; Stengel, Calori, and Giannoudis 2008). As emphasized by Edward Tufte (2001), professor emeritus at Yale University and prominent scholar of graphing quantitative information, effective communicators should "show data variation, not design variation" (61). Thus, the graphs you design should resemble the design of the other graphs in your target journal so that the audience, being familiar with the design, will focus on the data rather than on your presentation style.

Although recent articles will show you the design elements (such as point, line, and bar styles; fonts; and colors) for graphs published in your target journal, published graphs often contain errors of data presentation that remain

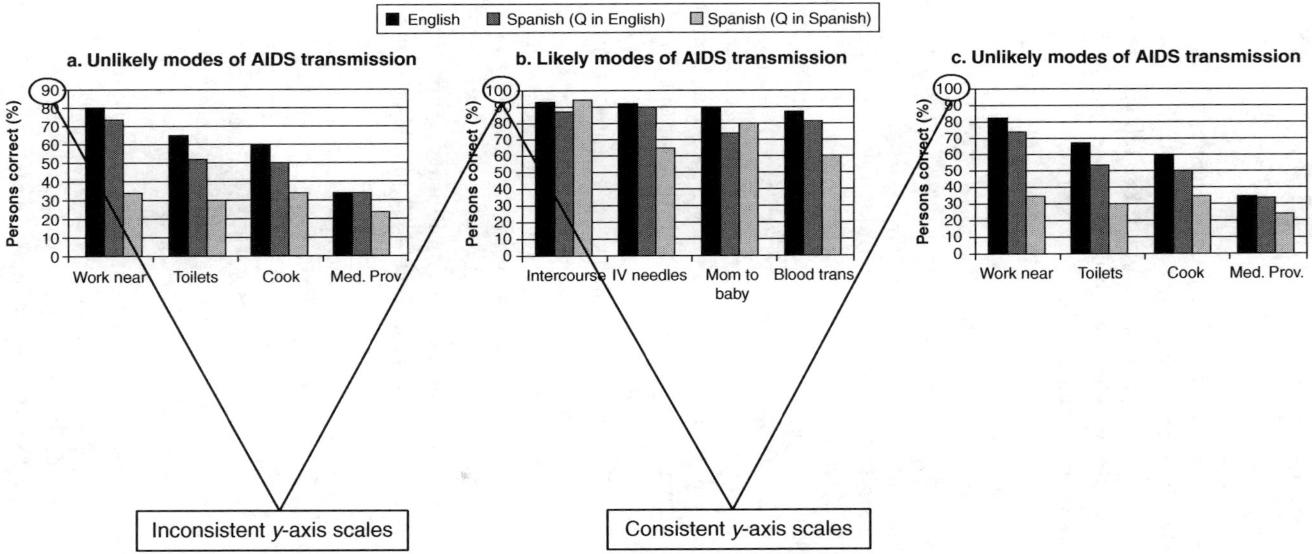

FIGURE 7.12 Example of a poorly designed multipanel graph in which the scales are different. You should maintain consistency in scales and orders of groups/variables to enable comparison between graphs.

Source: Jane Miller (2004).

unresolved even after the peer-review and editing process (Cooper, Schriger, and Tashman 2001; Schriger et al. 2006). To create clear, data-rich graphs you should:

1. **Use high figure-ground contrast for points, lines, and bars** (Gross 2007; Kostelnick 2007). In general, data points, lines, and bars should appear in black ink against a plain, white background. Use full contrast (black and white regions) to distinguish multiple points, lines, and bars. Often this means using solid and outlined points (i.e., solid circles and outlined circles), dashed lines, and hatch mark patterns in bars rather than color or shading (Allen, Erhardt, and Calhoun 2012; Bryan 1995; Ng and Peh 2009a; Puhan et al. 2006; Stengel, Calori, and Giannoudis 2008). However, looking at published articles may reveal accepted design variations in a journal or discourse community.
2. **Maximize the data-to-ink ratio and eliminate any nondata ink (ink that can be erased without losing meaning).** For example, Tufte (2001) argued that the graph in Figure 7.13 maximized the data-to-ink ratio with only 10% to 20% of the ink of the graph used for "non-data-ink" (96). In this example, Tufte (2001) noted that only the grid and tick marks could be eliminated before beginning to lose informative content from the graph because the tick marks are redundant with the unit measurements and the grid line serves to frame the graph. However, the grid lines and tick marks help define the graph space and are design conventions. Therefore, in practice, you would not actually remove these items—just keep them to a minimum. Common types of nondata ink that you should eliminate from graphs include grids and rule lines within the data rectangle and background colors and patterns.
3. **Ensure that the graph is visible in the published format** (Ng and Peh 2009a; Stengel, Calori, and Giannoudis 2008). To do this, you should obtain a copy of the printed journal and ensure that each graph you plan fits into the width of a column or page. Journal publishers will fit graphs to neatly fit within the column(s) of the page, and the height of the graph will be proportional to the width. Once adjusted for the column or page width, ensure that the data symbols and lines are clearly distinguished from one another (Puhan et al. 2006). If necessary, you may adjust the scale of the graph to improve the resolution of data points or lines.

In SCM, audiences are interested in the data patterns or trends, which you must represent clearly. As Tufte (2001) noted, if you do not have any "interesting" data values to show, then you should use a table (or possibly emphasize those values in the text). As an effective communicator, you should not resort to graphical design features to "attract" an audience. Specifically, you should not add false or empty dimensions, such as using a three-dimensional graph when you are only showing two variables (Allen, Erhardt, and Calhoun 2012; Tufte 2001).

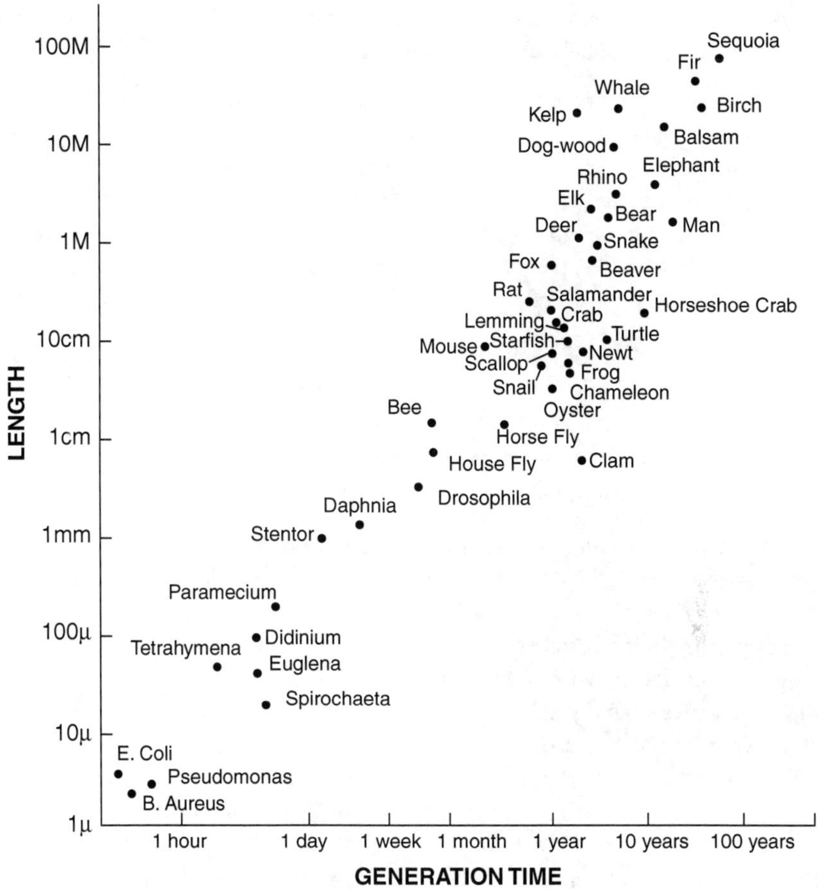

FIGURE 7.13 Comparison of the reproduction (generation) time to the length of an organism in *Size and Cycle: An Essay on the Structure of Biology* by John Tyler Bonner (1965) in Tufte (2001). Tufte emphasized the high data-ink ratio in this graph.

Integrating Graphs into the Text

Similar to tables, the each graph should be self-explanatory without requiring the reader to refer to the text. Specifically, the graph should (1) be clearly labeled, (2) include an informative title, and (3) include a descriptive caption that summarizes the trends of the data.

You should integrate the graph into the text of your article by following the same approach that we covered for tables. A few conventions regarding the labeling and numbering of graphs are worth mentioning. First, any visual that is not a table is labeled as a "figure." Thus, graphs are called figures, but so are photographs and diagrams. Second, figures are numbered independently from tables. Thus, manuscripts may contain both a "Table 1" and a "Figure 1."

Below are the most common errors of graphs found in SMC (Ng and Peh 2009a; Puhan et al. 2006), which you should avoid:

- Using the wrong type of graph to represent the data,
- Misusing three-dimensional graphs to add a pseudo or false third dimension,
- Not plotting graph to scale or distorting the scale to exaggerate an effect,
- Not defining axes, data symbols, or abbreviations,
- Not including measurements,
- Not including a descriptive title, caption, or legend, and
- Not crediting the source of previously published data.

After you have created the graphs, use this list to ensure accuracy.

Statistics

Statistics are calculated values from large data sets that provide information about the data. Statistics fall into two broad categories:

- DESCRIPTIVE STATISTICS—summarize (or "describe') the data set, or
- INFERENTIAL STATISTICS—extrapolate from a sample to a larger population.

Although the choice of statistics is beyond the scope of our discussion and often requires consultation with a statistician, we will briefly cover the general categories of statistics so that you have a sense of the information you are presenting.

Descriptive statistics are the basic calculations that provide the "average" value of a data set, which is properly referred to as a measure of CENTRAL TENDENCY. Descriptive statistics is the category of summary that students often ask instructors following an exam—such as: What was the "average" grade? and What was the "range"? Descriptive statistics provide a summary of the entire set of values (Sonnad 2002) and include:

1. CENTRAL TENDENCY—the best representation of the group of values based on the distribution, which is either the MEAN (arithmetic average), MEDIAN (midpoint value in which half of the values lie above and half lie below), or MODE (most common value).
2. VARIABILITY—the degree to which the data values are close together or spread apart.
3. DISTRIBUTION—the pattern of the data values (as "normal" bell shape, skewed in one direction or the other, or more complex patterns).
4. RANGE—the difference between the largest and smallest data values.

As Seema Sonnad (2002), from the Department of Surgery at the University of Michigan Medical Center, noted, "An adequate summary of a data set requires both a measure of center [central tendency] and a measure of variability" (624). The most commonly reported descriptive measures are the number of measurements (the *n*), the mean (for central tendency), and the standard deviation (for variability; De Muth 2008; Sonnad 2002). However, the most appropriate descriptive statistic for distribution depends on the data format and pattern of distribution (Bailar and Mosteller 1988; De Muth 2008; Overholser and Sowinski 2007; Sonnad 2002).

Inferential statistics are more complex than descriptive statistics. As James De Muth (2008), professor in the School of Pharmacy at the University of Wisconsin, stated, "Inferential statistics represent what most people think of as statistics" (2359). Essentially, computing inferential statistics involves making calculations of the data collected from a small group of individuals (a SAMPLE) and extending those measures to a larger POPULATION (the entire set of people whom the sample represent). A common form of inferential statistics is a survey poll for a political election, which predicts the percentage of votes for each candidate. Prior to voting day, small surveys of representative voters are used to predict how the entire population of eligible voters will vote on election day. Inferential statistics must include a measure of UNCERTAINTY, which describes the accuracy or precision of the statistic. Table 7.4 summarizes the symbols used for the most commonly reported inferential statistics in samples and populations.

Many journals (Bailar and Mosteller 1988; Vandenbroucke et al. 2007), the American Psychological Association (2010), and the CONSORT guidelines (Moher et al. 2010) recommend that authors use the confidence interval (CI) to indicate precision. *P* values are less informative than CI, do not convey the effect size, and may be misleading (Bailar and Mosteller 1988; International Committee of Medical Journal Editors 2014; Matthews and Altman 1996). Exact *P* values may be reported in addition to CI, but results should not be reported only as *P* values (Moher et al. 2010). When reporting *P* values, you should always report the exact *P* value to one or two decimal places (such as "$P = 0.03$" or "$P = 0.22$") rather than inequalities (such as "$P < 0.05$") or imprecise values (such as "*P* not significant") so that readers can evaluate the *P* value and make comparisons between studies (Bailar and Mosteller 1988; Brand and Huiskes 2001; Lang and Altman 2014).

With inferential statistics, in particular, you should consult a statistician to ensure that you are reporting appropriate measures because these decisions are

TABLE 7.4 Symbols used for common inferential statistics for samples and populations. Adapted from De Muth (2008).

	Sample	Population
Sample size	n	N
Mean	$\overline{x}$	μ
Standard deviation	S	σ

complex and are even debated within the field of statistics (Altman, Goodman, and Schroter 2002; De Muth 2008; Gardner, Machin, and Campbell 1986; Herbison 2007; Mike and Good 1977; Morton 2009; Nuzzo 2014; Overholser and Sowinski 2008; Saracino, Jennings, and Hasse 2013). The editors of *Nature Cell Biology* (2009) note that the statistics are often inappropriately used. They recommend that writers in SMC who are reporting studies with fewer than five independent repeats of an experiment avoid using inferential statistics and instead report the entire data set. Furthermore, writers should "be careful not to discard 'outlier' data points on a whim, as they may well be as relevant as clustered measurements" (*Nature Cell Biology* 2009, 667).

If you do find statistically significant results that were unplanned in the original study design, you should tell the reader that these results were an unplanned evaluation of the data (referring to the data as "post hoc analysis" or "exploratory" rather than "a priori") to appropriately contextualize the information (Iverson et al. 2007). Results from comparisons planned during the experimental design phase are typically more reliable than analyses following data collection (American Psychological Association 2010; Moher et al. 2010; Tukey 1977). The *AMA Manual of Style* states that

> performing significance testing on the baseline differences between groups is controversial. (Even with perfect random assignment, an average of 1 in every 20 comparisons will appear to be "significant" at the .05 level by chance alone; such random findings illustrate the dangers of post hoc analyses.) Furthermore, in small studies, large differences may nonetheless be statistically nonsignificant due to limited statistical power.
>
> *(Iverson et al. 2007, 839)*

Additionally, selective reporting of particular subgroups may bias the study (Moher et al. 2010). As John Tukey, professor of statistics and science at Princeton University, stated in an article published in *Science*, "Once multiple questions are asked, there will be pressures, some ethical in nature, to concentrate upon those questions for which the results appear most favorable" (681).

Photographs and Instrument Output

Photographs and instrument output are similar in that these images provide source evidence of a phenomenon (Gross 2007; Ng and Peh 2009b). The purpose of providing photographs or instrument output is to document a phenomenon and serve as evidence for future claims to be made by the researchers (Gross 2007; T. Miller 1998). Often these types of visuals show data that is neither quantitative nor easily or adequately described in words. Photographs and instrument output are "medium-use" visuals, which are less common than graphs and diagrams but more common than tables and equations (Mogull and Stanfield 2015).

In some situations, particularly in medicine, you should clarify if an image shows a typical, extreme, or unusual case (Siontis et al. 2010). When presenting photographs (see Table 7.5) or instrument output (see Table 7.6), you should adhere to the following guidelines:

- Crop the image to illustrate the point being made in the manuscript but include enough context to provide sufficient orientation of the image,
- Use a callout (e.g., arrows, labels, or circles) to highlight the important features in the image,
- Include a scale or familiar object to represent the size of the object (this is particularly important for microscopic images), and
- Maintain image quality by avoiding file compression formats. Often this means a TIFF or EPS format for files at a minimum resolution of 300 dpi.

Additionally, you might need to adjust the contrast or colors to optimize the image for the medium (print or screen) that will be used by the audience to visualize the image. In some cases, black and white photographs might provide greater contrast, which makes objects easier to visualize than color photographs. Consider viewing images in the actual medium to ensure that the images are easily visualized. If you make any adjustment to optimize the original image, you should describe the modifications in the caption (Rossner and Yamada 2004; Siontis et al. 2010).

While adjusting the contrast or colors of an image to highlight relevant features is acceptable, the general rule is that any adjustment must be made to an entire image rather than a selected portion of the image (Bertrand 2001; Rossner and Yamada 2004). Importantly, you may not erase extraneous marks on an image, splice different images or parts of an image together so that the resulting image appears to be a single image (such different images may be presented as

TABLE 7.5 Major categories of photographs.

Category	Description	Example
Patients and Specimens		
External signs	Shows external appearance of a clinical condition from a patient	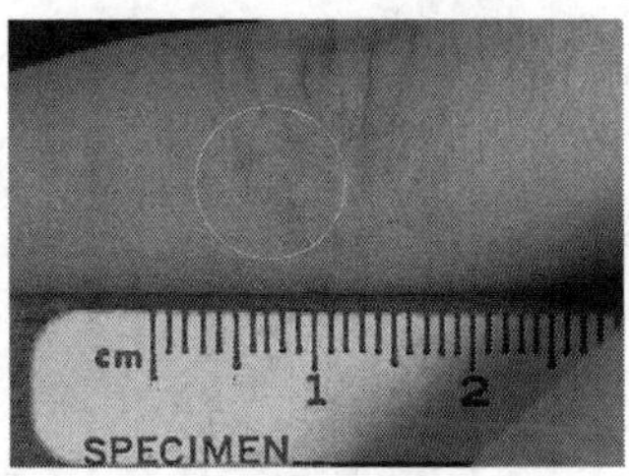 *Source*: Blanton et al. (2005).

Continued

Category	Description	Example
Internal signs	Shows internal appearance of a clinical condition in a patient from an endoscopic camera or radiology (such as CT, MRI, X-ray, or ultrasound)	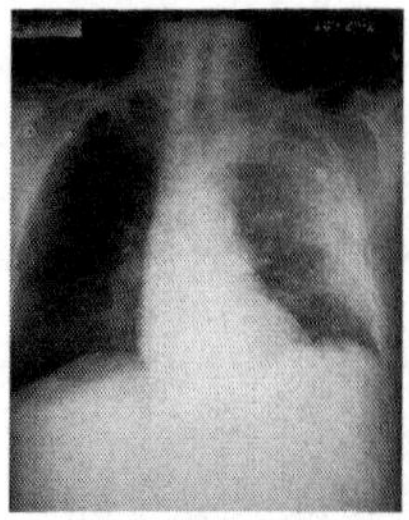 *Source*: Jernigan et al. (2001).
Microscopic images	Shows cells or tissue samples (often biopsy specimens) stained and imaged by a light or electron microscope	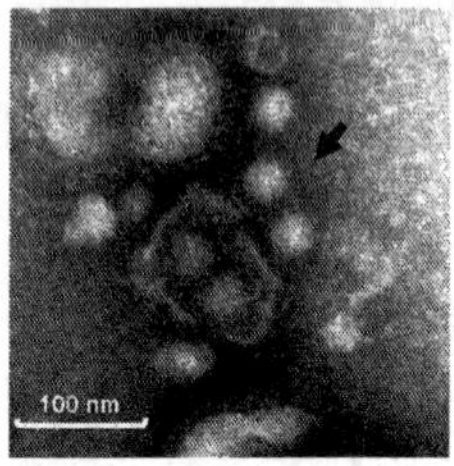 *Source*: Wang et al. (2005).
Detection of Macromolecules		
Gels and blots	Shows the presence or absence of nucleic acids or other molecules of a predicted size on gel electrophoresis or blots (respectively). Images must include a known standard for size reference (see lane 1). Unknown samples show either the presence (see lanes 4–7) or absence (see other lanes) of bands of a predicted size.	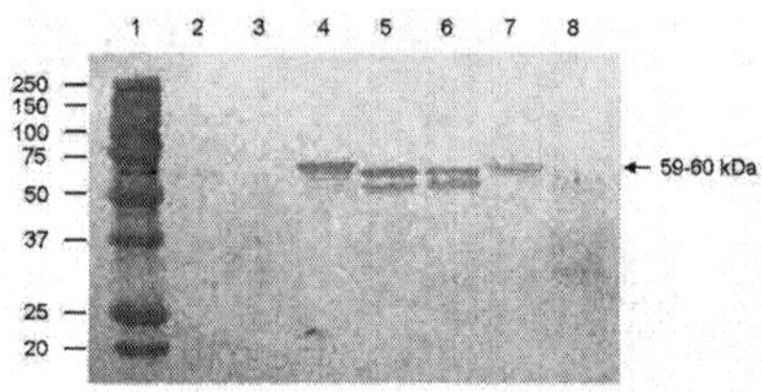 *Source*: Wang et al. (2005).

Sources:

1. Blanton, Jesse D., Nadine Y. Bowden, Millicent Eidson, Jeffrey D. Wyatt, and Cathleen A. Hanlon. 2005. "Rabies Postexposure Prophylaxis, New York, 1995–2000." *Emerging Infectious Diseases* 11 (12): 1921–7. Jernigan et al. (2001).
2. Jernigan, John A., David S. Stephens, David A. Ashford, Carlos Omenaca, Martin S. Topiel, Mark Galbraith, Michael Tapper, Tamara L. Fisk, Sherif R. Zaki, Tanja Popovic, Richard F. Meyer, Conrad P. Quinn, Scott A. Harper, Scott K. Fridkin, James J. Sejvar, Colin W. Shepard, Michelle McConnell, Jeannette Guarner, Wun-Ju Shieh, Jean M. Malecki, Julie L. Gerberding, James M. Hughes, Bradley A. Perkins, and members of the Anthrax Bioterrorism Investigation Team. 2001. "Bioterrorism-Related Inhalational Anthrax: The First 10 Cases Reported in the United States." *Emerging Infectious Diseases* 7 (6): 933–44.
3. Wang, Qiu-Hong, Myung Guk Han, Sonia Cheetham, Menira Souza, Julie A. Funk, and Linda J. Saif, 2005. "Porcine Noroviruses Related to Human Noroviruses." *Emerging Infectious Diseases* 11 (12): 1874–81.
4. Wang, Qiu-Hong, Myung Guk Han, Sonia Cheetham, Menira Souza, Julie A. Funk, and Linda J. Saif, 2005. "Porcine Noroviruses Related to Human Noroviruses." *Emerging Infectious Diseases* 11 (12): 1874–81.

TABLE 7.6 Major categories of instrument output.

Category	Common Type	Example
Physiological signal tracings	Electrocardiogram (ECG) or electroencephalogram (EEG)	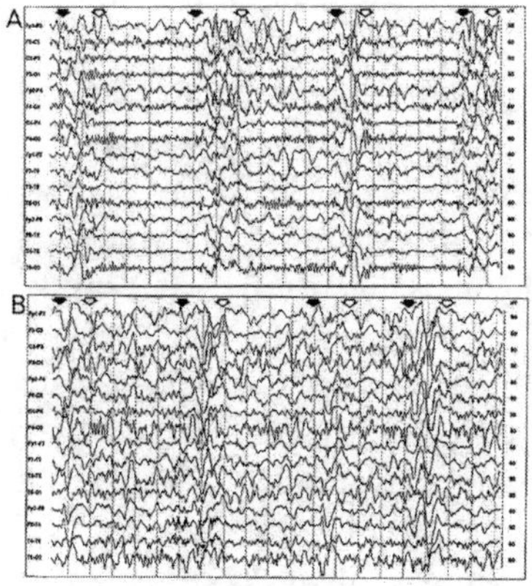 *Source*: Bonthius, Stanek, and Grose (2000).
Laboratory graphs	Chromatogram, electropherogram (DNA sequence), or karyogram	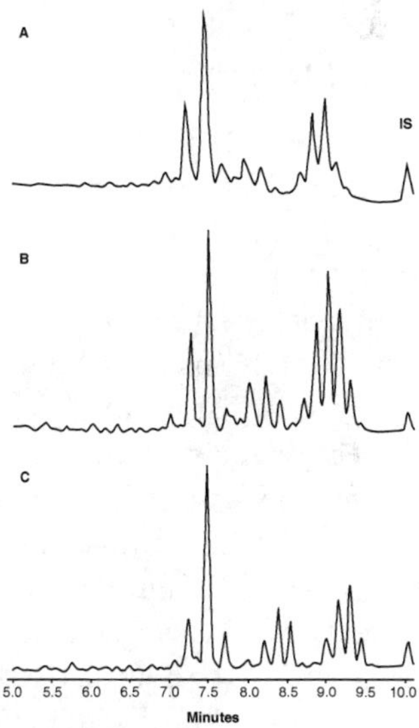 *Source*: Piersimoni et al. (2004).

Sources:

1. Bonthius, Daniel J., Nicholas Stanek, and Charles Grose. 2000. "Subacute Sclerosing Panencephalitis, a Measles Complication, in an Internationally Adopted Child." *Emerging Infectious Diseases* 6 (4): 377–81. Piersimoni et al. (2004).
2. Piersimoni, Claudio, Piergiorgio Zitti, Gianna Mazzarelli, Alessandro Mariottini, Domenico Nista, and Diego Zallocco. 2004. "Mycobacterium triplex Pulmonary Disease in Immunocompetent Host." *Emerging Infectious Diseases* 10 (10): 1859–62.

multipanel images), nor adjust the contrast to remove imperfections because such practices are unacceptable and considered data manipulation (Rossner and Yamada 2004).

Finally, you should ensure that you either have permission to use photographs of a patient or have adequately cropped or masked the images to ensure an individual's anonymity (Hill 2006; Ng and Peh 2009b). The standard practice when showing a person's face is to mask the identity of the individual by placing a black box over the individual's eyes (Ng and Peh 2009b), although such practice is often insufficient to actually protect a person's identity (Lang 2009). When using photographs that reveal one's identify, you must collect written permission from the patient or guardian (Hill 2006; Ng and Peh 2009b).

Despite the benefits of documenting actual evidence, photographs and instrument output are not always the best option for visuals. These images often contain excess detail, which may make the relevant features difficult to distinguish. Historically these images have been difficult and expensive to reproduce in some media although that is less of a problem with contemporary publishing. For these reasons, however, effective communicators use drawings and diagrams as substitutes for more detailed images.

Drawings and Diagrams

Diagrams (which include drawings as well as maps) are the second most prevalent form of visuals in scientific communication (Mogull and Stanfield 2015). Drawings and diagrams are similar to photographs in that they show the appearance of an object although drawings and diagrams are simplified representations of the object that lack the extraneous detail on object surfaces (Frankel and DePace 2012; Mishra 2004). In many situations, diagrams are used to represent a phenomenon or process that is not directly observable, illustrate the inner workings of an object, or show a proposed model. As such, diagrams are a tentative extrapolation of other data and, unlike photographs, are not direct evidence (see Figure 7.14; Northcut 2011; Richards 2003).

In drawings and diagrams, illustrators select the level of abstraction from highly representative, showing the actual appearance of an object, to highly abstract, using symbols or nomenclature to represent components of the object (see Figure 7.15). The level of abstraction depends on the relevant information being emphasized and may use standard representations to illustrate a concept (see Figure 7.16). Different types of drawings and diagrams can be used to accomplish different purposes (see Table 7.7).

The guidelines for designing effective drawings and diagrams are similar to those for designing effective photographs. Drawings and diagrams should focus on the feature of interest and include sufficient detail to provide adequate orientation of an object. The contrast, resolution, and use of colors should be appropriate for the duplication and presentation medium. Although color has become less

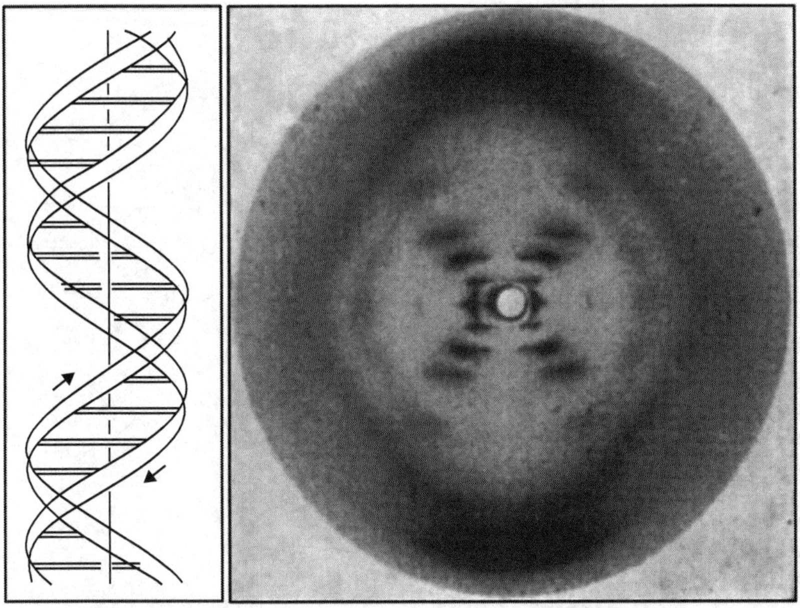

FIGURE 7.14 Graphical representations of the structure of DNA (a) proposed by Watson and Crick and (b) captured by Franklin and Gosling in an X-ray image.

Sources: Watson, J. D., and F. H. C. Crick. 1953. "Molecular Structure of Nucleic Acids: A Structure for Deoxyribose Nucleic Acid." *Nature* 171 (4356): 737–8.

Franklin, Rosalind E., and R. G. Gosling. 1953. "Molecular Configuration in Sodium Thymonucleate." *Nature* 171 (4356): 740–1.

expensive to reproduce, you should use color selectively so that the color highlights an area of interest but avoid excessive color use throughout the rest of the image to avoid distracting from the object of interest. Often black and white, or a single spot color, may provide greater emphasis than full-color diagrams and drawings.

Maps

Finally, maps are used to show location or geographic differences (see Figure 7.17). Maps should only include the necessary detail to show the object of interest and to provide orientation for readers. As in all images, unnecessary details should be omitted from maps. For example, to show the distribution of rainfall over a geographical region, you might use a basic map of the region with color or shading to represent the rainfall. A more complex map that contains additional unnecessary detail such as altitude will distract the readers from focusing on the relevant information. By convention, maps are plotted so that North is at the top of the image and the scale is provided.

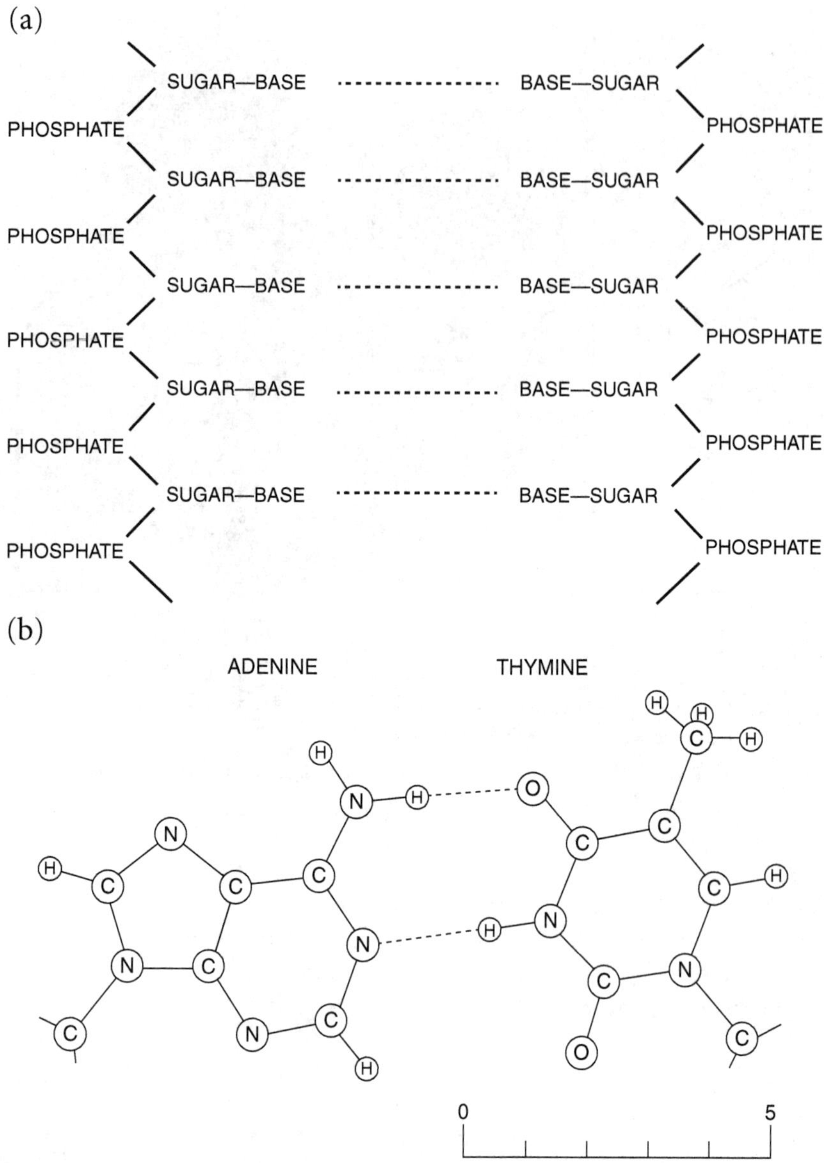

FIGURE 7.15 Various level of abstraction or detail as illustrated by Watson and Crick: (a) representing bonding of bases in DNA and (b) showing in increased detail interactions between atoms.

Source: Watson, J. D., and F. H. C. Crick. 1953. "Genetical Implications of the Structure of Deoxyribonucleic Acid." *Nature* 171 (4361): 964–7.

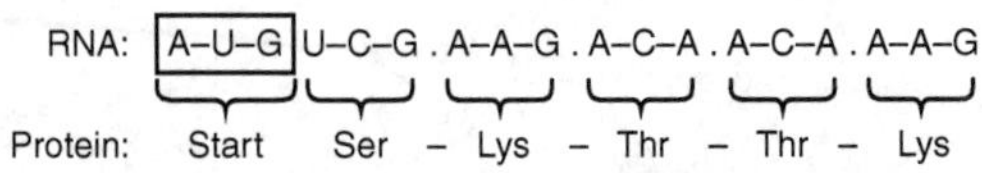

FIGURE 7.16 Initial genetic sequence and amino acids of bacteriophage MS2 replicase gene, part of the first genome sequenced. This figure illustrates the use of standard representations of nucleic acids and amino acids to represent a concept.

Adapted from Fiers, W., R. Contreras, F. Duerinck, G. Haegeman, D. Iserentant, J. Merregaert, W. Min Jou, F. Molemans, A. Raeymaekers, A. Van den Berghe, G. Volckaert, and M. Ysebaert. 1976. "Complete Nucleotide Sequence of Bacteriophage MS2 RNA: Primary and Secondary Structure of the Replicase Gene." *Nature* 260 (5551): 500–7.

TABLE 7.7 Major categories of drawings and diagrams.

Category	Description
Representative drawing	Shows the appearance and the relevant features of an object
Translucent (or cutaway) drawing	Shows the inner structure of an object
Action-view drawing	Shows movement or process by showing stepwise actions in a series of panels or a series of steps in a single image
Symbolic diagram (or schematic)	Illustrates a phenomenon or process using graphical representations
Flowchart	Illustrates a process in arrows and words

Concluding Points

In this chapter, we covered the content, organization, and style of the Results section. The Results section provides the evidence for the claims made in the manuscript. The content of the Results section varies somewhat by discourse community, journal, and article type (Kanoksilapatham 2005; Swales 1990). One major difference is the extent to which the Results section includes a summary of methods or initial interpretation. Some discourse communities insist that the Results sections only contain data whereas other discourse communities include a brief overview of the experimental procedure and some interpretation of the data. Regardless of the article type—experimental (or hypothesis-) testing or exploratory investigations—you should ensure that the content flows logically from one data set to the next. When writing the Results section, you should approach the content as a reader—where you highlight the results (or relevant information) rather than simply list the data (or details). Additionally, you should take the time to carefully organize the results into

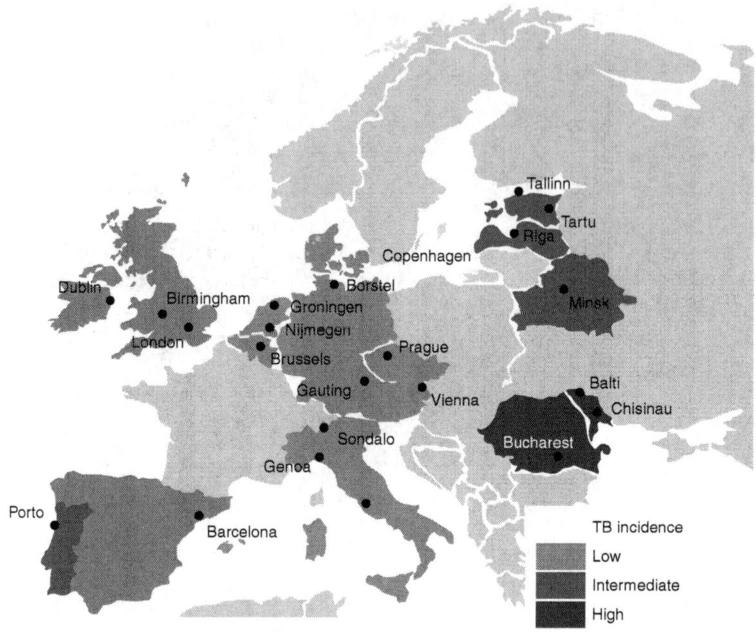

FIGURE 7.17 Map illustrating the rate of multidrug-resistant tuberculosis in European countries during 2010–2011. Note shading is used to focus attention and only one variable (TB incidence) is illustrated.

Source: Günther, Gunar, Frank van Leth, Sofia Alexandru, Neus Altet, Korkut Avsar, Didi Bang, Raisa Barbuta, Graham Bothamley, Ana Ciobanu, Valeriu Crudu, Manfred Davilovits, Martin J. Dedicoat, Raquel Duarte, Gina Gualano, Heinke Kunst, Wiel de Lange, Vaira Leimane, Cecile Magis-Escurra, Anne-Marie McLaughlin, Inge Muylle, Veronika Polcová, Emanuele Pontali, Christina Popa, Rudolf Rumetshofer, Alena Skrahina, Varvara Solodovnikova, Victor Spinu, Simon Tiberi, Piret Viiklepp, Christoph Lange, and TBNET. 2015. "Multidrug-Resistant Tuberculosis in Europe, 2010–2011." *Emerging Infectious Diseases* 21 (3): 409–16.

a coherent story or narrative before you begin writing. Organization of the Results section is important because it influences the organization of content in the Methods and Discussion sections.

Note

1 Many journals, the American Psychological Association, and the CONSORT guidelines recommend that authors use the confidence interval (CI) to indicate precision (American Psychological Association 2010; Bailar and Mosteller 1988; Moher et al. 2010; Vandenbroucke et al. 2007). For more information, refer to the section on "Statistics" in this chapter.

8

THE DISCUSSION SECTION

In this chapter, you will learn to:

- Select and organize content for the Discussion section
- Write in a scientific style and edit style problems that commonly arise in the Discussion section

Overview

The Discussion section, much like the closing argument of a legal case, is the place scientific writers present an explicit argument for the meaning they inferred from the data presented in the Results section and the contribution that the research makes to the field of study (Annesley 2010; Skelton and Edwards 2000). Highly influential journal articles are often an outcome of the argument presented in the Discussion section, which includes: (1) the theoretical significance of the data and an explanation of phenomena better than other studies, (2) the practical significance or useful implications of the findings, and (3) the value or implications of the study for future research (Sternberg and Gordeeva 1996). In addition, highly influential journal articles clearly address the research problem and have a logical flow and organization of ideas, which are particularly important in convincing the audience of the authors' conclusions.

What to Expect

The Discussion section, as the name implies, is the part of the manuscript where the authors explain the findings and implications of their research

(Skelton and Edwards 2000). In the Discussion section, you will use the evidence presented in the Results section to infer a greater meaning or principle about the subject studied (Docherty and Smith 1999; Skelton and Edwards 2000). Richard Horton (1995), editor in chief of *The Lancet*, explained the origins of this presentation:

> The format of research papers published in this and other journals conforms to classical ideas of rhetorical presentation. Aristotle distinguished four elements that make up successful oratory: introduction, narration, proof, and epilogue. The historical link to the familiar "IMRAD" format of a scientific paper—introduction, methods, results, and discussion—is self evident.
>
> To the extent that science is a search for the reason that lies behind an observation, the study of rhetoric is part of the scientific tradition.
>
> *(987)*

As you will see in the examples presented in this chapter, the Discussion section should not simply reiterate the data from the Results section; rather, it should provide a synthesis and interpretation of the data within the context of the published research (Brand and Huiskes 2001; Liumbruno et al. 2013). Ann Penrose and Steven Katz (2010), scholars of scientific rhetoric, emphasized that:

> research authors do not leave it up to readers to draw conclusions about the data. They tell readers their conclusions, the ones the researchers believe the data support, and they provide enough information about the data and the conditions under which the data were generated to enable readers to determine whether their conclusions are justified.
>
> *(113)*

The findings posed in the Discussion section fill the research niche set forth in the Introduction section (Motsitsi 2008; Nwogu 1997). The Discussion section is not the place to repeat information presented previously in the Introduction or Results sections, but to move beyond the data and background meaningfully (Cetin and Hackam 2005; International Committee of Medical Journal Editors 2014; Pique-Angordans and Posteguillo 2006; Skelton and Edwards 2000). As Jim Thornton (2003), editor in chief of *BJOG: an International Journal of Obstetrics and Gynaecology*, pointed out, "The discussion section is designed to help the reader see how the results fit in with previous work and to indicate the implications for future practice and research" (449).

Within the Discussion section, you will move from specific to broad detail on the research topic, which to some extent, is the inverse of the Introduction section (Annesley 2010; Cals and Kotz 2013b; Plaxco 2010; Swales 1990). In the Discussion section, you will (1) present the principle findings of the Results, (2) place these findings within the context of other research in the field, and

(3) explore the significance of the research (Berkenkotter and Huckin 1995; Swales 1990; Thornton 2003). The Discussion section often constitutes approximately one-third to one-quarter of the length of the entire paper (Brand and Huiskes 2001; Driscoll 1997). In this chapter, we will examine the content and the style of the Discussion section.

How to Prepare

The Discussion section should be the last major section of the IMRAD paper to be written because you must synthesize the information from all the other sections of the manuscript. Therefore, you should have drafted all the other major sections (Introduction, Methods, and Results) before you begin writing this section. As you will see, the storyline developed in the Discussion section should be a logical extension of the research lineage presented in the Introduction section. Additionally, you will need to compare the Results of the present study to previously published research. Therefore you should review the journal articles summarized previously in the Introduction section and conduct another search of the scientific literature to identify any recently published articles. You should not only compare the Results and Methods of these prior studies to contextualize the present research; but you should use the Discussion sections of these articles as models for writing your Discussion section. In an editorial, Peter Thrower (2010), former editor in chief of the journal *Carbon*, reflected on the importance of reading other studies:

> One thing that never ceases to amaze me is the fact that many submissions are prepared as if the writer had never read a good scientific paper, even though 50 may be cited! One sometimes wonders whether the author has ever read any of them carefully. As with many things in life, we can learn most from the experience of others and/or trying to do it ourselves, rather than from textbooks. You learn to write papers by reading other people's papers, and by writing them yourself.
>
> *(2676)*

Organization and Content of the Discussion and Conclusions Sections

In SMC, the Discussion section has fairly conventional content and structure (Peacock 2002; Skelton and Edwards 2000). Although the Discussion section is not formally divided into subsections like the Methods and Results sections, the Discussion section does contain standard content, referred to as "moves" (Cals and Kotz 2013b; Kanoksilapatham 2005; Motsitsi 2008; Nwogu 1997; Peacock 2002; Skelton 1994; Swales 1990, 2004; Wolka and Cutler 2008). The number and order of the moves vary somewhat between fields, although the differences appear more in the categorization (or labeling) of moves rather than in the content (Swales 2004). In a prototypical Discussion section, you should include the following:

1. Move 1: Present the principal claim,
2. Move 2: Contextualize research findings by comparing the Results of the current study to previously published research,
3. Move 3: Analyze the research approach (Methods), and
4. Move 4: Provide the Conclusions, by:
    a. Summarizing the principal claim,
    b. Presenting the implications of the research, and
    c. Providing specific recommendations for future research.

In the following subsections, we will examine each move in detail.

Move 1: Present the Principal Claim

As shown in Examples 8.1, 8.2, and 8.3, the first paragraph of the Discussion section typically begins with a sentence summarizing the PRINCIPAL CLAIM, or primary findings of the study (Docherty and Smith 1999; Kanoksilapatham 2005; Motsitsi 2008). In Move 1 of the Discussion, the principal claim directly addresses the primary research objective or answers the research question presented in the Introduction section (Motsitsi 2008; Nwogu 1997). This summary statement must synthesize all of the data in the article and account for all observations (Brand and Huiskes 2001).

In Example 8.1, the principal claim is from a journal article of basic research. In Examples 8.2 and 8.3, the principal claims are from medical journal articles. As illustrated in these examples, the principal claim is followed by a brief mention of key results, which serves to firmly ground the summary statement in the data. In the Discussion section, you should only reference key data from the Results section rather than describe data in detail or introduce new data (Cals and Kotz 2013b; International Committee of Medical Journal Editors 2014). At the end of the first paragraph, authors often introduce a proposed explanation and value of the observations (see Examples 8.1 and 8.2; International Committee of Medical Journal Editors 2014).

When examining the principal claim statement in Examples 8.1, 8.2, and 8.3, you should note that a general principle emerges from the specific data or observations. This form of logic, called INDUCTIVE REASONING, has an inherent degree of uncertainty. Therefore the principal claim of your Discussion must be (1) based in the evidence presented in the Results section, (2) appropriately tempered to avoid taking the interpretation too far beyond the data, and (3) presented carefully and precisely to avoid overstating the certainty of the statement (covered later in this

EXAMPLE 8.1 PRESENTATION OF THE PRINCIPAL CLAIM IN THE FIRST PARAGRAPH OF A DISCUSSION SECTION IN A BASIC RESEARCH ARTICLE.

dksA Is Required for Intercellular Spread of *Shigella flexneri* via an RpoS-Independent Mechanism

Text	Annotation
DISCUSSION	Subheading
Here, we identify DksA as a protein necessary for intercellular spread of *S. flexneri.*	Principal claim
DksA appears to influence or function in a variety of pathways in enteric bacteria. The phenotypes include a failure to properly localize IcsA, sensitivity to environmental stress, and cell elongation or filamentation. Two-dimensional gel analysis of the *S. flexneri dksA* mutant showed that expression of several proteins is altered compared to that for the wild type. Because the *dksA* mutant exhibited phenotypic deficiencies in several assays and exhibited altered protein expression relative to that of the wild type,	Briefly ground the summary in the key Results
we hypothesize that DksA may function in a central pathway within the cell.	Speculative explanation and value of the observations

Adapted from Mogull, Scott A., Laura J. Runyen-Janecky, Mei Hong, and Shelley M. Payne. 2001. "*dksA* Is Required for Intercellular Spread of *Shigella flexneri* via an RpoS-Independent Mechanism." *Infection and Immunity* 69: 5742–51.

EXAMPLE 8.2 PRESENTATION OF THE PRINCIPAL CLAIM IN THE FIRST PARAGRAPH OF A DISCUSSION SECTION IN A MEDICAL ARTICLE.

Coffee Drinking and Cutaneous Melanoma Risk in the NIH-AARP Diet and Health Study

Text	Annotation
DISCUSSION	Subheading
In this large prospective study, coffee drinking was inversely associated with malignant melanoma.	Principal claim
We found that as compared with non-coffee drinkers, those who drank the most coffee (≥4 cups/day) had a 20% lower risk of malignant melanoma but not of melanoma in situ,	Briefly ground the summary in the key results
which may indicate different disease etiologies or an inhibitory role of coffee consumption in disease progression.	Speculative explanation and value of the observations

Statistically significant inverse associations were also found for caffeinated coffee intake and malignant melanoma but not for decaffeinated coffee intake.	Minor or ancillary findings may also be reported

Source: Loftfield, Erikka, Neal D. Freedman, Barry I. Graubard, Albert R. Hollenbeck, Fatma M. Shebl, Susan T. Mayne, and Rashmi Sinha. 2015. "Coffee Drinking and Cutaneous Melanoma Risk in the NIH-AARP Diet and Health Study." *Journal of the National Cancer Institute* 107 (2): dju421.

EXAMPLE 8.3 PRESENTATION OF THE PRINCIPAL CLAIM IN THE FIRST PARAGRAPH OF A DISCUSSION SECTION IN A MEDICAL ARTICLE.

Prospective Study of Peripregnancy Consumption of Peanuts or Tree Nuts by Mothers and the Risk of Peanut or Tree Nut Allergy in Their Offspring

DISCUSSION	Subheading
In a large national cohort of mothers and their offspring, with extensive efforts taken to validate all cases of physician-diagnosed food allergy,	
we found that frequent peripregnancy consumption of peanut or tree nut by mothers was associated with a decreased risk of peanut or tree nut allergy in their offspring.	Principal claim
The inverse association was driven by the large majority of mothers without peanut or tree nut allergy. Among mothers with peanut or tree nut allergy, a nonsignificant positive association was found, and the statistical interaction of maternal peripregnancy peanut or tree nut consumption and maternal peanut or tree nut allergy status was statistically significant ($P_{interaction}$ = .004).	Briefly ground the summary in the key results

(Continued in Example 8.4)

Adapted from Frazier, A. Lindsay, Carlos A. Camargo, Susan Malspeis, Walter C. Willett, and Michael C. Young. 2014. "Prospective Study of Peripregnancy Consumption of Peanuts or Tree Nuts by Mothers and the Risk of Peanut or Tree Nut Allergy in their Offspring." *JAMA Pediatrics* 168 (2): 156–62.

chapter). As P. K. Nair (2005), editor in chief of *Agroforestry Systems*, stated, "The line between 'optimum' and 'excess' [interpretation] is often faint; the scientist has to do some balancing act to separate rote from reasoning" (xiii). Nair (2005) illustrated this point of keeping interpretation limited to the evidence by telling the classical joke of Sherlock Holmes and the mystery of stars:

> Mr. Sherlock Holmes and his sidekick Dr. John Watson went on a camping trip. They pitched their tent under the stars and went to sleep. During the night, Holmes awoke and nudged his friend "Watson, look up at the stars, and tell me what you deduce." Watson replied, "I see millions and millions of stars. Astronomically, it tells me that there are millions of galaxies and potentially billions of planets. Astrologically, I observe that Saturn is in Leo. Theologically, I can see that God is all powerful and that we are small and insignificant. Meteorologically, I suspect that we will have a beautiful day tomorrow. What does it tell you?" Holmes went silent for a moment, then said: "Watson, you idiot, somebody has stolen our tent."
>
> *(xiii)*

As illustrated in the story, you must exercise good judgment and common sense based on the available evidence. Some of the most common errors made by scientific authors when developing the principal claim are listed in Table 8.1.

Move 2: Contextualize Research Findings

The longest and least structured subsection of the Discussion section is a critical evaluation of the research in context of other studies (Peacock 2002). In this subsection, you should compare the results of the current study to previously published research results. According to Kathleen Fahy (2008) of the University

TABLE 8.1 Common errors in developing the principal claim.

- Presenting a cause-and-effect relationship when only a correlation or association is established
- Conflating statistical significance with clinically or practically relevant findings
- Failing to account for all data, in particular ignoring data that do not fit the model or statistically nonsignificant findings
- Extrapolating beyond the animal or model system to suggest human clinical significance
- Solving a problem far beyond the scope of the data
- Failing to account for alternative explanations of the study results
- Preventing room for criticism or differences of opinion on the part of the reader
- Speculating on economic benefits or costs when the research does not address these issues

Sources: Cetin and Hackam (2005); Drotar (2009); Hess (2004); International Committee of Medical Journal Editors (2014); Liumbruno et al. (2013); Nair (2005); Pique-Angordans and Posteguillo (2006); Skelton and Edwards (2000).

of Newcastle, "You should explain how the evidence provided in the paper supports your thesis and undermines alternative explanations" (116). When writing this subsection, you should explicitly tell readers how the principal claim and supporting data of your research fit with previously published theories and data (Brand and Huiskes 2001; Drotar 2009; Fahy 2008; International Committee of Medical Journal Editors 2014; Pique-Angordans and Posteguillo 2006). In this move, you should include a brief review of relevant previously reported findings, which includes both consistent and inconsistent data (Annesley 2010; Liumbruno et al. 2013; Motsitsi 2008). In this analysis, you should focus objectively on the data and methodological comparisons without personally criticizing the authors of previous studies (Hess 2004). Furthermore, any discrepancies in your own data, which includes inconsistencies either internally or externally with other studies, should be mentioned and explained when possible (Brand and Huiskes 2001).

The comparison of your Results to previously published research in the Discussion section provides a more detailed examination of the research findings and methodological approach than is provided in the Introduction section. Here, you should explain how your new Results contravenes, agrees, extends, or conflicts with previous studies as shown in Example 8.4 (Cetin and Hackam 2005; Harmon and Gross 1996; Nair 2005; Swales 1990). The review of previous literature in the Discussion section is at a greater level of detail than in the Introduction section. Specifically, in the Discussion section, you will critically evaluate the Methods and Results from previous research studies whereas, in the Introduction section, you only present a general summary of the principal claim (cf. Examples 8.4 and 8.5).

EXAMPLE 8.4 MOVE 2 CONTEXTUALIZES THE PRESENT RESEARCH BY COMPARING IT TO FINDINGS FROM PRIOR STUDIES.

The Introduction and Discussion sections both contain a review of the literature but with distinct purposes and different levels of detail.

(Continued from Example 8.3)

Although animal studies have suggested a benefit from frequent exposure to dietary antigens during pregnancy and lactation, prior human studies have important methodological limitations and have shown either no risk or increased risk of peanut or tree nut allergy in offspring; to our knowledge, no prior study in humans has shown that maternal peripregnancy exposure to peanut or tree nut is associated with a reduced risk of peanut or tree nut allergy in their offspring. We review the prior studies here.	Move 2 of the Discussion begins with an overview of problems with prior research and gap in knowledge (which corresponds to Move 3 in the Introduction)

Animal studies consistently demonstrate a protective effect of maternal exposure to dietary antigens during pregnancy and lactation.[27–30] However, most animal models use egg albumin and bovine serum albumin as the allergen; only 1 animal model has examined peanut allergy using peanut-sensitized mice.[30] In this mouse study, the offspring of mothers fed peanut were protected from peanut sensitization; maximum protection occurred in offspring of mothers fed peanut with the adjuvant.

Detailed comparison of Results and Methods of prior research studies

Human studies examining the impact of antenatal and perinatal maternal diets on peanut allergy show less consistent results. A prospective US study[13,31] showed no benefit of maternal and early childhood avoidance of milk, egg, or peanut in children up to age 7 years in preventing food allergies. Likewise, 2 population-based UK studies[18,19] showed that maternal peanut consumption during pregnancy and lactation was not associated with risk of peanut allergy in offspring. In contrast, a Dutch birth cohort study[14] found that frequent maternal peanut or tree nut consumption during pregnancy increased the diagnosis of asthma but not the diagnosis of peanut allergy in the offspring at age 8 years. A questionnaire-based UK study by Fox et al.[6] showed no association between maternal peanut consumption during pregnancy or lactation and the development of peanut allergy in the offspring. A 2010 meta-analysis[5] of 6 human studies[10,15,18,32–34] found no evidence that mothers' avoidance or exposure to peanut during the antenatal or perinatal period had an impact on sensitization or allergy to peanut in their offspring.

Note the detailed analysis of Methods and Conclusions for studies #18 and 19, #14, and #5 (respectively, for comparison with the Introduction in Example 8.5)

In contrast, other studies suggest an increased risk of peanut allergy with maternal peanut consumption during pregnancy and the perinatal period. A retrospective UK study by Hourihane et al.[16] evaluated 622 subjects with peanut allergy by questionnaire and found that subjects younger than 6 years were more likely to have mothers who ate peanuts during the pregnancy or lactation period. In South Africa, Frank et al.[15] conducted a retrospective case-control study examining 25 children with peanut allergy and found an association with maternal peanut ingestion during pregnancy, but statistical significance was not reached. In the United States, Sicherer et al.[17] performed a retrospective study examining 503 infants with eczema, milk allergy,

Note the detailed analysis of the Methods and Results from three previous studies correspond to the summary of the principle claim from these studies in the Introduction section (see Example 8.4)

or egg allergy who were at high risk for developing peanut allergy. They found a dose-dependent association between reported maternal peanut ingestion during the third trimester of pregnancy and peanut sensitization. A retrospective case-control study from Montreal of 403 infants with history of clinical reactions to peanut and positive peanut-specific IgE also found that mothers reported greater maternal consumption of peanut during pregnancy and lactation.[35] All these retrospective studies are limited by likely recall bias. Accordingly, the issue of maternal dietary modification during pregnancy and its effect on development of peanut or tree nut allergy in offspring remains controversial. Our novel finding of reduced peanut or tree nut allergy with increased maternal peanut or tree nut consumption in the peripregnancy period supports more recent studies showing that early food allergen exposure promotes food allergen tolerance and reduced risk of food allergy.[20–24]

(Continued in Example 8.6)

Adapted from Frazier, A. Lindsay, Carlos A. Camargo, Susan Malspeis, Walter C. Willett, and Michael C. Young. 2014. "Prospective Study of Peripregnancy Consumption of Peanuts or Tree Nuts by Mothers and the Risk of Peanut or Tree Nut Allergy in their Offspring." *JAMA Pediatrics* 168 (2): 156–62.

EXAMPLE 8.5 THE INTRODUCTION SECTION CONTEXTUALIZES THE PRESENT RESEARCH BY SUMMARIZING FINDINGS FROM PRIOR STUDIES.

The Introduction and Discussion sections both contain a review of the literature but with distinct purposes and different levels of detail.

Prospective Study of Peripregnancy Consumption of Peanuts or Tree Nuts by Mothers and the Risk of Peanut or Tree Nut Allergy in Their Offspring

INTRODUCTION

Peanut allergy affects 1% to 2% of the population in most Western countries,[1–3] and in the United States, the prevalence of childhood peanut allergy has more than

Move 1, Step 1: Establish the importance of the research topic tripled, from 0.4% in 1997 to 1.4% in 2010.[4] Typically, the onset of peanut allergy is in early childhood; 70% of reactions occur during the first known exposure.[5] These IgE-mediated hypersensitivity reactions require prior allergen exposure and sensitization, implying that prior exposure to peanut had already occurred in utero or through unknown exposures in the diet or environment, such as through skin or respiratory routes.[6] Because of frequent overlap between peanut allergy and tree nut allergy and their similar natural history, with 80% to 90% persistence of the food allergy into adulthood,[7] these 2 allergies are often considered together as peanut or tree nut allergy.

Move 1, Step 2: Summarize previous research to contextualize the present study

For many years, pediatric guidelines have recommended the avoidance of peanut and tree nuts for at least the first 3 years of life, with some experts also recommending that peanut and tree nuts be avoided during pregnancy.[8] These recommendations were rescinded recently when literature reviews showed little support for them.[9,10] For decades, many investigators have posited that modifications of the maternal diet during pregnancy might prevent food allergies.[11,14] However, some studies on maternal avoidance of peanut during pregnancy actually demonstrated an increase in peanut sensitization in the child,[15–17] while other studies found no association.[5,14,18,19] In related research, early exposure to allergenic foods in infant diets may decrease sensitization and increase oral tolerance to those foods.[20–24]

Move 2: Identify a gap or inconsistency in the previous research to establish a niche for the present study

Note the general summary of studies #5, 14, 18, and 19 in the Introduction

Note that the Introduction provides a general summary of the principal claim whereas the Discussion provides a detailed analysis (see Example 8.3)

Given the lack of clarity in the current literature, an important quandary exists: should the pregnant mother include or exclude peanut or tree nuts in her diet? The goal of our investigation was to clarify the association between peripregnancy consumption of peanut or tree nuts by mothers and the subsequent development of peanut or tree nuts allergy in their offspring.

Move 3: State the objective of the present study, which occupies the research niche

Adapted from Frazier, A. Lindsay, Carlos A. Camargo, Susan Malspeis, Walter C. Willett, and Michael C. Young. 2014. "Prospective Study of Peripregnancy Consumption of Peanuts or Tree Nuts by Mothers and the Risk of Peanut or Tree Nut Allergy in their Offspring." *JAMA Pediatrics* 168 (2): 156–62.

Move 3: Analyze the Research Approach

The next major move in the Discussion section is to provide an analysis of your research approach. When analyzing the research approach, you should include a balanced discussion of both the strengths and limitations of the research design (Cals and Kotz 2013b; Chiswick 2002; Docherty and Smith 1999; Skelton and

Edwards 2000). As explained in an article coauthored by Richard Smith, editor of *BMJ*, "The task here is not to show how your study is better than previous studies but rather to compare strengths and weaknesses" (Docherty and Smith 1999, 1224). A balanced discussion of the research approach will help readers to appropriately contextualize the principal claims.

In poorly developed Discussions, the analysis of the research approach is limited to a trite and superficial mention of generic limitations to the research design, such as a statement that the sample was of limited size. Such statements lack the depth of analysis that is the goal of this section. Rather, you should analyze the general research approach and specific methods used in the study to identify sources of meaningful limitations that may influence the quality of the data or analysis, such as methodological imprecision, systematic bias of instrumentation, and statistical significance versus clinical importance (Ioannidis 2007; Moher et al. 2010; Puhan et al. 2012).

In the following subsections, we will examine the strengths and limitations presented in the Discussion section of the medical article on peripregnancy consumption of peanuts or tree nuts and allergy (continued from the previous examples).

Strengths of the Methods Used

The next subsection of the Discussion section provides a detailed analysis of the strengths of the research methods in the current study. Importantly, writers provide an argument that the Methods were selected appropriately to answer the research question. Specifically, you should cite previous literature showing that the research methods and resulting data were valid to show the intended phenomena. Moreover, you should establish the extent that the Results of the study are appropriately generalized to other circumstances.

In Example 8.6, a summary of the research method and the advantages of the approach are provided.

EXAMPLE 8.6 MOVE 3 EXPLAINS THE STRENGTHS OF THE RESEARCH DESIGN AND ADVANTAGES TO THE APPROACH.

(Continued from Example 8.4)

Our study has several strengths. The maternal history of peanut or tree nut consumption was obtained during or within 6 months of the pregnancy, giving more accurate quantitative results than other studies that collected dietary histories retrospectively many	Strengths of the study design are listed along with the explicitly stating the advantages of each design

years after pregnancy. Recruitment of subjects	Summary of method
was not based on atopic history or on a genetically high-risk population, so generalizability is enhanced.	Advantage
We reviewed all self-reported cases of physician-diagnosed peanut or tree nut allergy by having 2 pediatricians independently examine the medical records, allergy skin test results, and specific IgE data and assign confirmation codes based on the strength of evidence.	Summary of method
Using this approach, many potentially false-positive cases of food allergy were removed (e.g., children with lactose intolerance).	Advantage
Our conclusions are derived from the cases with the highest likelihood of being a true food allergy as opposed to food sensitization. Our results did not change when we restricted the analysis to only peanut or tree nut cases with the highest level of confirmation (eTable 3 in Supplement).	Summary argument for the strength of this research design compared to limitations of prior research (identified in Example 8.4)

(Continued in Example 8.7)

Adapted from Frazier, A. Lindsay, Carlos A. Camargo, Susan Malspeis, Walter C. Willett, and Michael C. Young. 2014. "Prospective Study of Peripregnancy Consumption of Peanuts or Tree Nuts by Mothers and the Risk of Peanut or Tree Nut Allergy in their Offspring." *JAMA Pediatrics* 168 (2): 156–62.

Limitations of the Methods Used

Although all research studies have limitations, the goal of this section is to convince the reader that the limitations of the study design do not undermine the conclusions or applications of the research (Brand and Huiskes 2001; Drotar 2009; Hess 2004). A frank and insightful analysis of reasonable methodological limitations should not jeopardize the publishing of a journal article (Puhan et al. 2012). Meaningful limitations of methods worthy of discussing may include weaknesses or biases in the design, implementation, analysis, interpretation, or limited practical applications of the work (Ioannidis 2007; Veness 2010; Wolka and Cutler 2008). When discussing the limitations of the methods, you might include your reasons for selecting the approach to emphasize the benefits of the method (Driscoll 1997). Often the limitations that you present can be developed into suggestions for future research studies (Kurmis 2003; Puhan et al. 2012). Example 8.7 provides a detailed analysis of the methodological limitations in the medical study.

Move 4: Conclusions

The conclusions of a journal article are summarized in the last paragraph of the Discussion section, which may be a separate section in some journals (Nair 2005;

EXAMPLE 8.7 MOVE 3 ALSO EXPLAINS THE LIMITATIONS OF THE RESEARCH DESIGN.

(Continued from Example 8.6)

Text	Annotation
Our study also has potential limitations. The dietary questionnaires were not specific for the actual dates of the pregnancy but were chosen as the one completed closest to the child's date of birth. Consequently, only 45% of the dietary questionnaires were completed during the pregnancy; 76% were within 1 year of the pregnancy. Because the dietary assessment was completed long before the start of this food allergy study,	Methodological limitation
the data are less subject to recall bias.	Impact on study
Lastly, our study only examined maternal history,	Methodological limitation
so any paternal influences on the development of peanut or tree nut allergy were not investigated.	Impact on study
Another potential limitation of our study is confounding by other dietary and behavioral differences among mothers with frequent peanut or tree nut consumption. For example, these mothers also were more likely to consume fruits and vegetables, which are high in antioxidants and may lower the risks of allergic sensitization and asthma.[36,37] Likewise, they were more likely to introduce peanut or tree nut into the child's diet at a younger age, which may promote oral tolerance.[23,24]	Methodological limitation
Thus, there was a theoretical possibility that higher in utero antioxidant exposure (not higher peanut or tree nut exposure) was responsible for incident food allergy in offspring.	Theoretical impact on study
To address this possibility, we controlled for these factors and they had no material effect on our results (eTable 2 in Supplement).	Measures taken to account for limitation

(Continued in Example 8.8)

Adapted from Frazier, A. Lindsay, Carlos A. Camargo, Susan Malspeis, Walter C. Willett, and Michael C. Young. 2014. "Prospective Study of Peripregnancy Consumption of Peanuts or Tree Nuts by Mothers and the Risk of Peanut or Tree Nut Allergy in their Offspring." *JAMA Pediatrics* 168 (2): 156–62.

Veness 2010). The conclusions provide your final statement on the research in the article (Wolka and Cutler 2008) and consist of the following:

1. Summary of the principal claim,
2. Implications of the research, and
3. Specific recommendations for future research.

Although the conclusions section (or paragraph) contains these three elements, the order may vary (as in Example 8.8).

Typically, the conclusions paragraph (or section) begins with the "take-home message" (Annesley 2010, 1672; Hess 2004, 1240; Veness 2010, 398) from the research and should provide a "citable statement" (Cals and Kotz 2013b, 1064) for other researchers and practitioners to use to summarize the journal article in future studies. This summary statement should be a single sentence that transforms the research findings into a GENERALIZED CLAIM about the research topic (Harmon and Gross 1996; Liumbruno et al. 2013; Pique-Angordans and Posteguillo 2006; Veness 2010). Much like the principal claim that was presented previously, the generalized claim should occupy the research niche defined in the Introduction and set the boundary of extrapolation to the research study (Fahy 2008; Harmon and Gross 1996).

Following the summary of the generalized claim, you should identify the implications or relevance of the current work and its contribution to existing knowledge on the topic (Kurmis 2003). Although some writers speculate too far on the possible implications of the research, you should try to keep claims

EXAMPLE 8.8 CONCLUSIONS OF A MEDICAL ARTICLE.

The conclusions may be in a separate section (as in this example) or the last paragraph of the Discussion section.

(Continued from Example 8.7)

CONCLUSIONS

We found a strong inverse association between peripregnancy peanut or tree nut consumption by mothers and risk of peanut or tree nut allergy in their offspring. The more peanut or tree nuts the mother consumed, the lower the risk was of her child developing physician-diagnosed peanut or tree nut allergy.	Summary of the research leading to a generalized claim
Our study supports the hypothesis that early allergen exposure increases the likelihood of tolerance and thereby lowers the risk of childhood food allergy.	Generalized claim (or "take-home message")
Additional prospective studies are needed to replicate this finding.	Specific recommendations for future research
In the meantime, our data support the recent decisions to rescind recommendations that all mothers avoid peanut or tree nuts during pregnancy and breastfeeding.[9,10]	Implications of the research

Adapted from Frazier, A. Lindsay, Carlos A. Camargo, Susan Malspeis, Walter C. Willett, and Michael C. Young. 2014. "Prospective Study of Peripregnancy Consumption of Peanuts or Tree Nuts by Mothers and the Risk of Peanut or Tree Nut Allergy in their Offspring." *JAMA Pediatrics* 168 (2): 156–62.

grounded in the data. As Dean Hess (2004) of Massachusetts General Hospital and Harvard Medical School stated:

> After all the hard work that goes into a study, it is easy to attribute unwarranted importance to study findings. We all want our study to make an important contribution that will be cited for generations to come. However, unwarranted inflation of the importance of the study results will disgust reviewers and readers. A measure of humility goes a long way.
>
> *(1240)*

Example 8.8 shows an appropriately tempered statement explaining the implications of the research. The International Committee of Medical Journal Editors (2014) specifically discourages writers from speculating on economic benefits or costs of the research findings if the research manuscript lacks the appropriate economic data or analyses.

Finally, at the end of the section, you should recommend specific future research studies that are needed to advance the topic or validate a change in practice (Annesley 2010; Harmon and Gross 1996; Liumbruno et al. 2013; Nair 2005; Thornton 2003). The recommendations for future research should be based on the outcomes of the current study and thoughtfully considered to further knowledge on the topic (Drotar 2009). In an editorial published in *Current Paediatrics*, Malcolm Chiswick (2002) of St. Mary's Hospital for Women and Children stated:

> Simply ending a paper with an isolated statement without context: "We conclude that future research is needed" is a journalistic hanging offence. Of course, future research is needed—we rarely read: "We conclude that our findings are the final say in this matter. No future research is needed."
>
> *(416)*

Writing Style in the Discussion Section

In this section, we will examine the writing style of the Discussion section. The primary style issues to consider when writing this section are verb tense, passive versus active voice, and hedging.

Verb Tense Revisited

In the Discussion section, the verb tense (past, past perfect, or present) may vary within the section and even each paragraph. The appropriate tense depends on the type of information being presented (Cargill and O'Connor 2013; Cetin and Hackam 2005; Fahy 2008; Hawes and Thomas 1997; *Nature* 2014; Penrose and Katz 2010). If the information refers to specific results or data that were found in the current study, the information is presented in past (or past perfect) tense.

Similarly, specific results or data that were collected from and attributed to previous studies are typically presented in past (or past perfect) tense since the data represent direct observations from a previous event (i.e., the research method). In contrast, claims of general knowledge are typically presented in present tense because these claims are currently accepted as a general explanation (for, which may include generalized claims from both the current and previous research studies). But claims from previous studies that have been disproven or are no longer accepted are typically presented in past (or past perfect) tense and clearly identified as outdated understandings.

Voice and Personal Pronouns Revisited

As discussed in Chapter 5, writers' use of active and passive voice varies by section of the journal article. Notably, the use of active voice constructs is highest in the Discussion section (see Table 5.7). As shown in the examples throughout this chapter, writers often attributed research findings, data, and methods to the researchers who performed the experiments. In Moves 1, 3, and 4 of the Discussion section, writers use personal pronouns to refer to the present research (see Examples 8.3, 8.6, 8.7, and 8.8) and the last names of researchers in the evaluation of previous studies (see Example 8.4). In Move 2, which is a detailed analysis of previous research data, writers often place the authors of the study at the beginning of a sentence (in the stress position) when referring to a specific journal article (see Example 8.9). This style reflects expert conversation, in which scientists refer to specific studies by the authors' names. Using this style in written discourse is the clearest and most efficient way to reference a specific journal article as well as contextualize the findings into the scientific method. Writers may also use third-person pronouns as further illustrated in Example 8.9 (Bernhardt 1985).

EXAMPLE 8.9 USE OF ACTIVE VOICE AND PERSONAL PRONOUNS. (THE USE OF THIRD-PERSON PRONOUNS IS ALSO ACCEPTED.)

In the United States, Sicherer et al[17] performed a retrospective study examining 503 infants with eczema, milk allergy, or egg allergy who were at high risk for developing peanut allergy. They found a dose-dependent association between reported maternal peanut ingestion during the third trimester of pregnancy and peanut sensitization.

Source: Frazier, A. Lindsay, Carlos A. Camargo, Susan Malspeis, Walter C. Willett, and Michael C. Young. 2014. "Prospective Study of Peripregnancy Consumption of Peanuts or Tree Nuts by Mothers and the Risk of Peanut or Tree Nut Allergy in Their Offspring." *JAMA Pediatrics* 168 (2): 156–62.

"A possible eureka."

Hedging

In the Discussion section, you should envision yourself in a dialogue with the reader in which your goal is to justify your analysis and interpretation of the Results (Cals and Kotz 2013b). This dialogue is a logic-based case in which you try to convince the reader that your reasoning is valid given the available evidence. In the Discussion section, your writing style must be tempered to avoid presenting your interpretations too forcefully (Cals and Kotz 2013b; Pique-Angordans and Posteguillo 2006; Skelton and Edwards 2000; Thornton 2003).

The tempered writing style used in SMC, often referred to as "HEDGING," shows that you, as the writer, are conscious of the limitations inherent in the scientific method and are allowing room for your readers and future researchers to disagree and modify your conclusions without negating your entire work (Hyland 1995, 1996; Myers 1992). Students (and novice scientific writers) often incorrectly present information too forcefully, as illustrated by their frequent inappropriate use of the word "prove," which is almost never used in published journal articles (Parkinson 2011). In contrast, experienced scientific authors employ much greater use of the word "suggest" than do novice student writers (Parkinson 2011). As explained by linguist Ken Hyland (1995), "Scientific knowledge involves the consensus of the research community and hedges are an important rhetorical device in acknowledging the reader's role in ratifying claims" (34). Table 8.2 summarizes the hedging words and phrases commonly used in scientific and medical writing. As shown in Table 8.3, the exact word or phrase should indicate an appropriate degree of caution given the evidence. Hedging words and phrases typically appear in clusters when writers are presenting a particular concept (Hyland 1995). In particular, hedging is used when analyzing the validity of the experimental methods, the data that arise from such methods, and generalized claims that the authors derive from interpreting the data (Hyland 1995; Myers 1992).

TABLE 8.2 Hedging words and phrases identified by Hyland (1995) that are commonly used in scientific and medical writing.

Category	Specific Function	Common Words and Phrases	Example
Lexical verbs	Used by writers as they infer a general statement or claim from specific data	Indicate Suggest Appear Propose	"The results presented here **suggest** that the protection conferred by Shanchol may be higher in the first 6 months than in the remaining 2 years after vaccination."[1]
Modal adjectives	Introduce a degree of uncertainty in interpretations or speculation as to the cause or reason for an observation	Likely Possible Most Consistent with	"In the pediatric population, this difference did not reach statistical significance, **likely** because of the small sample size and the wide range of observed values in the pediatric patients with nonfatal cases."[2]
Adverbial forms ("downtoners")	Decrease the force of a verb	Quite Almost Usually	"This difference is notable because children **usually** have slightly lower levels of total IgG than do adults."[2]
	Decrease the force of a statement or claim	Probably Generally Evidently	"This is **generally** compatible with trends reported in previous patient surveys in other countries regarding willingness to try placebo treatments."[3]

Continued

TABLE 8.2 (Continued)

Category	Specific Function	Common Words and Phrases	Example
Modal verbs	Introduce a degree of uncertainty in statements	Would May Could	"The low proportion of PCR-positive samples **could** be related to the small amount or the poor quality of biologic material."[1]
Modal nouns	Indicate the uncertainty of a statement or claim	Possibility Assumption Estimate Tendency	"This difference may also be explained in part by the **possibility** that some nonvaccinated participants in the study by Sur et al. had naturally acquired immunity because cholera is endemic in Kolkata."[1]

Sources:

[1] Luquero, Francisco J., Lise Grout, Iza Ciglenecki, Keita Sakoba, Bala Traore, Melat Heile, Alpha Amadou Diallo et al. 2014. "Use of *Vibrio cholerae* Vaccine in an Outbreak in Guinea." *The New England Journal of Medicine* 370 (22): 2111–20.

[2] McElroy, Anita K., Bobbie R. Erickson, Timothy D. Flietstra, Pierre E. Rollin, Stuart T. Nichol, Jonathan S. Towner, and Christina F. Spiropoulou. 2014. "Biomarker Correlates of Survival in Pediatric Patients with Ebola Virus Disease." *Emerging Infectious Diseases* 20 (10): 1683–90.

[3] Hull, Sara Chandros, Luana Colloca, Andrew Avins, Nancy P. Gordon, Carol P. Somkin, Ted J. Kaptchuk, and Franklin G. Miller. 2013. "Patients' Attitudes About the Use of Placebo Treatments: Telephone Survey." *BMJ* 346: f3757. doi: 10.1136/bmj.f3757.

TABLE 8.3 List of sample hedging terms and phrases that convey an appropriate degree of caution verses those that are too strong. These terms and phrases are typically used to complete a construct such as "The results of this study ___________ that."

Strength of Claim	Sample Terms and Phrases
Appropriately cautious	Suggest Indicate Show Demonstrate
Agree with other studies	Support Are consistent with

Strength of Claim	Sample Terms and Phrases
Disagree with other studies	Differ Are inconsistent with
Overly confident (too strong so avoid using)	Prove Confirm Substantiate Verify Mean

Concluding Points

Although the Discussion section is the last of the major sections of an IMRAD-structured journal article, and finishing this section represents a major milestone in the writing process, you still have a few more sections to write and prepare before submitting a manuscript to a journal for peer review. Specifically you will need to prepare the sections that summarize the content of the entire journal article, which are the Abstract and Title. Both the Abstract and Title are important sections because these are typically the two sections that potential readers review in database searches and use to determine if a journal article is worth accessing and reading in full. Thus, these summary sections are arguably the most important sections of the entire article. As a writer, you should wait to write these summary sections until the Introduction, Methods, Results, and Discussion are complete so that both the Title and Abstract accurately and thoroughly summarize and describe the content of the manuscript. In the following chapter, we will discuss writing these summary sections (the Title and Abstract). Additionally, you will need to prepare the list of references according to the journal citation style that is specified in the Instructions to Authors. After all of these sections are complete, you are ready to submit the manuscript to a journal for review. For the journal submission process, you will need to write a cover letter to the journal editor. We will discuss cover letters and the peer-review process in the last chapter of this section.

9

SUMMARY SECTIONS: TITLES AND ABSTRACTS

In this chapter, you will learn to:

- Write scientific Titles and Running Titles
- Write comprehensive Abstracts

Although appearing at the beginning of journal articles, Titles and Abstracts should be written last. Both Titles and Abstracts summarize the content of a journal article and enable potential readers to locate it through a database search and determine if the research topic is relevant and meets their needs (Harmon and Gross 2009; Liumbruno et al. 2013; Thrower 2007). To fully appreciate the importance of the Title and Abstract, you should think about your own experience of searching the scientific and medical literature (which we discussed in Chapter 3). Like you, potential readers will perform a keyword search of a database and obtain a list of Titles to examine. They will briefly scan each Title and evaluate it for relevance to their information needs. For Titles that appear to be most relevant, potential readers will click on a hyperlink and scan or read the Abstract. If the Abstract confirms the relevance of the topic and, moreover, provides particularly meaningful results, then potential readers will go through the steps (which sometime are tedious and somewhat expensive) to access the complete text of a journal article. Therefore as a courtesy to other researchers, you should develop a simple and clear, yet precise and complete Title and Abstract that accurately summarizes the content of your journal article. In this chapter, we will discuss writing such scientific Titles and Abstracts.

Writing Scientific Titles

Although seemingly simple and straightforward to prepare, the Title is the most prominent piece of a journal article. Scientific Titles should summarize the content of the manuscript in approximately 10 to 12 words (American Psychological Association 2010). Potential readers will determine whether or not they want to read the paper based on the Title. As Joseph Harmon and Alan Gross (2009), scholars of scientific communication, describe, "Scientific titles have evolved over time. Compared with titles from earlier times, modern ones are much more specific and technical, stripped of anything personal or openly literary" (455). Therefore, you should wait until you finish the manuscript to finalize the Title so that it will accurately and thoroughly reflect the content of the research.

Writing Effective Titles

The Title should be written as a clear, direct, and specific statement—providing a brief, yet complete and accurate summary of the research (Harmon and Gross 2009; Hartley 2005, 2007). Scientific Titles generally convey the following:

- The research topic, purpose, question, or hypothesis;
- The key subjects of the research investigation (such as organisms, chemicals, or genes);
- A brief summary or indication of the conclusion; and
- The experimental approach or methods used (especially if important to the findings or a methods paper).

When writing a Title, the goal is to maximize the informative content in the fewest words. Scientific Titles are fairly uniform in grammatical structure, particularly within the subgenres of experimental (or hypothesis) testing, exploratory research, and new research methods (see Example 9.1).

As Harmon and Gross (2009) recommend, you should construct Scientific Titles around the HEADER NOUN, which is the central object or topic of the journal article manuscript (see Example 9.2). The header noun should be placed at the beginning of the Title, which is the most prominent location. Next, modifying words or phrases may be placed to the left or right of the header noun. You should limit modifying words and phrases, particularly preceding the noun (to the left), because adding too many modifiers will bury the header noun and make the subject of the manuscript less clear. Additional content summarized in the Title often depends on the type of research study. For experimental or hypothesis testing, you should identify the dependent variable (see Example 9.2a) whereas for exploratory research, you should summarize the research findings (see Example 9.1b). For new research methods, you should identify the application or purpose

EXAMPLE 9.1 SCIENTIFIC TITLES OF THREE SUBGENRES.

(a) Experimental or Hypothesis Testing

Melatonin Secretion and the Incidence of Type 2 Diabetes

(b) Exploratory Research

dksA Is Required for Intercellular Spread of *Shigella flexneri* via an RpoS-Independent Mechanism

(c) New Research Methods

A Low Gastric pH Mouse Model to Evaluate Live Attenuated Bacterial Vaccines

Sources:

(a) McMullan, Ciaran J., Eva S. Schernhammer, Eric B. Rimm, Frank B. Hu, and John P. Forman. 2013. "Melatonin Secretion and the Incidence of Type 2 Diabetes." *JAMA* 309 (13): 1388–96.
(b) Mogull, Scott A., Laura J. Runyen-Janecky, Mei Hong, and Shelley M. Payne. 2001. "*dksA* Is Required for Intercellular Spread of *Shigella flexneri* via an RpoS-Independent Mechanism." *Infection and Immunity* 69: 5742–51.
(c) Brenneman, Karen E., Crystal Willingham, Jacquelyn A. Kilbourne, Roy Curtiss 3rd, and Kenneth L. Roland. 2014. "A Low Gastric pH Mouse Model to Evaluate Live Attenuated Bacterial Vaccines." *PLoS One* 9: e87411.

of the method (see Example 9.1c). For all of these Titles, you should ensure that the Title states the principal claim of the journal article so that potential readers will know how your research fits in the field.

Many researchers argue that subtitles are overused in scientific Titles and often result in a cumbersome Title that could easily be simplified (Thrower 2007). Yet subtitles do have some recommended uses. For example, subtitles separated by colons are often used in medical papers to separate a subtitle that indicates the experimental approach, such as "a randomized controlled trial" or "a systematic review and meta-analysis" (Iverson et al. 2007; see Example 9.2). Alternatively, colons may be used to precede the part number for each part of a multipart article (Thrower 2007). In such cases, you should keep the main title the same and then indicate each part of the article (such as "Part 1" and "Part 2").

Developing Titles from a Purpose Statement

If you are having difficulty writing a Title, you might find it useful to start with a purpose statement, and then edit the purpose statement into a brief summary. In the following subsections, I will demonstrate the process.

EXAMPLE 9.2 ORGANIZATION OF CONTENT IN SCIENTIFIC TITLES.

☑ **Recommended: constructing a scientific title around the header noun.**

Low-Dose Hydrochlorothiazide and Preservation of Bone Mineral Density in Older Adults: A Randomized, Double-Blind, Placebo-Controlled Trial

Main Title:

- Modifying words
- Header noun
- Dependent (measured) variable
- Population

Subtitle:

- Experimental approach

☒ **Not recommended: burying a header noun so that the subject matter of the research is less prominent.**

Efficacy of High-Dose Versus Standard-Dose Influenza Vaccine in Older Adults

- Dependent (measured) variable
- Modifying words
- Buried Header Noun
- Population

Sources:

(a) LaCroix, Andrea Z., Susan M. Ott, Laura Ichikawa, Delia Scholes, and William E. Barlow. 2000. "Low-Dose Hydrochlorothiazide and Preservation of Bone Mineral Density in Older Adults: A Randomized, Double-Blind, Placebo-Controlled Trial." *Annals of Internal Medicine* 133 (7): 516–26.

(b) DiazGranados, Carlos A., Andrew J. Dunning, Murray Kimmel, Daniel Kirby, John Treanor, Avi Collins, Richard Pollak, Janet Christoff, John Earl, Victoria Landolfi, Earl Martin, Sanjay Gurunathan, Richard Nathan, David P. Greenberg, Nadia G. Tornieporth, Michael D. Decker, and H. Keipp Talbot. 2014. "Efficacy of High-Dose versus Standard-Dose Influenza Vaccine in Older Adults." *The New England Journal of Medicine* 371: 635–45.

Step 1: Identify the Specific Audience

Defining the purpose and Title of a manuscript is an evaluative process in which you highlight a particular application or angle of the research. To determine the main purpose of the research, you should ask the following questions:

- What is important about this information?
- What is the benefit of this information (to the audience)?
- How will other scientists or medical practitioners use this information?
- Why is this research interesting to others?

As the author, you must visualize the typical audience member and determine the type information that will be of use to that reader. In doing this audience analysis, you determine the information to emphasize.

Also, you may use the audience of the target journal to help guide your writing of the Title. As an example, I selected *Infection and Immunity* as the target journal for one study that I referenced in this text. Even if you, as an author, are unfamiliar with a particular journal, you can still prepare an article for submission to it. To determine the focus of information in a journal, you can visit the journal's website or read a recent edition. Journals usually publish a statement that describes the scope of the content and a general profile of the readers. These descriptions provide a useful starting point for your audience analysis and Title development.

After locating the description of the journal content and audience, you can focus on the pertinent details that apply to your research. In the following description from *Infection and Immunity* (2015) website, I have highlighted (in bold) the parts of the description that are most relevant to the example (see Example 9.1b):

> *Infection and Immunity* (IAI) reports key discoveries that help **microbiologists**, immunologists, epidemiologists, pathologists, and clinicians to gain new insights into the **underlying mechanisms of host-pathogen interactions** and develop novel strategies to prevent or treat infectious diseases. Topics include infections caused by **pathogenic bacteria**, fungi, and parasites; **molecular mechanisms of pathogenicity**; interactions of virulence factors with host cells; innate and adaptive immunity to infection; development of vaccines against nonviral pathogens; and genomes of pathogenic bacteria.

Using this description of the audience and scope of the journal, I have identified the specific audience (microbiologists conducting basic research) and highlighted the key subjects and issues to identify:

- Organism (pathogenic bacteria): *Shigella flexneri*
- Relevant issue (underlying mechanism of host-pathogen interaction): *Deficiency of intercellular spread in epithelial cells*
- Feature or part of the issue that is relevant (molecular mechanisms of pathogenicity): *Function of dksA and rpoS genes*

An effective Title will identify these specific details.

Step 2: Write a Purpose Statement to Summarize the Research

After identifying the audience and key subjects and issues, the next step in Title development is to compose a purpose statement for the research with specific attention to your audience and scope. The purpose statement is a good place to begin because it is written as complete sentences and you are not limited by word count. Alternatively, if you have the purpose statement already composed, review it to confirm that you have included the specific details mentioned in step 1.

To develop the purpose statement, once again consider including the major points that are being made in the paper:

- The question or hypothesis evaluated: *Function of dksA in the intercellular spread of Shigella flexneri*
- The major subject(s) of the investigation (such as organisms, chemicals, or genes): *Shigella flexneri, dksA, and rpoS*
- A brief summary of the conclusions: *dksA is required for intercellular spread in a cell culture of Henle cells*
- The experimental approach or methods used (if important to the findings): *Transposon mutagenesis followed by tissue culture analysis.*

Once you have identified the important details, you can generate a purpose statement. Here is the purpose statement for our example:

> My purpose is to communicate to primary researchers (specifically microbiologists) that the *dksA* gene is required for the intercellular spread of *Shigella flexneri* in Henle cells, potentially by altering cell phenotype and IcsA localization. Although the exact mechanism is unknown, we know it works via an RpoS-independent mechanism.

Here I generated a purpose statement that is several sentences, which required additional details that I took from the journal article. Similarily may use several sentences to clarify the purpose of this research at this point. Here, your objective is to identify the most important concepts and key components of the research—we will make the statement more concise in the next step. To generate a purpose statement from your research, you should first gather a few summary statements and write a brief passage.

In this example, a significant portion of the experimental approach was excluded from the preceding purpose statement because I determined that the experimental techniques were not important to the research investigation. (Based on the audience description from the target journal, I determined that the audience would be more interested in the scientific phenomenon rather than the methods I used.) To illustrate this point, note fact that the main point of the article was investigating a *dksA* mutant, not that transposon mutagenesis, which was the technique used to create the mutant. A *dksA* mutant could have been generated by any number of techniques, such as ultraviolet radiation, with the same result (a nonfunctional *dksA* gene). Although the details of this example might be overly technical, the point is that you will need to make decisions about the information that you emphasize and eliminate. To do this effectively, you must have a comprehensive understanding of your own research as well as the primary audience and target journal.

Step 3: Edit the Purpose Statement into a Concise Title

The last step in developing an effective Title is to carefully edit the purpose statement into a concise, direct, and specific statement (which is often a sentence fragment). Table 9.1 provides a list of words and phrases that are unnecessary in Titles of journal articles and can be eliminated without any loss of meaning. Consider removing any obvious, redundant, or general information, as in the following example:

> ~~My purpose is to communicate to primary researchers (specifically microbiologists) that the~~ *dksA* ~~gene~~ is required for ~~the~~ intercellular spread of *Shigella flexneri* ~~in Henle cells, potentially by altering cell phenotype and IcsA localization. Although the exact mechanism is unknown, we know it works~~ via an RpoS-independent mechanism.

Using this process of heavy editing down to key concepts and one primary concept, you can develop an effective Title for your research from your purpose statement.

Writing Running Titles (Headers or Footers)

RUNNING TITLES (either as HEADERS or FOOTERS) are abbreviated Titles that appear on the top or bottom, respectively, of each internal page of the manuscript and published journal article. Running Titles identify the pages comprising the manuscript that you submit to a journal and the published journal article. The Running Title is usually a condensed version of the Title that typically includes the header noun and a few other key terms from the Title (see Table 9.2 for example

TABLE 9.1 Words and phrases to eliminate in Scientific Titles.

Word or Phrase	Reason to Eliminate
"Analysis of..." **"Investigations of..."** **"Observations of..."** **"Report of..."** **"Study of..."**	Such statements are obvious due to the genre (journal articles)
"New" or "novel"	Only new or novel research findings are published
"Careful" or "detailed" (when describing the methods or analyses)	Scientists are expected to be detailed and careful
Beginning a title with "a," "an," or "the"	Including an article at the beginning of a title is unnecessary
Abbreviations, jargon, and trade names	Such terms may confuse potential readers and impede database indexing and retrieval

Sources: Day and Gastel (2006); Iverson et al. (2007); Nair (2005); Thrower (2007).

TABLE 9.2 Running Titles for journal articles published in different journals.

Journal Article Title	Number of Words in Article Title	Running Title	Characters in Running Title (With/without Spaces)
Melatonin Secretion and the Incidence of Type 2 Diabetes	9	Melatonin Secretion and the Incidence of Type 2 Diabetes	48/57
dksA Is Required for Intercellular Spread of *Shigella flexneri* via an RpoS-Independent Mechanism	13	*dksA* Is Required for *S. flexneri* Intercellular Spread	46/53
A Low Gastric pH Mouse Model to Evaluate Live Attenuated Bacterial Vaccines	12	A Murine Low Gastric pH Model	24/29

Sources:

(a) McMullan, Ciaran J., Eva S. Schernhammer, Eric B. Rimm, Frank B. Hu, and John P. Forman. 2013. "Melatonin Secretion and the Incidence of Type 2 Diabetes." *JAMA* 309 (13): 1388–96.

(b) Mogull, Scott A., Laura J. Runyen-Janecky, Mei Hong, and Shelley M. Payne. 2001. "*dksA* Is Required for Intercellular Spread of *Shigella flexneri* via an RpoS-Independent Mechanism." *Infection and Immunity* 69: 5742–51.

(c) Brenneman, Karen E., Crystal Willingham, Jacquelyn A. Kilbourne, Roy Curtiss 3rd, and Kenneth L. Roland. 2014. "A Low Gastric pH Mouse Model to Evaluate Live Attenuated Bacterial Vaccines." *PLoS One* 9: e87411.

Titles and Running Titles of published articles). If the original Title is short, as in the first example in Table 9.2, the complete Title may be used as the Running Title. For articles with longer Titles, as in the other examples, a condensed Title is used as the Running Title for internal pages of the manuscript and journal article. The length of a Running Title varies and is usually based on the space limitations of the printed page in the journal. Therefore, Running Titles usually have specific requirements, such as a maximum of 60 characters including spaces. The exact length of a Running Title differs by journal and is typically specified in the Instructions to Authors.

Writing Comprehensive Abstracts

Abstracts provide a comprehensive, yet concise, summary of a research manuscript (Kurmis 2003; Plaxco 2010; Van Way 2007). Abstracts are read more than any other content section and thus are arguably the most important section of a journal article

(Chiswick 2002; Driscoll 1997; Pitkin and Branagan 1998; Plaxco 2010; Winker 1999). Often, Abstracts are read online without access to the full content of the journal article. Additionally, many databases, such as MEDLINE, search only the Title and Abstract of journal articles rather than the complete text. Since Abstracts are often read and searched independently of the complete journal article, you must thoroughly summarize the content in the Abstract. Moreover, the Abstract should be written as a self-contained communication that includes all relevant and necessary information without referring to any content that is not included in the Abstract, such as citations to sources listed in the References section (Falagas and Vergidis 2004; Pitkin, Branagan, and Burmeister 1999; Weinert 2010). In the following subsections, we will discuss writing a complete Abstract that accurately summarizes a journal article.

Including Complete Information

Conventionally, Abstracts have been poorly structured and lacked essential information (Ad Hoc Working Group for Critical Appraisal of the Medical Literature 1987; Falagas and Vergidis 2004; Rennie and Glass 1991; Salager-Meyer 1990, 1994, 2006). As a result, journal editors, initially in clinical medicine, implemented STRUCTURED ABSTRACTS, which are Abstracts with specific subheadings, to provide authors and readers with a framework for the content to include in an Abstract (Ad Hoc Working Group for Critical Appraisal of the Medical Literature 1987; Huth 1987; Lock 1988; Rennie and Glass 1991; Salager-Meyer 2006). Example 9.3 compares an unstructured and structured Abstract for a clinical medical study. Example 9.4 compares Abstracts for basic research. The primary difference between structured and conventional (or unstructured) Abstracts is that structured Abstracts contain subheadings that parallel the IMRAD sections and provide one to three sentences of content for defined subsections. In contrast, conventional Abstracts lack the subheadings and are formatted into a paragraph with each subsection leading directly into the next. As a result, structured Abstracts are:

- Higher quality (contain more information and are more likely to include relevant information),
- More accessible (easier to search and read), and
- Easier to write (provide a framework for writers) (Hartley and Benjamin 1998; Hartley and Sydes 1997; Hartley, Sydes, and Blurton 1996; Taddio et al. 1994).

Due to their many advantages over conventional abstracts, structured Abstracts in journal articles indexed in MEDLINE increased from 2.5% of articles published in 1992 to 20.3% of articles published in 2005 (Ripple et al. 2011).

Although theoretically the content of structured Abstracts should be the same as unstructured Abstracts, this is not the case. Therefore, when writing either type of Abstract, you can use the subheadings and content guidelines of structured Abstracts to ensure that you include all of the relevant information. So even if your target

EXAMPLE 9.3 UNSTRUCTURED AND STRUCTURED ABSTRACTS FOR A CLINICAL MEDICAL STUDY ORIGINALLY PUBLISHED IN *ANNALS OF INTERNAL MEDICINE.*

☒ Not Recommended: Unstructured Abstract

Twelve patients with refractory rheumatoid arthritis were treated with weekly pulse methotrexate in a double-blind, placebo-controlled, crossover study. After 13 weeks of therapy, patients receiving methotrexate showed greater improvement, judged by degree of joint swelling and tenderness, duration of morning stiffness, and subjective assessments of clinical condition, compared with those receiving placebo ($p \leq 0.002$). This improvement was associated with a decrease in sedimentation rate and decreases in levels of IgG, IgM, and IgA; no changes were seen in serum rheumatoid factor titer or complement protein levels. Proportions of mononuclear cell subsets that were abnormal before treatment (decreased percentage of total T cells, increased percentage of monocytes) improved toward normal after therapy with methotrexate. However, no changes were seen in elevated pretreatment Leu-3/Leu-2 ratios, in in-vitro proliferative responses of lymphocytes to mitogens, or in immunoglobulin secretory responses to pokeweed mitogen. Weekly pulse methotrexate is effective in the short-term treatment of refractory rheumatoid arthritis. Little evidence for cellular immune suppression was associated with this clinical benefit.

☑ Recommended: Structured Abstract

Study Objective: To determine the efficacy of weekly pulse methotrexate in refractory rheumatoid arthritis.

Design: Randomized, double-blind, placebo-controlled, crossover trial with 13-week treatment periods.

Setting: Referral-based rheumatology clinics at two army medical centers.

Patients: Sequential sample of 15 patients with active definite or classical rheumatoid arthritis and previous treatment failure. Twelve patients (80%) completed the study; 1 patient removed because of drug toxicity (pancytopenia).

Interventions: Nonsteroidal anti-inflammatory drugs and prednisone were continued. Methotrexate 5 mg intramuscular test dose was given at week 1, increased in 5-mg steps to a maximum of 25 mg/wk if clinically needed. Intramuscular saline placebo given in control periods.

Measurements and Main Results: The following results (statistically significant findings, $p < 0.05$) were in favor of methotrexate compared with placebo: number of swollen joints 6.9 (5.2, SD) with methotrexate and 19.4 (12.1) with placebo; number of tender joints 12.6 (14.2) and 26.2 (17.0); minutes of morning stiffness 78 (117.8) and 242 (131.6); joint pain (scale 0-10) 1.1 (2.1) and 4.8 (3.1); 50-foot walk (seconds) 16.1 (10.0) and 23.1 (16.3). Laboratory tests showing differences ($p < 0.05$) favoring methotrexate included erythrocyte sedimentation rate and IgG. Other tests of physical and laboratory function, including immunologic tests, did not show important differences.

Conclusions: Weekly pulse methotrexate results in improvement of multiple measures of disease activity in refractory rheumatoid arthritis. The mechanism of methotrexate action is uncertain with little evidence of short-term cellular immune suppression. Larger and longer trials are needed to assess the safety of methotrexate for refractory rheumatoid arthritis.

Adapted from Ad Hoc Working Group for Critical Appraisal of the Medical Literature (1987).

EXAMPLE 9.4 UNSTRUCTURED AND STRUCTURED ABSTRACTS FOR A JOURNAL ARTICLE OF BASIC RESEARCH ORIGINALLY PUBLISHED IN *HORMONES AND BEHAVIOR*.

☒ Not Recommended: Unstructured Abstract

The male offspring of mice stressed by crowding during the final third of pregnancy showed reductions in sexual behavior and fertility. When paired with receptive females, their latencies to mount and to achieve intromission and ejaculation were greater than controls, and 30% of them failed to ejaculate in the 100-min test. When housed continuously for four days with females, 31% of them failed to impregnate their partners, compared with 4% of controls. The sexual receptivity of the untreated females with prenatally stressed males was not affected. Resting testosterone levels of prenatally stressed males did not differ from those of controls, and the pattern of rise and fall of testosterone during a 60-min interaction with a female showed only minor differences. The results suggest a central, rather than a peripheral, mediation of the behavioral effects of prenatal stress.

☑ Recommended: Structured Abstract

Background: A substantial number of studies have reported that prenatal stress in mice affects the sexual behavior of male offspring. However, these studies have all relied on information obtained over brief periods of time.

Aims: The aims of this study were (i) to see whether "demasculinized" males were capable of impregnating females, given sufficient time, and (ii) to consider the nature of the alteration to demasculinized males.

Methods: Five experiments are reported. In Experiment 1 the male offspring of mice stressed by crowding during the final third of pregnancy and control mice, 10–12 weeks of age, were individually observed for 100 min of interaction with a receptive female. In Experiment 2 prenatally stressed mice and controls were re-housed individually at 12 weeks and, at 15 weeks, each male was paired with a virgin 3- to 6-month-old female. In Experiment 3 this procedure was repeated with sexually experienced instead of naive experimental and control males. In Experiment 4 the sexual responses of experienced estrous females to naive prenatally stressed males and controls was tested and, in Experiment 5, the effect of prenatal stress on plasma testosterone concentrations was examined, both in resting males and after exposure to a female.

Results: The prenatally stressed mice showed significant reductions in sexual behavior and fertility compared with the controls. In the 100-min test, their latencies to mount and to achieve intromission and ejaculation with receptive females were significantly greater than controls, and 30% of them failed to ejaculate. In Experiments 2 and 3, 31% of them failed to impregnate their partners, compared with 4% of the controls. Observations made in Experiment 4 showed that the sexual receptivity of the untreated females with prenatally stressed males was not affected. Experiment 5 showed that the resting testosterone levels of prenatally stressed males did not differ from those of controls, and that the pattern of rise and fall of testosterone during a 60-min interaction with a female showed only minor differences.

Comment: These results confirm and extend the earlier findings obtained with briefer time spans. They suggest a central, rather than a peripheral, mediation of the behavioral effects of prenatal stress.

Adapted from Hartley (1998).

journal does not use structure Abstracts, you may use the subheadings in Examples 9.3 and 9.4 to write an Abstract to ensure that you cover all of the important content. You can later remove the subheadings and reformat a structured Abstract into a single paragraph for journals that have unstructured Abstracts. Additionally, Table 9.3 provides the latest list of subheadings in clinical medical studies, which you should use for medical studies and may adapt for other research. While this table provides you with prototypical structured Abstract categories, journal editors and authors adapt the categories as appropriate for different types of research. For the appropriate content of structured abstracts in your target journal, you should refer to the Instructions to Authors and recently published articles in the specific journal. Refer to Appendix C for a glossary of terms to describe medical research studies.

Consistency with the Article

Although structured Abstracts help guide writers to include relevant information, the accuracy and consistency between the Abstract and main text of published journal articles remains a major problem. For example, 18% to 68% of conventional Abstracts in six major medical journals (*Annals of Internal Medicine*, *BMJ*, *CMAJ*, *JAMA*, *Lancet*, and *The New England Journal of Medicine*) from the late 1990s were inconsistent with the main text of the journal articles—with different data in the Abstract and main text, data in the Abstract not appearing in the main text, or both (Pitkin, Branagan, and Burmeister 1999). Alarmingly, the most common error was providing different data in the Abstract and in the body of the journal article, which means that the content of the Abstract not carefully cross-checked with the main text of the article by either the writers, reviewers, or

TABLE 9.3 Subheadings and content of structured Abstract for journal articles in medical journals.

Subheading	Content
Importance	Briefly summarize the reasons and relevance for the study
Objective	Clearly state the research question or purpose of the study
Design, setting, and participants	Detail the basic design of the study along with the relevant setting and features of the study participants
Interventions	Specify the experimental interventions and explain relevant features such as method and duration of administration
Main outcomes and measures	Identify the primary outcomes of the study
Results	Quantify the main outcomes of the study and include relevant statistics
Conclusions and relevance	Provide conclusions of the study that are directly supported by the results
Trial registration	For clinical trials, include the name of the trial registry, registration number, and URL of the registry

editors (Pitkin, Branagan, and Burmeister 1999). As summarized in Table 9.4, a more recent analysis of structured Abstracts revealed continued problems with such content discrepancies (Fontelo, Gavino, and Sarmiento 2013). Often the discrepancy arises because writers forget to update the Abstract when they revise the manuscript. Therefore, the point of this section is to remind you to check the content of your Abstract and ensure that it matches the journal article exactly. As an aside, this also demonstrates another reason why you cannot rely on the information and data from Abstracts alone in citations.

Concluding Points

In this chapter, we examined scientific Titles and Abstracts. As covered in this chapter, you should think of both Titles and Abstracts as complete summaries of journal articles. Both the Title and Abstract should be informative and fully comprehensible to the reader as an isolated communication because each is

TABLE 9.4 Comparison of data accuracy in structured Abstracts versus the main text of the journal article. A corpus of 60 "most-read," "most-cited," or "most e-mailed" articles published in 2011 in six high-impact medical journals (*American Journal of Emergency Medicine, BMJ, JAMA, Lancet, New England Journal of Medicine,* and *Obstetrics and Gynaecology*).

Section of Journal Article	Discrepancy (%)	No Discrepancy (%)
Introduction	6.7	93.3
Methods	13.3	86.7
Results	45.0	55.0
Discussion	3.3	96.7

Adapted from Fontelo, Gavino, and Sarmiento (2013).

most often read without the accompanying full-text journal article. The Title should convey the primary purpose of the research and should not incorporate any creative or literary references because such expressions are often confusing (especially to international audiences). The Abstract, whether structured or conventional (unstructured), should be a comprehensive and accurate summary of the entire, full text of a journal article. You may consider the Title, Abstract, and journal article as being increasingly more comprehensive and detailed communications of the same information.

10

PREPARING A MANUSCRIPT FOR SUBMISSION

Cover Letters, Publication Ethics, and the Peer-Review Process

In this chapter, you will learn to:

- Adhere to the ethical practices of publishing scientific and medical research
- Write a cover letter to accompany a manuscript submission
- Submit a manuscript to a journal for peer review

Journal editors and publishers expect that everyone involved in scientific research and publication adhere to the established ethical practices of scientific communication (Committee on Science, Engineering, and Public Policy 2009; CSE Editorial Policy Committee 2012). Yet with significant pressure to publish, many scientific authors resort to unethical scientific communication practices (Kirby and Houle 2004). For example, in a survey of 3,247 scientists, 33% of respondents admitted that they had engaged in at least one sanctionable offense (Martinson, Anderson, and de Vries 2005). Alarmingly, much of the misconduct in all of science involves issues related to scientific communication (see Table 10.1, Mogull and Katz 2012). In a similar study, 39% of physicists reported personal knowledge of one or more violation of science communication ethics within the first three years after getting their doctoral degree (see Figure 10.1, Kirby and Houle 2004). Cases of scientific communication misconduct span across scientific disciplines and throughout the history of science (LaFollette 2000; Swazey, Anderson, and Seashore Louis 1993). Therefore, journal editors expect that authors to confirm

TABLE 10.1 Top 16 ethical issues reported in scientific communication and research. This table lists the ethical misconduct from most to least frequent and provides the percent of scientists in early- and mid-career who admitted to being engaged in an activity deemed "misconduct" within the previous three years of the 2002 survey. The ethical issues related to scientific communication are contrasted to those that are exclusively research process or methods.

Top 15 reported activities of misconduct reported by scientists[1]	Percent of scientists admitting to engagement of offense ($n = 3,247$)	Ethical Issue	
		Related to **Scientific Communication**	***Exclusively*** **Research Process or Methods**
1. *Inadequate record keeping* related to research projects	27.5	✓	
2. Changing the design, methodology, *or results of a study* in response to pressure from a funding source	15.5	✓	
3. *Dropping observations or data points from analyses* based on a gut feeling that they were inaccurate	15.3	✓	
4. Using inadequate or inappropriate research designs	13.5		✓
5. *Overlooking others' use of flawed data or questionable interpretation of data*	12.5	✓	
6. *Withholding details of methodology or results in papers or proposal*	10.8	✓	
7. *Inappropriately assigning authorship credit*	10.0	✓	
8. Circumventing certain minor aspects of human-subject requirements	7.6		✓

Continued

TABLE 10.1 Continued

Top 15 reported activities of misconduct reported by scientists[1]	Percent of scientists admitting to engagement of offense ($n = 3{,}247$)	Ethical Issue	
		Related to **Scientific Communication**	***Exclusively*** **Research Process or Methods**
9. *Failing to present data that contradict one's own previous research*	6.0	✓	
10. *Publishing the same data or results in two or more publications*	4.7	✓	
11. *Unauthorized use of confidential information* in connection with one's own research	1.7	✓	
12. *Using another's ideas without obtaining permission or giving due credit*	1.4	✓	
13. Relationships with students, research subjects, or clients that may be interpreted as questionable	1.4		✓
14. *Not properly disclosing involvement in firms* whose products are based on one's own research	0.3	✓	
15. Ignoring major aspects of human-subject requirements	0.3	✓	✓
16. *Falsifying or 'cooking' research data*	0.3	✓	

Adapted from the analysis by Mogull and Katz (2012).

[1] Survey data from Martinson, Anderson, and deVries (2005).

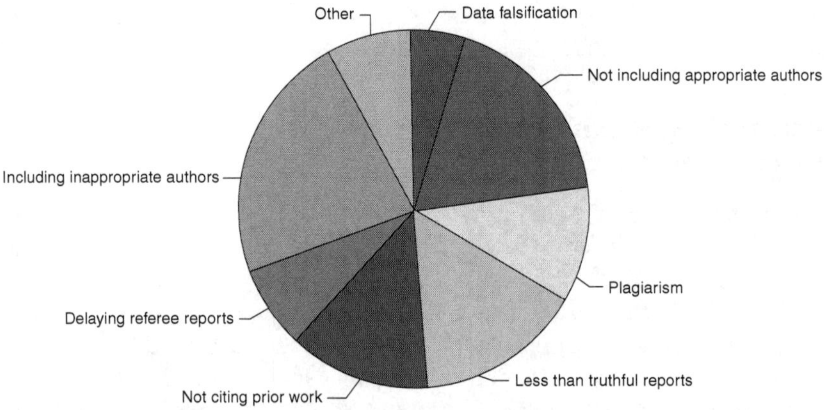

FIGURE 10.1 Violation of science communication ethics reported by physicists.
Source: Kirby and Houle (2004).

their adherence to the ethical and legal publication issues in the cover letter or submission documentation that accompanies the manuscript. In this chapter, we will examine the issues to address in the cover letter and the process of submitting a manuscript to a journal for peer review.

Writing a Cover Letter

Journal editors and professional organizations strongly recommend that authors include a COVER LETTER when submitting a manuscript to a journal for review (Benson and Silver 2013; Chipperfield et al. 2010). The cover letter provides authors the opportunity to establish a direct dialogue with the journal editor and address issues, particularly ethical and legal issues, that are of concern to the editor. Communicating through a cover letter builds your rapport with the editor and shows that you are professional and committed to the publishing process (something that might become important later if your manuscript requires significant revision and a journal editor needs to decide whether to reject your paper or work with you through an extensive revision process). The main outcome of an effective cover letter is that a manuscript will more quickly pass through the initial screening by the journal editors and begin the peer-review process (Benson and Silver 2013; Roederer et al. 2013).

Although different sources recommend varying lengths and content for cover letters, many editors and professional organizations suggest that cover letters should be approximately one page and include information explaining the submission and some precisely worded statements confirming certain legal and ethical issues related to publication if those statements are not included in the online submission system (American Psychological Association 2010; Benson and Silver

2013; Chipperfield et al. 2010; Day and Gastel 2006; International Committee of Medical Journal Editors 2014).

Introduce the Manuscript and Explain the Relevance to the Journal's Audience

The primary message to convey in the cover letter is to show the editor that the research is an appropriate fit for the audience and scope of a journal, and that the manuscript conforms to the journal's style and formatting requirements (Benson and Silver 2013). In a sense, the cover letter is your opportunity to "sell" the paper to the journal editors (Cargill and O'Connor 2013) or to tell the editors why your research is important, how it differs from other studies, and why it belongs in that particular journal (Hofmann 2014; Wachs 1996; Welch 2006; Welch 2007). According to Stephen Welch (2007, 1593), executive editor of *CHEST*, "The cover letter is a crucial part of any manuscript submission. . . . With increasing competition for limited space in medical journals, authors should take every opportunity to clarify the importance of their research." Specifically, journal editors are interested in articles about important topics that fit within the defined scope of the publication, as covered in the Instructions to Authors, and therefore would be interesting and relevant to the readers (Ortinau 2011; Radford et al. 1999). For example, the *Bulletin of the World Health Organization* states that the top reason for rejection is that the content of the manuscript is not suitable for the scope of the journal (World Health Organization n.d.). In initial screening of topic relevance or fit of the research to a specific journal, 40% to 60% of manuscripts may be rejected (Huler 1990). Furthermore, according to Lynn Worsham (2008), editor of *JAC*, "Submitting an article that is clearly inappropriate for the journal will result in a swift rejection, and may well cause the editor to be wary of your future submissions."

Cover letters should begin by summarizing the submission—identifying the manuscript (by title and, if available, ID number) as a submission of a particular article type (such as a research article, case study, or letter) as defined by the target journal. Since the rationale for submitting the manuscript to the journal is the primary message of the letter, you should cover this information in the first paragraph by briefly summarizing the content of the manuscript and then explaining the relevance and significance of the research to the audience of the journal (Chipperfield et al. 2010). Essentially, in the cover letter you are presenting an argument to the editor in which you explain why that journal is the most appropriate place to publish the article (Hofmann 2014; Matthews, Bowen, and Matthews 2000; Roederer et al. 2013). Ideally, this argument should focus on the value of the information to the journal's audience and explain the benefits of new or innovative information or application since editors are most interested in publishing novel and relevant research (Benson and Silver 2013; Triadafilopoulos 2006; Wachs 1996). Yet as suggested by medical writer Thomas Lang (2010), authors should not overstate the originality of their research. He recommends that authors include statements such as "*To the best of our knowledge*, our report is the first to report..." (271). In other words, claims of novelty and innovation of the research should be appropriately tempered and consistent with the hedging of claims in scientific communication (see Chapter 8).

Provide Details of the Submission and Address Copyright Issues

After you introduce the topic of the manuscript and value to the journal's audience, you should mention that you have followed the Instructions to Authors and provide a summary of the manuscript elements. By stating the word count of the manuscript and the number of figures and tables, you make it easy for the editorial staff to confirm that they have received the entire submission (or recognize if they are missing part of it) and to estimate the space requirements for the article.

If your manuscript includes the reprinting of figures, tables, data, or excerpts (typically longer than a few hundred words) from another source, you should obtain and provide copyright permission for reprinting the work in addition to appropriately citing the original source of information (American Psychological Association 2010; Coghill and Garson 2006; CSE Editorial Policy Committee 2012). Often copyright has been transferred from the authors of the study to the journal, so you should contact the journal's or publisher's copyright permissions office, which is usually referenced on the publication's website. Many journals hold and provide copyright clearance to facilitate the distribution and sharing of information (American Psychological Association 2010; Friedman 1981). In some cases, journals may not hold the copyright for the content but rather have been granted permission to publish the material. Sometimes the authors retain copyright for articles published in open-access journals, so you may need to contact the author directly for copyright clearance (International Committee of Medical Journal Editors 2014). (You can usually find author contact information included with published articles.) In certain cases the material may not be subject to copyright because the work was written under US government contract, which specifies that the copyright belongs to the government and may be reproduced by others. Once you are granted permission to reproduce material, the grantor will specify the exact wording for the credit line to be included in your manuscript.

Likewise, if your manuscript is accepted for publication, you will probably be requested to transfer copyright or permissions to the journal. Often, journals have a form that they ask authors to complete during the manuscript-submission process or occasionally later after the manuscript has been accepted. Journals holding copyright permission act on behalf of authors to facilitate copyright permissions in order to have their work shared widely and appropriately acknowledged (Lagowski 1981; Weil 1981). The copyright transfer agreement for the American Chemical Society, for example, states that the "author retains the right to revise, adapt, prepare derivative works, present orally, or distribute the work, provided that such use is for [the author's] personal noncommercial benefit" (Lagowski 1981, 72). Therefore, authors might need to obtain copyright clearance to republish their own work if copyright of the original material has been transferred to the publisher.

Next you should also address any issue directly related to the publication of the manuscript. For example you may request to have the manuscript-submission fees or page charges waived, which is often considered for authors from developing nations or students who lack financial support from their institutions (Lang 2010).

For journals that request the names of possible reviewers but do not provide a separate form, you should provide their names, titles, institutions, and contact information in the cover letter (Boellstorff 2011; Chipperfield et al. 2010). In some cases, you may also request to exclude certain individuals as reviewers if they may not be able to objectively evaluate the manuscript (Boellstorff 2011; Korner 2008).

Acknowledge Compliance with Scientific Publication Ethics

In the scientific publication process, authors are expected to follow specific ethical standards that help maintain the quality and integrity of the science. As a checkpoint in the process, authors confirm that they followed proper ethical practices when they submit the manuscript to a journal. In the cover letter, you should confirm that the research has not been published previously or accepted for publication elsewhere (referred to as DUPLICATE or REDUNDANT PUBLICATION). Duplicate publication is specifically defined as a report with content that is at least 10% identical (or highly similar) to a previous publication (Iverson et al. 2007). The INGELFINGER RULE states that journals will not publish scientific findings that were published elsewhere (Angell and Kassirer 1991; International Committee of Medical Journal Editors 2014; Iverson et al. 2007; Langdon-Neuner 2009).

You should also state that the current manuscript is not being simultaneously considered for publication by any other journal or publisher (referred to as DUPLICATE SUBMISSION). Duplicate publication and duplicate submission of research in which identical data lead to the same conclusion are considered unethical practices that violate copyright and misuse journal resources that should be used for new information (Audisio et al. 2009; CSE Editorial Policy Committee 2012; Huston and Moher 1996; International Committee of Medical Journal Editors 2014; Von Elm et al. 2004). Duplicate publications may also be considered SELF-PLAGIARISM, which is plagiarism of one's own work (Cicutto 2008).

Studies of duplicate publication in the scientific and medical literature report that duplicate publication occurs in 2% to 20% of articles (Bailey 2002; Bloemenkamp et al. 1999; Schein and Paladugu 2001). Further, 40% of authors of review articles identified some form of duplicate publication (Von Elm et al. 2004). Approximately 20% of duplicate publications occur by authors inappropriately dividing the results from a single study into multiple papers to increase the number of publications, which is pejoratively referred to as "SALAMI SLICING" (Huth 1986). A similarly unethical publication practice is to add data to one's own previously reported research without reaching new conclusions, which is pejoratively referred to as "MEAT EXTENDERS" (Huth 1986).

A few notable exceptions to the rule against prior publication allow for prior presentation of research at a scientific meeting (as a scientific poster, an oral presentation, and publication of the conference abstract). Conference genres, discussed in Section III, are considered tentative or work in progress unless the research is being published in the conference proceedings (Chipperfield et al. 2010; International Committee of Medical Journal Editors 2014; Langdon-Neuner 2009). Additionally, it is a legal and ethical requirement of clinical trial transparency for data to be posted in clinical trial registries (such as ClinicalTrials.gov, which is sponsored by the US National Library of Medicine), so such posting is *not* considered prior publication (DeAngelis et al. 2004; International Committee of Medical Journal Editors 2014; Tse, Williams, and Zarin 2009; Zarin et al. 2011). Indeed, journals may reject papers if the disclosure to clinical trial registry is incomplete or inaccurate (DeAngelis et al. 2005; Irwin 2007). Any previous announcements of the research, including the exceptions mentioned above, should be stated and specifically identified (as to which data and the format, including the conference or database name and year, of prior disclosure) in the cover letter.

Finally, news reporting of scientific findings prior to publication in a primary journal should be avoided because such premature public announcements undermines important quality controls of the scientific process and may violate the policy of many journals (International Committee of Medical Journal Editors 2014). (For a discussion on communicating with the news media, see Chapter 13.)

Identify Funding Sources and Possible Conflicts of Interest

In the cover letter, you should identify any funding sources and potential conflicts of interest if the information is not included in the manuscript or on a separate form included with the submission (Chipperfield et al. 2010; International Committee of Medical Journal Editors 2014). Often the funding sources are identified on the first page of the article or in the Acknowledgments section.

Particularly in the medical literature, editors and professional organizations scrutinize any potential conflict of interest (Irwin 2009). Conflict of interest is often considered any form of financial incentive that leads to either biased research design or PUBLICATION BIAS, such as selectively communicating only favorable data about a product, failing to publish unfavorable data, delaying publication of unfavorable data, duplicating publication of positive data, or exaggerating interpretations that emphasize benefits or minimize risks (Miller and Brody 2005). Writers should expect to have funding sources published along with the article (CSE Editorial Policy Committee 2012; Iverson et al. 2007).

Confirm Ethical Research Conduct

The cover letter is also the place for authors to confirm that they followed ethical research practices and have received the necessary legal approval (Chipperfield et al. 2010; CSE Editorial Policy Committee 2012; International Committee of Medical Journal Editors 2014). This may include confirming that:

- Research on human subjects received the appropriate approval from an institutional review board (IRB) or complied with the Declaration of Helsinki;
- Good clinical practice requirements were met, such as by providing registration details for clinical trials (or details about how you followed the CONSORT checklist for randomized, controlled trials);
- Research involving animals was approved by a committee on animal care and use and appropriate research protocols for animal experimentation were used; and
- The authentication methods were performed on cell lines to ensure that the cell lines were properly identified.

In many cases, verification of these standard practices is integrated into the electronic (online) submission process and forms. In such cases, you may not need to repeat this information in a cover letter.

Conclude with Correspondence Information

At the end of the cover letter, you should express your appreciation to the editor for considering the manuscript and provide contact information for the CORRESPONDING AUTHOR, the individual who will receive correspondence about the article (International Committee of Medical Journal Editors 2014). If you are working with a group of authors, you should confirm that the corresponding author takes responsibility for keeping all authors informed of the communications throughout the peer-review process.

For manuscripts with multiple authors, you should confirm that all individuals listed as authors have reviewed and approved the manuscript for submission to the journal (Benson and Silver 2013; Chipperfield et al. 2010; International Committee of Medical Journal Editors 2014; Matthews, Bowen, and Matthews 2000). In addition, you should confirm that all individuals listed as authors meet the requirements for authorship, which often does not include professional communicators (Chipperfield et al. 2010; Matthews, Bowen, and Matthews 2000; Welch 2007). Professional communicators are generally mentioned in in the Acknowledgments section rather that listed as an author (Woolley 2006). (For more information on authors and professional writers in SMC, see Chapter 1.)

Manuscript Submission

The logistics of manuscript submissions are specified in the Instructions to Authors (Schriger, Arora, and Altman 2006). Many journals use an online SUBMISSION-MANAGEMENT SYSTEM, which is a content-management system designed to accept manuscripts electronically (ELECTRONIC SUBMISSIONS) and facilitate the peer-review process (Bishop 2007; Honeyman-Buck 2003a, 2003b; Proto 2007; Thomson Reuters n.d.; Welch 2006). Many journals, such as *Technical Communication Quarterly*, use ScholarOne Manuscripts (formerly called Manuscript Central; see Figure 10.2). Each management system is customized for the specific journal, but the submission process for authors is similar among all these systems.

In submission-management systems, authors (or AGENTS, who submit manuscripts on behalf of the authors) will upload a copy of the manuscript (typically as a single Microsoft Word document) and supplemental documents such as figures. Many systems require that each supplemental item is uploaded in an individual file, so you may have to place each figure in a separate document. Tables and appendices are often placed at the end of the manuscript. These are just general conventions, but may not be the case for your particular target journal. Therefore you will need to review the Instructions to Authors for your target journal as you prepare the final files for submission.

These submission-management systems are also configured to require authors to input short data fields, such as the title, running head, Abstract, and figure captions. These data-field entries are configured to only accept the exact character or word limits based on the journal's publication standards. The precise limits for these fields are stated in the Instructions to Authors and differ between journals (see Chapter 9). For example, *Nature* requires article titles to be limited to 75 characters, which include spaces. Many journals provide submission or manuscript checklists that repeat these requirements. If a checklist is not available, you should examine the manuscript-submission system early in the writing process to identify particular word or character limits that are not specified in the Instructions to Authors.

Additionally, many journals have authors identify KEYWORDS from a defined menu during the manuscript-submission process. Journal editors often use keywords to categorize articles and determine appropriate reviewers within the submission-management system, so these keywords are not necessarily the INDEX TERMS used to categorize the article in databases once published (Proto 2007; Roederer et al. 2013; Welch 2006). (For more information on selecting keywords and index terms, see Chapter 3.)

One of the most common errors in manuscript submission is incomplete manuscript blinding (Oxford University Press n.d.). BLINDING (or "ANONYMIZING") a manuscript involves removing the authors' names and identifying information from the entire submission—including manuscript, figures, and tables—so that the reviewers do not know the authors identity (Boellstorff 2011). To blind a manuscript, you need to:

1. Remove the author names and institutional affiliations from the title page (Boellstorff 2011; *Nature* n.d.; Oxford University Press n.d.).
2. Remove the names and institutional affiliations from the metadata of all files you submit (the metadata content is usually found under "File" and then "Properties" in software programs; *Nature* n.d.).

(a)

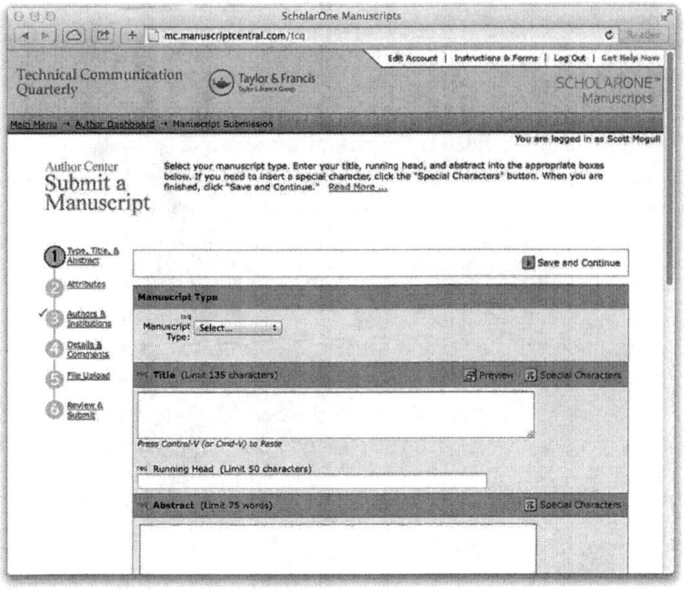

(b)

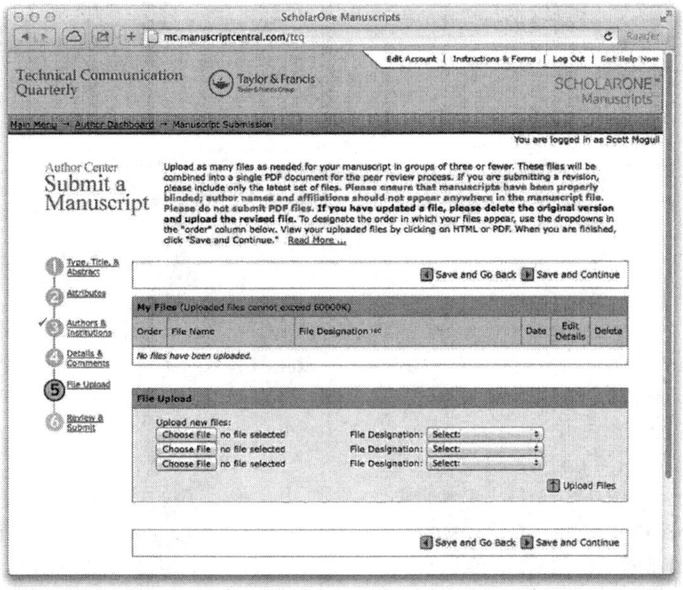

FIGURE 10.2 The author's interface in the ScholarOne Manuscripts submission-management system for the journal *Technical Communication Quarterly*. Note that character and word counts are stipulated for specific sections such as the Title (a) and must be adhered to in order for the system to accept a submission. The manuscript and figures are uploaded as individual documents(b).

3. Exclude the "Acknowledgements" section from the manuscript; some journals, such as *Nature* (n.d.) recommend that authors add this information to the cover letter whereas other journals will let authors add this information later in the publication process (Boellstorff 2011).
4. Remove any self-identifying self-citation in the manuscript:

    a. Only refer to published work (*Nature* n.d.; Oxford University Press n.d.);
    b. Cite your own publications the same way you cite the work of others (i.e., do not use any personal pronouns in the manuscript to indicate that a citation was your own work; *Nature* n.d.; Oxford University Press n.d.);
    c. Avoid self-identifying references to your prior research Methods in a description of the methodology (rather you should include the relevant details in the Methods section of the current manuscript; Oxford University Press n.d.); or
    d. If you must refer to your own work and doing so would reveal your identity, you should substitute "Author Citation" in the text and in the Reference list along with the year of publication (or, where appropriate, "Unpublished," "In press," or "Forthcoming"; Oxford University Press n.d.). Do not include additional information in the Reference list, such as the Title of the article, which may be used to identify the authors of the work.

Following these steps will help ensure that your manuscript is properly blinded for review.

In the submission-management system, the manuscript is configured into a proof (as a PDF or HTML document) that is distributed to editors and peer reviewers (we will discuss the peer-review process next). You should review the proof documents to ensure that the manuscript is converted accurately and completely—especially that the figures and tables are legible and that special characters and symbols have been properly converted (Welch 2006).

The Peer-Review Process

PEER REVIEW is a centuries-old tradition by which the scientific content of a manuscript is evaluated by REFEREES, or other researchers in the same field (Campanario 1998a, 1998b; Suls and Martin 2009). Referees evaluate each manuscript for scientific originality, reliability, relevance, and appropriateness to the audience and scope of a particular journal (Goldbeck-Wood 1999; Triadafilopoulos 2006). After reviewing a manuscript, referees are asked to provide detailed comments and suggestions to improve the manuscript. Along with this feedback, each reviewer provides a publication recommendation, usually one of the following: accept, accept with revision, revise and resubmit, or reject. If the reviewers are evenly divided on whether to accept or reject a manuscript, the journal editor may make thc final decision or request a review by another referee (Benson and Silver 2013). Ultimately the journal editor makes the final publication decision and communicates the decision and the reviewers' comments (or a summary of those comments) to the authors (Benson and Silver 2013).

Variations in and Limitations to the Peer-Review Process

The peer-review process may differ slightly from one journal to another, but the general steps are as follows. First, the journal editor conducts an initial review of a manuscript to ensure that the manuscript is appropriate for the journal and that the authors adhered to the journal style. Following this preliminary assessment, the journal editor assigns the manuscript to two or more referees to independently review the scientific content of the manuscript. This review is often either a BLIND (or MASKED) REVIEW, in which the referees' identities are not disclosed to the authors but the authors' identities are revealed to the referees, or a DOUBLE-BLIND REVIEW, in which the referees' identities are not disclosed to the authors and the authors' identities are not disclosed to the referees. The process of blinding the authors' identities to the referees is more problematic than vice versa because authors generally publish on continuing lines of research and usually heavily cite their own work; therefore, they might inadvertently reveal their identity or institution through text allusions or, to those familiar with their work, by the style or voice used in the manuscript (Cho et al. 1998; Matsuda and Tardy 2007; Yankauer 1991). (Refer to the previous section on manuscript submission for guidelines on blinding a manuscript.) Table 10.2 summarizes referees' primary reasons in favor and in opposition to double-blind review. The peer-review system is recognized as an imperfect process, but modifications that have been implemented were also problematic and journal editors have not devised an adequate substitution (Benson and Silver 2013; Block 1998; Davidoff 1998; Goldbeck-Wood 1999; Rothwell and Martyn 2000; Triadafilopoulous 2006). At this time, peer review is considered the best option for quality control of the scientific literature.

Criteria Used by Referees to Evaluate Manuscripts

In the peer-review process, the quality of the scientific research is the primary basis for evaluation, although the quality and clarity of communication also affect the overall effectiveness of the manuscript and may influence the editor's final publication decision (Caulfield et al. 2008; Jefferson, Wager, and Davidoff 2002; Radford et al. 1999; Triadafilopoulous 2006; Van Rooyen et al. 1999). Table 10.3 provides a list of seven factors ranked by journal editors based on the influence of each on the final publication decision. Further, because reviewers

TABLE 10.2 Referees primary reason for favoring or opposing double-blind review.

Reasons	Number of Respondents (n = 216)
Favoring blind review	
Objectivity: judge content, not author	66.2% (143/216)
Comfort: less pressure, less hesitation to criticize	6.5% (14/216)
Helps beginning researcher if name is unknown	5.6% (12/216)
Total Favoring	**78.2% (169/216)**
Opposing Blind Review	
Blinding does not work	8.3% (18/216)
Names do not influence my judgment	6.9% (15/216)
Knowing author's background and track record contributes to judging paper	6.5% (14/216)
Total Opposing	**21.8% (47/216)**
TOTAL	**100% (216/216)**

Adapted from Yankauer (1991).

are expected to evaluate only the information provided in the manuscript, the peer-review process does not typically reveal scientific fraud or misconduct (Triadafilopoulous 2006). Such scientific integrity is entrusted to the authors of the manuscript.

The editorial and peer-review process takes weeks to months, depending on the journal and reviewers. Only rarely are manuscripts accepted without some sort of revision (Chipperfield et al. 2010). Thus, the initial publication decision is typically either a provisional acceptance of the manuscript, which is contingent on revisions specified by the referees, or an outright rejection (Chipperfield et al. 2010; Kotz and Cals 2014). The common reasons for manuscript rejections are summarized in Table 10.4.

Journal editors strongly encourage authors to use the editorial decision and feedback from reviewers as a means to strengthen the science and improve the manuscript (Benson and Silver 2013; Chipperfield et al. 2010; Kotz and Cals 2014). For manuscripts with an option for revision, you should address each issue summarized by the editor or referee. Addressing each issue often means providing a more detailed or clear analysis, adding more data, or accounting for

TABLE 10.3 List of factors ranked ordered (1 to 7) by journal editors based on the influence of each on the final publication decision (to accept or reject). Factors are shown in order of importance by ranking.

Factor	Mean/Median Rank	Rank Code*
The significance and importance of the research work	1.8/1	A
The manuscript is appropriate and of interest to the reader: it complies with the stated aim of the journal	2.1/1	A
The experimental and statistical method is valid	2.9/3	B
The literature review, discussion, and conclusions are relevant and correctly interpreted	4.2/4	C
The manuscript is understandable and readable	4.5/5	C
The presentation of the manuscript has been undertaken with care (e.g., spelling mistakes, minor grammatical errors)	6.2/6.5	D
The author has closely followed the Instructions to Authors (e.g., reference style, numbering of figures and tables, the submission of the correct number of double-spaced manuscripts)	6.3/6	D

* Factors with different codes (A, B, C, D) showed statistically significant differences in ranking ($p < 0.01$). Factors with the same code were not significantly different.

Source: Radford et al. (1999).

alterative perspectives to your interpretation of the findings (Benson and Silver 2013). You should not only address each issue in the manuscript but provide a point-by-point response to each issue in an updated cover letter accompanying the resubmission of your revised manuscript (Benson and Silver 2013; Kotz and Cals 2014). You are not required to make all the changes that were identified by the editor or referee, although you should address each issue in your resubmission cover letter. If you disagree with a recommendation for revision, explain your reasons clearly in the resubmission cover letter and provide data or references to support your argument for not revising the manuscript (Kotz and Cals 2014). Throughout all your correspondence with the editor, take a respectful

TABLE 10.4 Common reasons for manuscripts rejection following review.

Category	Translation
Science is not novel	Something similar (or identical) has been published
Information is not new enough	The science does not represent enough of an advance to warrant publication
Arguments are unsound	Data are poorly analyzed or incorrectly interpreted
Arguments are weak or the evidence presented is unconvincing	Conclusions made on the basis of these data are unsupported or the importance of the data is exaggerated
Data or other evidence are missing	Certain experiments were not done or previously published findings were not taken into account
Important arguments were not addressed, or seminal work was ignored or not cited properly	Again, some findings (your own or that of other researchers) have not been taken into account in building your argument
Paper is biased	In a scientific question with two or more possible hypotheses or interpretations, the paper highlights one side of the argument rather than presents a balanced view
Arguments are unclear and confusing	The research or conclusions may be valid but are not clearly explained; alternatively, there may be something wrong with the way the basic premise, the approach, the results, or the conclusions have been presented
Paper is uninteresting	The editor may feel that the paper is boring, will not enhance the journal, or will not be of interest to readers; unfortunately, papers describing negative results may be included in this category even though the resulting publication bias is a major problem in scientific publishing
The paper was not prepared according to the guidelines set out in the instructions to authors	If the manuscript clearly doesn't fit the criteria listed in the instructions, and this fact was missed at the initial review, it may mistakenly be sent to reviewers; if the editor feels that it would be impossible to change the paper sufficiently to conform to the journal criteria, it will be rejected

Adapted from Benson and Silver (2013).

tone and adopt the perspective that the editors and reviewers are interested in improving the manuscript. A journal editor once told me that authors and editors both have the same goal—to make each manuscript the best. A strong manuscript benefits the editor, authors, and readers.

Concluding Points

After you have finished writing and editing the manuscript, you will immediately want to submit the research to a journal for review. But taking the time to develop a comprehensive cover letter will initiate the dialogue with a journal editor that, under the best case, extends throughout the peer-reviewing, copyediting, and proofreading stages. The cover letter is the opportunity that authors have to demonstrate both their professionalism and their adherence to publication and research ethics. Regardless of the outcome of the review process (accept, revise, or reject), the feedback should be used to strengthen your research. If rejected, you should evaluate and address the comments and consider submitting a revised manuscript to another journal. Ultimately those who persevere have the greatest success in publishing research.

SECTION III

Presenting Research at Conferences

11

INTRODUCTION TO SCIENTIFIC CONFERENCES

In this chapter, you will learn to:

- Select a scientific conference to present research
- Apply to present your research at a conference

Scientific conferences, or meetings, range in size from large, international meetings of tens of thousands of attendees (see Table 11.1) to small, regional, or departmental meetings that may have a few dozen attendees. Many of the major scientific and medical associations have conferences that range from 10,000 to 20,000 attendees (Healthcare Convention and Exhibitors Association 2015). For example, the 2015 American Chemical Society national meeting had nearly 14,000 participants and 11,000 presentations (American Chemical Society 2015).

Regardless of the size, SCIENTIFIC CONFERENCES are the primary means through which researchers share research in progress. In addition to sharing research with others, presenters personally benefit from presenting research at scientific conferences (see Table 11.2).

The major presentation genres at scientific conferences are scientific posters, oral presentations, and workshops. Of these various genres, scientific posters (see Chapter 12) are the most common presentation format, particularly for junior researchers, faculty, and fellows (McNutt 2015; Wood and Morrison 2011; Woolsey 1989). SCIENTIFIC POSTERS are the primary format for presenting research in progress at conferences because the format facilitates dialogue (particularly

TABLE 11.1 Top seven largest scientific and medical conferences (meetings) in 2013.

Association	Total Attendees	Percent of Professional Attendees (Not Sponsors)
1. FIME Intermed USA/Federation of International Medical Equipment Suppliers	57,000	92.8%
2. Greater New York Dental Meeting	54,629	84.7%
3. Radiological Society of North America	54,008	60.5%
4. HIMSS: Healthcare Information & Management Systems Society	34,834	41.1%
5. American Society of Clinical Oncology (Annual Meeting)	32,200	82.0%
6. American Academy of Orthopaedic Surgeons	31,643	55.3%
7. Society for Neuroscience (Annual Meeting)	30,469	81.5%

Source: Healthcare Convention and Exhibitors Association (2015). Regardless of size, scientific conferences are the primary means for researchers to share research in progress.

TABLE 11.2 Personal benefits from presenting research at scientific conferences.

- Developing your national reputation
- Associating your name with a topic
- Finding opportunities to network and collaborate
- Contributing to job promotion
- Finding new jobs
- Helping with manuscript preparation by:
  - Forcing organization of your thoughts
  - Giving you a deadline
  - Giving you feedback before manuscript submission to shape analyses, interpretations, and future research directions

Adapted from Wood and Morrison (2011).

feedback, networking, and collaboration) between presenters and attendees (Tulsky and Kouides 1998).

In this chapter, we will explore the steps involved in applying (or proposing) to present research at scientific conferences. Since each conference is different, you should use this information as a general guide and refer to the specific information about each conference provided online by the meeting organizers.

Planning to Present Research at Conferences

When considering to present research at a conference, you should follow the following three general steps:

1. Select an appropriate conference for the topic,
2. Gather the requirements and deadline for submissions, and
3. Write a CONFERENCE ABSTRACT, which is the brief proposal that describes the research you intend to present.

Step 1: Selecting an Appropriate Conference for the Topic

The first step for planning to present at a conference is to select an appropriate venue for your research. Often researchers do not think consciously about selecting the conference because they work in a lab or department with other, more senior researchers who regularly present at specific conferences, so applying to present research at a particular conference is the norm. But consciously considering the reasons for selecting an appropriate conference will make you an informed presenter and, furthermore, help you shape the content of your presentation. Moreover, being able to select appropriate conferences will enable you to identify other conferences and help you to broaden the reach of your research to new audiences.

Above all else, you should present at conferences at which the audience will be interested in the research topic (Wood and Morrison 2011). Scientific conferences are often sponsored by professional societies, which will define the intended group of researchers. These professional societies state the scope and purpose of the organization on the organization's website (see Example 11.1), which summarizes the interests of researchers that will be attending the conference sponsored by the society. Additionally, some conferences may have particular meeting themes, or areas of focus, that are more specialized, which are described on the conference website (see Example 11.2). If you are considering proposing research to present at a specific meeting, then your research findings should have direct implications for the meeting attendees—either to further their own research or to apply the research findings as practitioners. The more distant your research

EXAMPLE 11.1 DESCRIPTION OF A PROFESSIONAL SOCIETY AVAILABLE ONLINE.

The American Society for Microbiology describes the scope and specific research interests of the field. Such information should be used to determine if your research is appropriate for a general conference hosted by a professional organization.

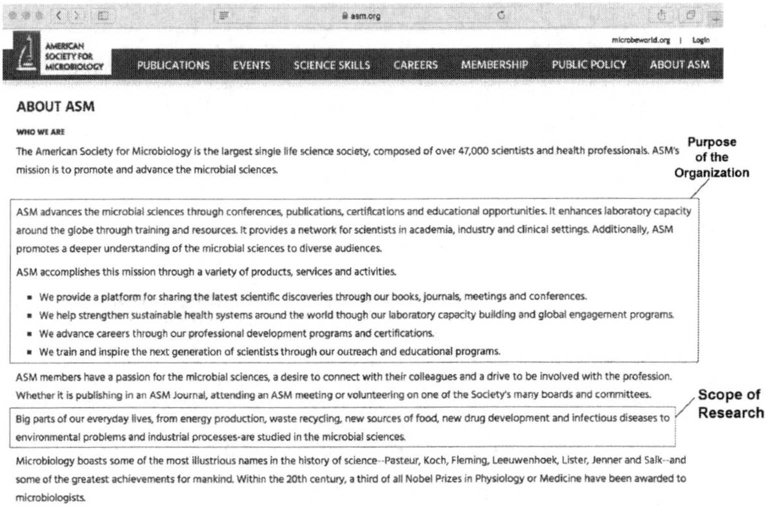

Source: American Society for Microbiology. 2017. *About ASM*. Accessed June 21, 2017. http://www.asm.org/index.php/about-asm2.

topic is from the meeting scope and purpose, or the more you have to think about creating ways to make your research useful to the attendees, the less likely that particular conference is appropriate for your research.

Step 2: Gathering the Requirements and Deadline for Submissions

The primary point of step 2, gathering the conference requirements and deadline for submissions of Conference Abstracts, is to ensure that you have sufficient time to prepare the materials to propose a conference presentation. Conference organizers and presenters often begin organizing and planning conference

EXAMPLE 11.2 SPECIALIZED CONFERENCE WEBSITE AVAILABLE ONLINE.

The American Society for Microbiology Biothreats conference is a specialized conference for a defined audience and topic. Such conference websites state the audience and scope of the meeting, which should be used to determine if your research is appropriate for the venue.

Source: American Society for Microbiology. 2017. 2017 ASM Biothreats. Accessed June 21, 2017. http://www.asm.org/index.php/biothreats-2017.

presentations a year in advance, and the initial deadline to submit a Conference Abstract may be 6 to 12 months in advance of the conference. Most conference organizers have strict requirements and deadlines for Conference Abstracts, which is a specific genre for proposing a research presentation at conferences (Pierson 2004; Wood and Morrison 2011). For example, the American Society for Microbiology (2015) has a 2,000-character limit for Conference Abstracts and a firm deadline for submission (see Example 11.3). For this annual conference, like most, potential presenters must follow the submission guidelines and deadlines precisely. Online submission forms usually prevent authors from submitting a Conference Abstract that is incomplete or longer than the specified number of characters or words. Furthermore, partly because of the size of these conferences and the amount of work involved in managing the on-time submissions, many conferences have a policy to not accept late Conference Abstracts (McNamara et al. 2001).

To ensure meeting the requirements of a conference, you should begin thinking about a conference 8 to 12 months in advance and review the conference website to find out the requirements and timeline, which are often available in the INSTRUCTIONS FOR PRESENTERS, which summarizes the requirements and process for each conference. In particular, you will need to check the requirements and submission deadline for Conference Abstracts because that is the critical first step when preparing to present research at conferences. As stated by the American Society for Microbiology (2015), "If you prepare an abstract that follows the rules and guidelines for submission, then your abstract stands an excellent

EXAMPLE 11.3 EVENTS AND TIMELINE FOR AN ANNUAL SCIENTIFIC CONFERENCE.

The American Society for Microbiology requires submission of Conference Abstracts by the end of January for the annual conference at the end of May. As you can see by this timeline, the conference organizers and peer reviewers must examine thousands of Conference Abstracts during the month of February. Due to the large number of submissions and tight time frame, the American Society for Microbiology (like most conferences organizers) will not accept late Abstracts.

Event	Responsibility	Date/Deadline
Conference Abstract submissions (due)	Researchers	January 27 11:59 p.m. EST
Review of Conference Abstracts	Conference organizers and peer reviewers	February
Notification of decision (Acceptance or rejection)	Conference organizers	Week of March 2, 2015
Conference		May 30 to June 2, 2015

Source: American Society for Microbiology. 2015. *About ASM*. Accessed June 16, 2015.

chance of being accepted. If you do not follow the rules and guidelines, it will be rejected."

Step 3: Writing a Conference Abstract (the Presentation Proposal)

Conference Abstracts are a short genre that summarizes research that is proposed for presentation at a scientific conference. Conference Abstracts are similar to Abstracts of journal articles in that both are based on the IMRAD (Introduction, Methods, Results, and Discussion) format. Conference Abstracts differ from journal article Abstracts in that Conference Abstracts constitute a complete, stand-alone genre that is written before the longer presentation is developed.

A Conference Abstract should be succinct and clearly focused on the key research questions (McNamara et al. 2001). The audiences of Conference Abstracts are usually looking for answers to the following questions (Pierson 2004):

- Why did you start the research?
- What did you do?
- What did you find?
- Why does it matter?

These questions are based on the IMRAD organization, with each question referring to one of the primary sections of a journal article (Introduction, Methods, Results, and Discussion). Therefore, you might use the structured Abstract guidelines presented in Chapter 9 to write a Conference Abstract.

The selection committee or peer reviewers will decide to accept or reject a presentation based entirely on the content included in the Abstract (Pierson 2004). Additionally, if your Conference Abstract is accepted, it is often published in the conference proceedings and will be used by attendees to decide whether or not to attend your presentation (Pierson 2004).

Often the data presentation that is proposed in a Conference Abstract is part of ongoing research that may eventually be included in a journal article (Carroll et al. 2003; McNamara et al. 2001; Scherer Ugarte-Gil, Schmucker, and Meerpohl 2015; Sprague et al. 2003). However, the amount of data presented at a conference is less than what would be presented in a journal article. Many scientific conferences also specify that the research data proposed for presentation must not have been previously published in a journal article (Wood and Morrison 2011).

Similar to Abstracts of journal articles, Conference Abstracts rarely cite original research nor contain tables or figures. But these conventions differ by conference and field. Therefore, you should refer to the Instructions for Presenters, review other Conference Abstracts from the previous meeting of the particular conference, and set up an account for the online submission system (if available) in order to identify the content and norms of a particular meeting. The more carefully your Conference Abstract, Title, and other required content adhere to the conventions of the specific conference (see Example 11.4), the more likely your research will be accepted for presentation.

EXAMPLE 11.4 CONFERENCE ABSTRACT ACCEPTED FOR PRESENTATION AT THE SOCIETY FOR NEUROSCIENCE ANNUAL MEETING

Content required for submission is identified.

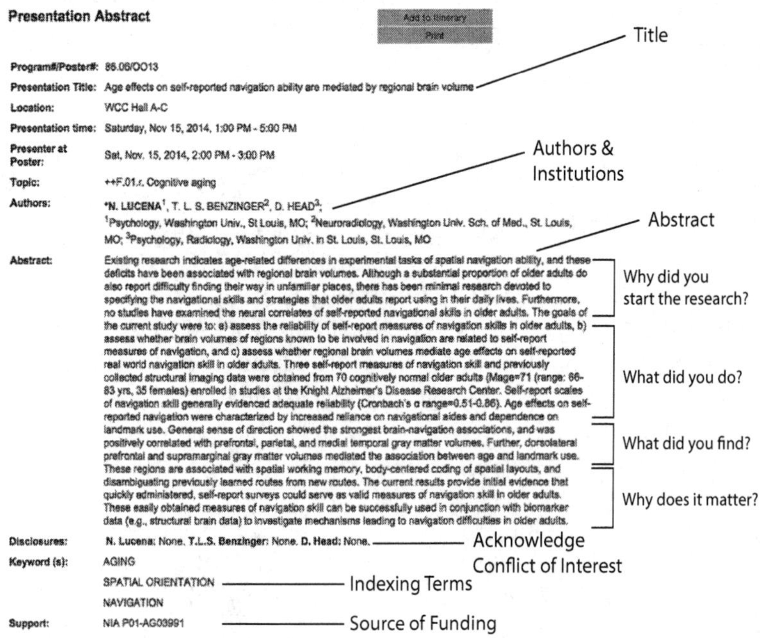

Presentation Abstract

Add to Itinerary
Print

Program#/Poster#:	86.06/OO13
Presentation Title:	Age effects on self-reported navigation ability are mediated by regional brain volume
Location:	WCC Hall A-C
Presentation time:	Saturday, Nov 15, 2014, 1:00 PM - 5:00 PM
Presenter at Poster:	Sat, Nov. 15, 2014, 2:00 PM - 3:00 PM
Topic:	++F.01.r. Cognitive aging
Authors:	***N. LUCENA**[1], T. L. S. BENZINGER[2], D. HEAD[3]; [1]Psychology, Washington Univ., St Louis, MO; [2]Neuroradiology, Washington Univ. Sch. of Med., St. Louis, MO; [3]Psychology, Radiology, Washington Univ. in St. Louis, St. Louis, MO
Abstract:	Existing research indicates age-related differences in experimental tasks of spatial navigation ability, and these deficits have been associated with regional brain volumes. Although a substantial proportion of older adults do also report difficulty finding their way in unfamiliar places, there has been minimal research devoted to specifying the navigational skills and strategies that older adults report using in their daily lives. Furthermore, no studies have examined the neural correlates of self-reported navigational skills in older adults. The goals of the current study were to: a) assess the reliability of self-report measures of navigation skills in older adults, b) assess whether brain volumes of regions known to be involved in navigation are related to self-report measures of navigation, and c) assess whether regional brain volumes mediate age effects on self-reported real world navigation skill in older adults. Three self-report measures of navigation skill and previously collected structural imaging data were obtained from 70 cognitively normal older adults (Mage=71 (range: 66-83 yrs, 35 females) enrolled in studies at the Knight Alzheimer's Disease Research Center. Self-report scales of navigation skill generally evidenced adequate reliability (Cronbach's α range=0.51-0.86). Age effects on self-reported navigation were characterized by increased reliance on navigational aides and dependence on landmark use. General sense of direction showed the strongest brain-navigation associations, and was positively correlated with prefrontal, parietal, and medial temporal gray matter volumes. Further, dorsolateral prefrontal and supramarginal gray matter volumes mediated the association between age and landmark use. These regions are associated with spatial working memory, body-centered coding of spatial layouts, and disambiguating previously learned routes from new routes. The current results provide initial evidence that quickly administered, self-report surveys could serve as valid measures of navigation skill in older adults. These easily obtained measures of navigation skill can be successfully used in conjunction with biomarker data (e.g., structural brain data) to investigate mechanisms leading to navigation difficulties in older adults.
Disclosures:	**N. Lucena:** None. **T.L.S. Benzinger:** None. **D. Head:** None.
Keyword (s):	AGING SPATIAL ORIENTATION NAVIGATION
Support:	NIA P01-AG03991

Source: Lucena, N., T. L. S. Benzinger, and D. Head. 2014. "Age Effects on Self-Reported Navigation Ability Are Mediated by Regional Brain Volume." *2014 Neuroscience Meeting Planner*. http://www.abstractsonline.com/plan/start.aspx?mkey=%7B54C85D94-6D69-4B09-AFAA-502C0E680CA7%7D.

Concluding Points

In this chapter, we covered the issues and considerations involved in planning research presentations for scientific conferences. The purpose of presenting at scientific conferences is to share research in progress with the scientific community before it is officially published in refereed journals. Additionally, presenting research at conferences provides researchers with many personal benefits, in particular developing the research prior to publication and networking with like-minded researchers. Because conference organizers frequently have to manage thousands of Conference Abstracts (or research proposals), you must follow the guidelines precisely and adhere to the submission deadlines. In the next chapter, we will examine the scientific poster genre, which is the major presentation format at scientific conferences.

12

CREATING SCIENTIFIC POSTERS

In this chapter, you will learn to:

- Plan and write content for a scientific poster
- Design and print a scientific poster

A SCIENTIFIC POSTER displays research data and is used as a visual aid at a POSTER SESSION, which is a popular presentation and discussion format at scientific conferences (Hess, Tosney, and Liegel 2009; Powell 2012; Purrington 2015; Shelledy 2004; Woolsey 1989). Scientific posters are a unique genre that emerged from oral presentations and still maintain many similarities. The contemporary conventions of scientific posters, which we will examine in this chapter, evolved from the display of oral presentation slides on a bulletin board (MacIntosh-Murray 2007; Woolsey 1989). In this chapter, we will discuss the current format and content of scientific posters necessary for an effective poster presentation.

Overview

A scientific poster may be considered a storyboard view of an oral presentation slide deck in which all the slides are visible on a large canvas that is about four-feet high by six-feet wide (see Figure 12.1), or approximately the size of a midsized classroom whiteboard (Butz, Kohr, and Jones 2004; Lang 2010; Matthews 1990; Shelledy 2004). In the past 20 years, scientific researchers have shifted from manually affixing these slides to bulletin boards at the conference site (see

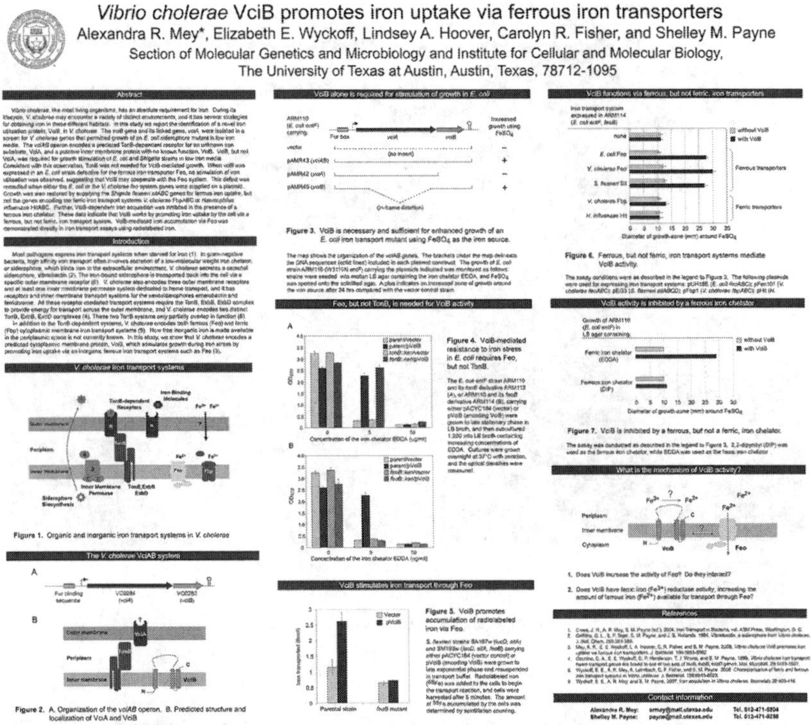

FIGURE 12.1 Scientific poster presented at a recent annual conference of the American Society for Microbiology.

Source: Drs. Alex Mey and Liz Wyckoff.

Figure 12.2) to predesigning conference posters on computers and printing as a large, full-color single sheet of paper to display on a presentation board at the conference venue.

Scientific poster sessions are typically part of a larger meeting and take place in a convention space with rows of blank poster-presentation boards. Presenters are assigned a period of time (usually a couple of hours) to stand with their posters at a numbered poster board and discuss the information with attendees who visit their poster (see Figure 12.3).

At a poster session, attendees may read the poster and examine the data figures from a distance of two to three feet, or they may broadly scan the entire poster from a slightly further distance. The most effective posters attract and engage attendees at both the close reading and global overview stages. In contrast, the least effective posters are off-putting at one or both levels—with text and visuals too small or difficult to read at a comfortable distance or with an overall design that lacks adequate aesthetic or professional appeal.

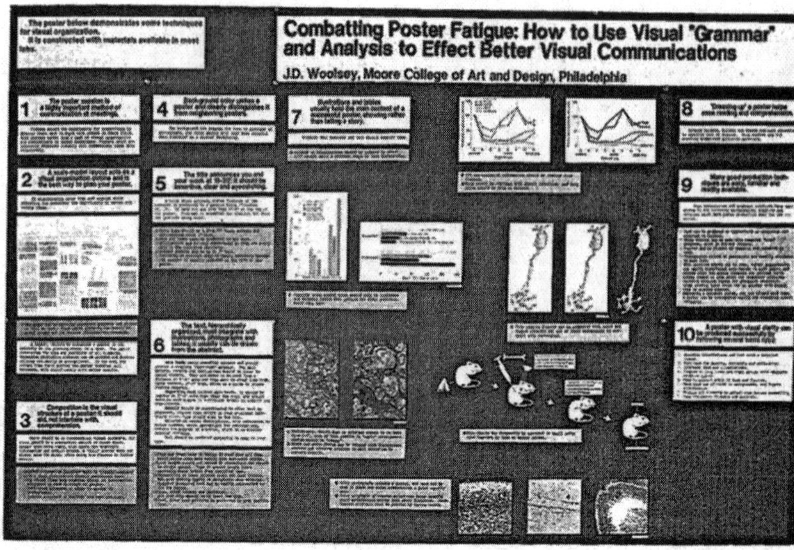

FIGURE 12.2 Historical scientific poster from the late 1980s based on oral presentation slides arranged on mat boards and assembled on a bulletin board at the conference site.

Source: Woolsey (1989).

FIGURE 12.3 Poster sessions encourage dialogue between researchers in a one-to-one or one-to-few discussion of research results.

Source: McNutt (2015).

The primary audience of scientific posters, other scientific researchers, is composed of two types: casual and highly interested. Casual attendees encounter a poster by browsing during a themed poster session in which presenters are grouped to display research on similar topics. Alternatively, highly interested attendees specifically seek out a particular poster based on the information (Title and Abstract) published in the CONFERENCE PROGRAM, the catalog of research presentations that is distributed online or in print at the beginning of the conference. Therefore, the audience is somewhat heterogeneous—including both researchers conducting similar investigations who are highly interested in your research and plan to visit your poster and other researchers of related fields who may only have a peripheral interest in the topic. Since your audience is varied in interest and expertise, you will need to adapt or tailor your short oral presentations (of approximately two to five minutes) as you explain the poster content to each individual or group of attendees (Block 1996; Hess, Tosney, and Liegel 2009; McNutt 2015).

In addition to narrating the poster content (as in a slide presentation), presenters at scientific poster sessions are expected to engage the attendees in a discussion of the research data, which often involves attendees asking questions probing into the data, methods, and theory (McNutt 2015; Powell 2012; Woolsey 1989). In some cases, attendees may prefer to read a scientific poster without an accompanying narrative from the presenter and then ask specific questions when they are done reading. Therefore, the scientific poster must be self-sufficient in that it includes adequate information, description, and transitions so that readers can understand the information without the accompanying narration by the presenter.

For the planning of the scientific poster content, you should envision that the scientific poster will function as a visual aid, much like the slides of an oral presentation (Powell 2012). The content should be visual—relying heavily on figures and simple tables—with minimal text that primarily serves to establish the context for each visual. The most common complaint of attendees is that scientific posters contain too much text (often because the authors inappropriately use the journal article or manuscript as a model for a scientific poster; Block 1996; Carter 2013; Lang 2010; Matthews 1990; Miller and Lewis 2008; Powell 2012). Steven Block (1996) of the Department of Molecular Biology at Princeton University provides this advice:

> DON'T write your poster as if it were a scientific paper. It's not. . . . And don't ever lift long sections of text directly from some manuscript and use these as a part of your poster. A poster is not a worked-over manuscript.
>
> *(3528)*

Elaine Miller and Katrina Lewis (2008) further explain in an article published in *Rehabilitation Nursing* that "Posters and manuscripts serve distinct purposes, target different audiences, and showcase content differently" (53). The key to an effective scientific poster is communicating visually (primarily through figures) in order to encourage dialogue with attendees (Carter 2013; Lawson 2005; McNutt 2015; Miller and Lewis 2008; Powell 2012; Woolsey 1989). As Rowena Murray, Morag Thow, and Rosanne Strachan (1998) from the University of Strathclyde and Glasgow Caledonian University emphasized, "A poster discussion can produce genuine two-way learning" (326). Therefore, you should take the time to plan and design a scientific poster carefully as this format of the content will directly influence how attendees will engage with you as the presenter and your poster.

Poster Design

The specifications of the scientific poster canvas, which include the dimensions and layout, should be established before beginning to create and fit the content. In this subsection we will discuss the initial design decisions that you need to make first in order to save you time and effort later in the process if you must redesign your poster.

Poster Dimensions

To determine the dimensions (width and height) of your scientific poster, you must answer the following questions:

1. *What are the maximum dimensions of the presentation space allotted to you?* (To answer this question, you should refer to the INSTRUCTIONS FOR PRESENTERS available from the conference website.) Obviously, the final poster canvas must fit within the allotted space and not encroach on adjacent presenters' space at the conference.
2. *Will your poster be horizontal (with the width greater than the height) or vertical (with the height greater than the width)?* The majority of scientific posters are designed horizontally, reflecting the dimensions of the allotted space for presentation and maximizing the content that is at a comfortable reading height for attendees without requiring them to stretch or kneel to read content that is posted too high or low, respectively. The optimal vertical placement of scientific posters is centered between waist level to a foot above eye level (Lang 2010).
3. *What software and printer will you use to design and print the poster? (And, more important, what physical limitations are introduced by the software and printer?)*

    a. **Selecting the software:** Selecting the software that you will use to create the poster depends on the software programs that you have that can accommodate the dimensions of the poster canvas and design elements (text, tables, and figures). Moreover, you should consider your own software skills in selecting a program that adequately meets your needs. The most commonly used software program for scientific posters is Microsoft PowerPoint. In fact, many PowerPoint templates for scientific posters are available from labs, departments, and online (search for "scientific poster template"). Individuals with more advanced software skills often prefer Adobe Illustrator, which provides greater design capabilities. Both PowerPoint and Illustrator accommodate custom canvas settings. Since PowerPoint is available and familiar to most individuals, I have included instructions to setup the canvas for scientific posters in Figure 12.4.
    b. **Selecting a printer:** Most scientific posters are printed on a single, continuous sheet of paper. The poster dimensions are limited by the width of the roll of paper used for printing, which determines the size of one side (often this

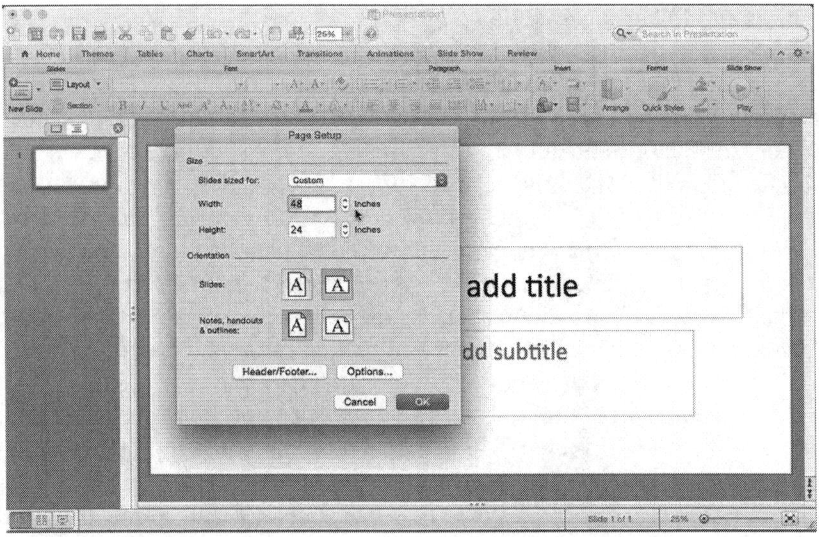

FIGURE 12.4 Setup of a scientific poster canvas in Microsoft PowerPoint. To set a custom canvas size, go to the File menu and select Page Setup. Although PowerPoint limits the size of a canvas to 56 inches, you may design larger posters by increasing the print scale from the Print menu when you are ready to print. For example, if you want a scientific poster to be 8-feet (96 inches) wide and 4-feet (48 inches) high, you can decrease the canvas size by 50% (48 inches wide by 24 inches high) and then set the printer scale to print at 200%.

turns out to be the height of a horizontal poster). If you plan to use a printer available in your department or organization, you should verify the width of the paper roll (usually either 36 or 48 inches). Copy shops and commercial printers, which may be found through an online search for "scientific poster printing," offer more sizes, but standard poster dimensions are usually less expensive and have faster production times than large, custom sizes.

After you have answered these questions, you will be able to specify the dimensions of your scientific poster canvas. This presentation space is analogous to the time limit of an oral presentation in that the content of your scientific poster (which we will cover later in this chapter) must be selected and designed to fit within this presentation space.

Poster Layout and Typography Decisions

The poster layout should be designed around the experience of attendees and optimized to facilitate communication and dialogue. To optimize the experience and minimize strain and confusion from the organization of information, you should follow a conventional magazine or newspaper layout design and employ a logical, consistent, readable, and uncluttered presentation format (Block 1996; Carter 2013; Matthews 1990). A typical layout of a scientific poster is shown in Figure 12.5. As illustrated in this figure, scientific posters include a banner for the title and authors that spans the entire width of the poster. You should use multiple columns (usually three to five, depending on the width of the poster) to provide a framework for vertically organizing content (Block 1996; Carter 2013). To create a clean, professional look, you should ensure that text blocks and visuals span the full width of each column. The organization of content vertically in columns follows the convention of printed documents (top to bottom within each column and then left to right between columns) and is a simple design that does not require attendees to interpret a more complex or unfamiliar organization (Carter 2013). Moreover, organizing content vertically in scientific posters is more effective than doing so horizontally in rows because vertical columns minimize the physical demand on attendees to walk across a poster (typically 6- to 8-feet wide) only once while they read content and also maximizes the number of attendees who can simultaneously read a poster without having to exchange places (Block 1996).

In addition to the layout of the scientific poster, you need to make typography decisions (such as typeface, type size, and alignment) that facilitate

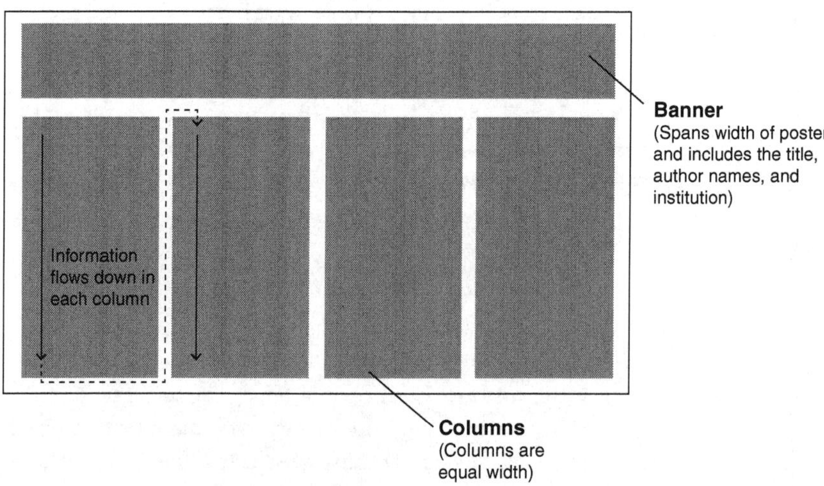

FIGURE 12.5 Layout of a typical scientific poster. The shaded regions illustrate the often-invisible framework of a poster used to align and fit content. The number of columns usually ranges from three to five, depending on the width of the poster. Columns should be of equal width and the empty space between columns should be consistent.

communication. TYPEFACES, which are commonly yet imprecisely called "fonts," are the names for the design of an alphabet and characters. (Technically, the term FONT is reserved for all of the typefaces of a single size.) When designing scientific posters, you should select legible typefaces and sizes, and use the size and weight of typefaces to reflect the hierarchy of information (with larger sizes and heavier weight typefaces for titles and headings).

In general, most scholars who have critiqued scientific posters recommend that you use standard typefaces (such as Times and Helvetica), which are familiar and easier to read, rather than decorative typefaces (Alred, Oliu, and Brusaw 1992; Block 1996; Carter 2013; Wong 2010). Donna Wong, who studied under Edward Tufte at Yale University and is the former graphics editor of the *New York Times* and graphics director of the *Wall Street Journal*, provides useful advice on selecting appropriate typefaces:

> Typography should not be center stage. The data is the focus. . . . Poor typography [selection] draws undue attention away from the underlying data, which carries the main message. The impulse to use type styles to spice up the chart [or poster] should be avoided at all costs.
>
> *(32)*

Although you have numerous typefaces to select, the most commonly used typefaces fall in two families: serif and sans serif. Serif typefaces (e.g., Times) include serifs, the small projections at the ends of letters. Sans serif typefaces (e.g., Helvetica) are those typefaces that do not include any projections at the ends of letters ("sans" is the French word meaning "without"). Often, designers use a typeface for the headings and title that is from the contrasting family that of the body text of the poster. Often, the body typeface is a serif typeface of normal weight whereas the title and headings are a sans serif typeface of a heavy weight (Alred, Oliu, and Brusaw 1992; Benson 1985; Ettenberg 1980; Pipes 2009).

In addition to selecting an appropriate typeface, you should use a type size for the main body text that is easily read at a distance of approximately four feet (Block 1996; Butz, Kohr, and Jones 2004). The optimal type size for the main body text is approximately 20 to 24 points (Block 1996; Lang 2010; Powell 2012; Shelledy 2004; Woolsey 1989). Using the body text as a foundation, you should make the section headings and poster title more prominent by using a larger typeface and heavier weight (see Table 12.1). In contrast, you should make the references and acknowledgments less prominent by using a smaller typeface (see Table 12.1). As summarized in Box 12.1, typeface sizes can help you estimate the size of the printed text. Since typeface sizes vary, you should always print a draft of your poster at actual size to ensure that it is legible. Box 12.2 explains TILED PRINTING, which will enable you to print a draft of your poster on a standard printer.

TABLE 12.1 Typography suggestions for scientific posters.

	Size[1] (Points)	Typeface	Style (Weight)	Alignment
Title	120	Sans serif (e.g., Helvetica)	Bold	Center
Authors	96	Sans serif (e.g., Helvetica)	Normal	Center
Institution	72	Sans serif (e.g., Helvetica)	Normal	Center
Section headings	36–48	Sans serif (e.g., Helvetica)	Bold	Centered or left justified
Body	20–4	Serif (e.g., Times)	Normal	Left justified (Ragged right margin)
References	18	Same as body	Normal	Left justified
Acknowledgments	18	Same as body	Normal	Left justified
Contact information	18	Same as body	Normal	Left justified

[1] The final size of printed letters varies slightly based on typeface (see Box 12.1). Therefore, you should print a sample poster to check that the text, particularly the body text, is clearly legible at four feet. To print a poster at actual size on a regular paper printer, see Box 12.2.

Box 12.1 Typeface Sizes

Estimating the printed size of text

Printing is a centuries-old industry that uses historical technologies for terms and is based on Imperial units for measure. The standard units of measure in printing are points, picas, and inches. The standard measure of a point is 1/72 inch and a pica is 1/6 inch (see Table 12.2).

TABLE 12.2 Conversion table of points to picas and inches.

	Points	Picas	Inches
One point	1 point	1/12 pica	1/72 inch
One pica	12 points	1 pica	1/6 inch
One inch	72 points	6 picas	1 inch

Although the conversion between points and inches is precise, the point sizes you select for typefaces will provide only an approximation of the final printed letter. Additionally, different typefaces of the same point size may appear to be different sizes when printed (see Figure 12.6). The reason for the inaccuracy between point size and printed letters is because points actually measure the theoretical size of the traditional metal letterpress type (see Figure 12.7). The point size of the metal type block must be the same for every character within the font and must accommodate the top of the tallest letter to the bottom of the lowest letter (see the dashed boxes in Figure 12.6 that approximate the placement of a letter on the face of a metal type character). Therefore, you should print your poster at actual size (see Box 12.2) to check that the actual size of the text is clearly visible at four feet.

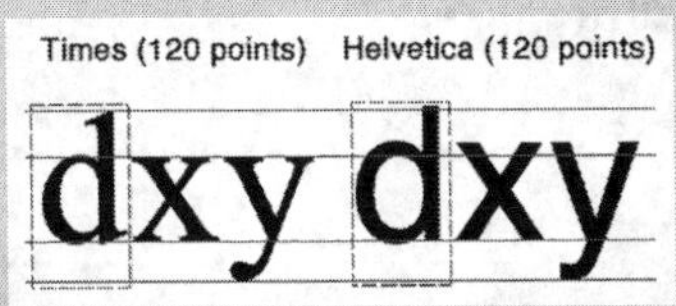

FIGURE 12.6 Comparison of different typefaces, both at 120 points (enlarged to show differences). Note that different typefaces of the same point size may appear to be different sizes when printed.

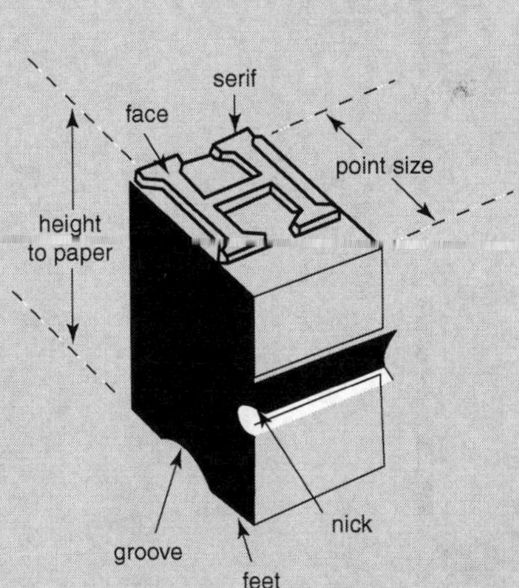

FIGURE 12.7 Block of text traditionally used for printing. Note that the point size is based on the face of the block, not the printed letter.

Box 12.2 Tiling a Scientific Poster

Printing a scientific poster at full size on a regular paper printer

When you TILE a large document (see Figure 12.8), a document of large dimensions is divided into frames that are the size of regular paper (e.g., US Letter paper at 8.5 × 11 inches) and each piece of the document is printed on a separate piece of paper. Then, the pages may be manually assembled to display the final document at actual size. If the tiling option is not available directly from a software program, you can export your poster as a PDF and then print from Adobe Acrobat. The "Tile Scale" (see arrow in Figure 12.8) should be the final poster dimension (usually 100%) unless you have reduced the canvas size as suggested in Figure 12.4. If you have reduced the canvas dimensions as suggested in Figure 12.4, then you need to compensate when printing (e.g., if you reduced the canvas size in PowerPoint to half, or 50%, of the final poster size, then you need to print the PDF through Adobe Acrobat at 200% to view at actual size).

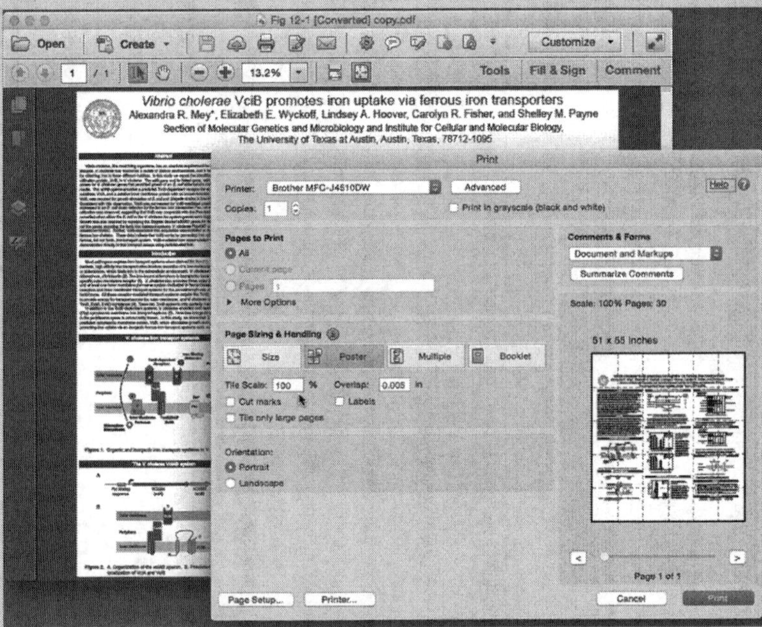

FIGURE 12.8 Tiling of a scientific poster in Adobe Acrobat prints the entire poster on the default paper in a printer.

Content

As with all genres of SMC, the content is the primary reason that your scientific poster will attract an audience at the poster session. Although the content of scientific posters is based on the prototypical IMRAD (Introduction, Methods, Results, and Discussion) organization, your scientific poster should emphasize the Results (the data) and have a highly abbreviated and focused Introduction, Methods, and Discussion. Scientists really just want the data—anything that comes before the Results delays and fatigues the audience. In fact, you may consider the scientific poster as a genre that presents the latest research data, with the other IMRAD content as ancillary to the Results.

As we discuss the content of scientific posters, you will find that the genre is more similar to the short, focused "report articles" in the journal *Science* than the information in the traditional IMRAD journal article we covered in Section II. Similar to report articles in *Science*, scientific posters should:

1. Focus on a single research question or problem,
2. Emphasize the Results (data) and have an abbreviated Introduction, a summarized Methods, and a focused Discussion section,
3. Show data and create a visual story,
4. Be concise with a total word count of less than 600 to 800 words (Carter 2013; Purrington 2015), and
5. Present information at a slightly more general level to interest and communicate to a slightly broader scientific audience than that of the conventional IMRAD journal article.

We will emphasize these themes as we explore the content for each section of scientific posters.

As Matt Carter (2013) assistant professor of biology at Williams College somewhat comically stated, "There is an inverse correlation between the amount of text on your poster and the probability that someone will actually read it" (319). A scientific poster should have a maximum of 600 to 800 words, which is 10% to 20% of the word count of a typical journal article (Carter 2013).

As we examine the content of scientific posters, you will notice that the genre is not as rigidly structured as journal articles. Often scientific posters use section headings that are more precise and descriptive than are the Introduction, Methods, Results, and Discussion headings in IMRAD journal articles. Additionally, scientific posters do not necessarily include or identify every IMRAD section (Block 1996). Since you are already familiar with the content of the IMRAD sections from Section II, we will use it as a framework for our discussion and focus on adapting content for the scientific poster genre.

Title and Authors

The scientific poster title, authors' names, and authors' institutional affiliation are included in the banner that spans the entire width of the top of the poster. As with the title of journal articles, the title is the most often-read piece of the poster and your opportunity to interest a casual audience (Lang 2010). The title is typically a sentence fragment that summarizes or describes the research findings (see Chapter 9). As mentioned earlier, the title should be most prominent text on the poster (largest font size and heaviest typeface). The title and authors should be consistent with the Conference Abstract previously submitted to the conference organizers (Lang 2010).

The authors of the poster are those who made substantial contributions to the research, either procedurally or theoretically, as discussed in Chapter 1. Usually the first author is also the PRESENTING AUTHOR, or the individual responsible for standing alongside the poster and discussing the content with attendees. Regardless of the order of the presenting author in the list of authors, the presenting author is usually identified with an asterisk. Additionally, the institutional affiliation (institution name and location) for each author is included. Department names are included for authors from academic institutions but are usually omitted from corporations. If the authors are from different institutions, superscript numbers are used to indicate the corresponding institution for each author. Often the primary institution where the research was conducted (or the institution of the presenting author) influences the visual design of the poster. In particular, the institution's logo is included in the banner and the brand colors and fonts of the institution are used throughout the poster design.

If you are presenting a scientific poster at a large conference, you may be assigned a poster number and required to display it prominently. Often the poster number is located in the top, left-hand corner of the banner although the exact location, if required, is specified in the Instructions for Presenters.

Abstract (Optional)

The inclusion of an Abstract on scientific posters is debated and may be required, optional, or discouraged by different conference organizers. The Conference Abstract is often printed in the conference program and thus may be omitted on the poster to avoid redundancy (Hofmann 2010). Since a scientific poster is itself a brief communication, it does not require an even shorter summary (Lang 2010). Moreover, the scientific poster is designed to facilitate conversation between the presenter and attendees, but an abstract, which is a paragraph of text, does not facilitate dialogue and uses space that may be better used presenting unique content. Some conferences organizers, however, require Abstracts to be reprinted on the poster. If included, the Abstract on the poster should be identical to the Conference Abstract submitted previously to the conference organizers (see Chapter 11) and published in the conference proceedings (Shelledy 2004). For a summary of writing Abstracts, see Chapter 9.

Introduction (Background and Aims)

In the Introduction, you should briefly (in 200 words or less) explain the rationale and importance of the research question or problem and cite a limited number (approximately three to seven) previous journal articles to establish the context for the present study (Butz, Kohr, and Jones 2004; Hofmann 2010; Shelledy 2004). In creating the Introduction for scientific posters, you should use the same moves that are covered in the Create a Research Space (CARS) model (Swales 1990; Swales 2004; Swales and Najjar 1987) which was presented in

1. Move 1: Establish a research territory by:
    a. Establishing the importance of the research topic, and
    b. Summarizing previous research to contextualize the present study.
2. Move 2: State the gap or inconsistency in the previous research to establish a niche for the present study.
3. Move 3: State the objective of the present study (which is to fill the research niche stated in Move 2).

As in *Science* report articles, scientific posters should provide just enough background so that the audience understands exactly what was known about the topic immediately prior to the present study (that is, what prompted the investigation).

You should consider presenting the Introduction content in a format that is conducive to an oral presentation. One popular approach is to present the content in two sections, the (1) Background and (2) Aims (or Hypothesis), rather than as a conventional Introduction section. In the Background section, authors often present the information in a bulleted list. In the Aims (or Hypothesis) section, authors often present Move 3 of the CARS model as a single, concise sentence. Another approach, as used in the scientific poster shown in Figure 12.1, is to present the information in a graphical model. In that poster, the authors presented the Introduction content in a short paragraph and summarized the known research and research question in the first illustration. Bulleted text and visual models are two effective approaches for presenting the Introduction content to facilitate dialogue between the presenter and attendees and thus enhance the presentation experience for all.

Methods

As in a *Science* report article, the Methods are often integrated into the Results with just enough information so that the audience can understand the data. You might integrate the Methods into the Results by providing a descriptive caption that summarizes each data collection method (we will cover this in the next subsection on Results).

If you decide to include a separate Methods section in your scientific poster, the section should provide a general overview of the experimental approach rather than a comprehensive description including all of the details necessary to repeat the experiment (Hofmann 2010). In this regard, the scientific poster differs from the journal article. Because the author is available to answer any specific procedural questions at the poster presentation, the scientific poster should provide only enough information so that attendees are able to understand the overall approach and judge the validity of the Methods as being appropriate to answer the research question (Shelledy 2004). Additionally, you may consider presenting the

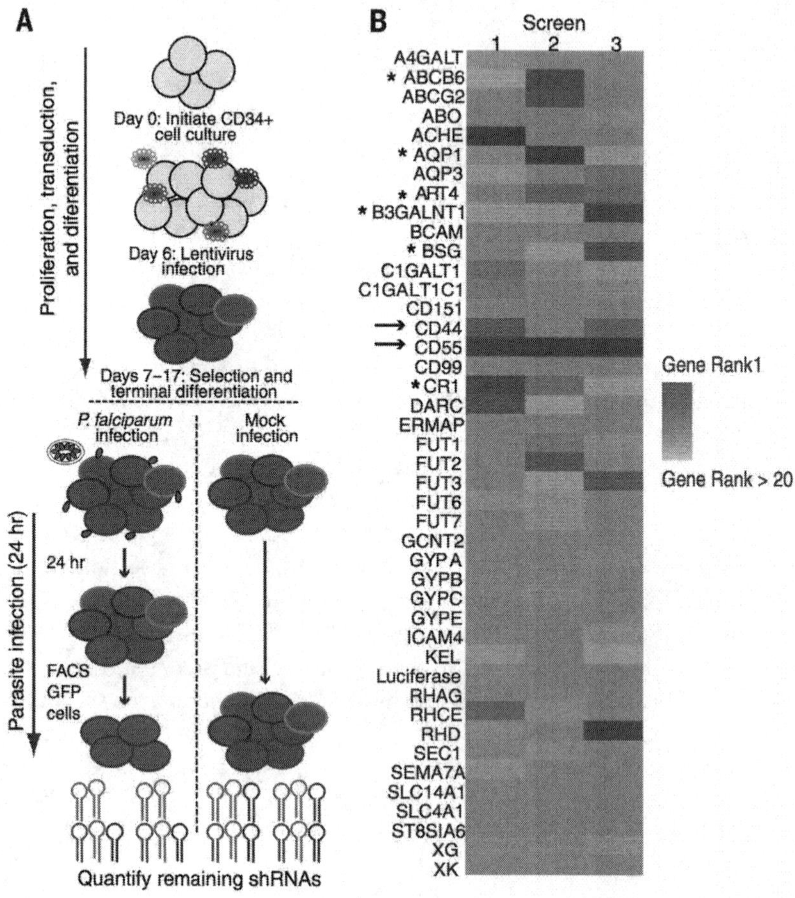

FIGURE 12.9 Diagram summarizing the research approach and descriptive caption from a *Science* report article. In this example, the Methods (A) is presented alongside the Results (B).

Source: Egan, Elizabeth S., Rays H. Y. Jiang, Mischka A. Moechtar, Natasha S. Barteneva, Michael P. Weekes, Luis V. Nobre, Steven P. Gygi, Joao A. Paulo, Charles Frantzreb, Yoshihiko Tani, Junko Takahashi, Seishi Watanabe, Jonathan Goldberg, Aditya S. Paul, Carlo Brugnara, David E. Root, Roger C. Wiegand, John G. Doench, and Manoj T. Duraisingh. 2015. "A Forward Genetic Screen Identifies Erythrocyte CD55 as Essential for Plasmodium Falciparum Invasion." *Science* 348 (6235): 711–4.

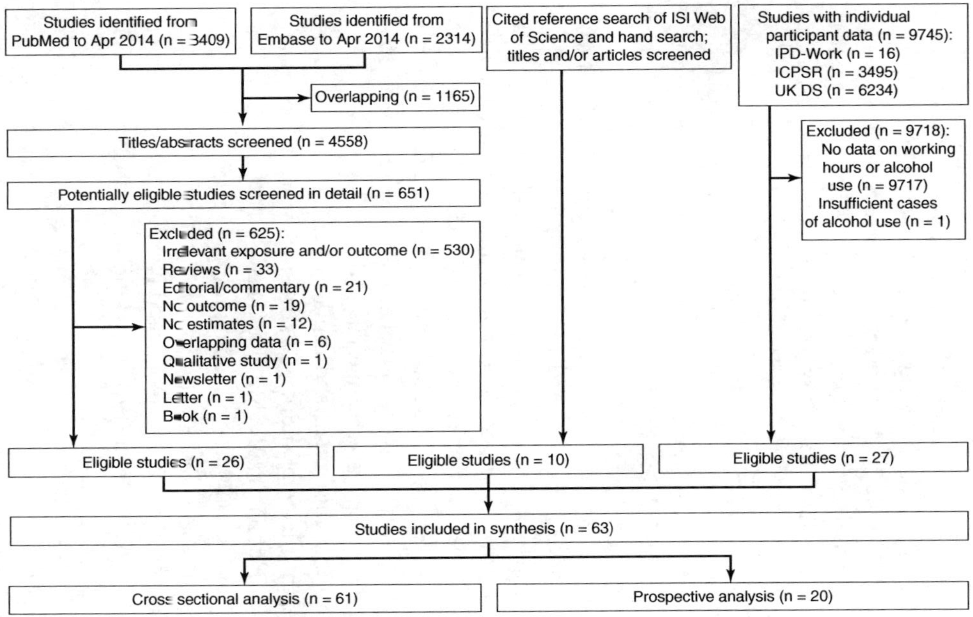

FIGURE 12.10 Flowchart showing experimental design from a literature review, which is a useful model for visually communicating the Methods in scientific posters. In this flowchart, the authors illustrate the research approach and provide an overview of the selection and exclusion process and criteria (along with the number of subjects).

Source: Virtanen, Marianna, Markus Jokela, Sola T. Nyberg, Ida EH Madsen, Tea Lallukka, Kirsi Ahola, Lars Alfredsson, G. David Batty, Jakob B. Bjorner, Marianne Borritz, Hermann Burr, Annalisa Casini, Els Clays, Dirk De Bacquer, Nico Dragano, Raimund Erbel, Jane E. Ferrie, Eleonor I. Fransson, Mark Hamer, Katriina Heikkilä, Karl-Heinz Jöckel, France Kittel, Anders Knutsson, Markku Koskenvuo, Karl-Heinz Ladwig, Thorsten Lunau, Maria Nordin, Tuula Oksanen, Jan H. Pejtersen, Jaana Pentti, Reiner Rugulies, Paula Salo, Jürgen Schupp, Johannes Siegrist, Archana Singh-Manoux, Andrew Steptoe, Sakari B. Suominen, Töres Theorell, Jussi Vahtera, Gert G. Wagner, Peter J.M. Westerholm, Hugo Westerlund, and Mika Kivimäki. 2015. "Long Working Hours and Alcohol Use: Systematic Review and Meta-Analysis of Published Studies and Unpublished Individual Participant Data." *BMJ* 350: g7772.

Methods visually as a diagram (see Figure 12.9A) or flowchart (see Example 7.2 in Chapter 7 and Figure 12.10) to facilitate the presentation and discussion format.

Results

The Results section is the core of the scientific poster. It is the most interesting section to attendees and the primary reason they are visiting your scientific poster. In a scientific poster, the data figures and tables should tell the research story visually rather than through large blocks of text (Hess, Tosney, and Liegel 2009; McNutt 2015; Rowe and Ilic 2011; Woosley 1989). The text should serve to augment the visuals (Woolsey 1989) rather than vice versa, which is more the style of a journal article. As in the scientific poster presented in Figure 12.1, the Results section should:

1. Occupy the majority of the poster space (Miller and Lewis 2008) and
2. Create a visual story in which the outcome from one data set leads directly to the next (Hofmann 2010; Miller 2007).

As in Figure 12.1, each visual is introduced by a summary heading and followed by a descriptive caption, which often references the Methods used.

You should present the data in a scientific poster as you would in the Results of any SMC—in a logical order but not necessarily in the order that the data were collected (see Chapter 7). The order of each visual will establish the order and flow of the presentation (Miller and Lewis 2008). You should carefully organize the data into a narrative or story that you construct around the visuals and communicate through the descriptive captions. In each caption, you should include sufficient information so that the visuals and captions stand alone and sufficiently explain the relevant Methods and key observations without having an author present to explain (Hofmann 2010; Miller and Lewis 2008). In a sense, the captions will provide the narration that you would provide to attendees.

Visuals are not only an effective means for communicating data; they also attract an audience from other posters with overused blocks of text (Hamilton 2004; Hofmann 2010; McNutt 2015). Thus, when presenting the data you should:

1. Show the data visually (see Chapter 7) rather than summarize them in paragraph form (Hofmann 2010; Miller 2007; Miller and Lewis 2008; Woolsey 1989);
2. Focus each visual on a single, primary message (Miller 2007);
3. Design visuals for the poster presentation format, which is simpler than visuals in journal articles (Hofmann 2010; McNutt 2015), by making them large, keeping the design basic (and uncluttered), clearly labeling the components and important features, larger text, and a using color purposefully (see Box 12.3);
4. Design tables for the poster presentation format, with fewer rows and columns than in tables in journal articles, larger text, and a focus on key data (Block 1996); and

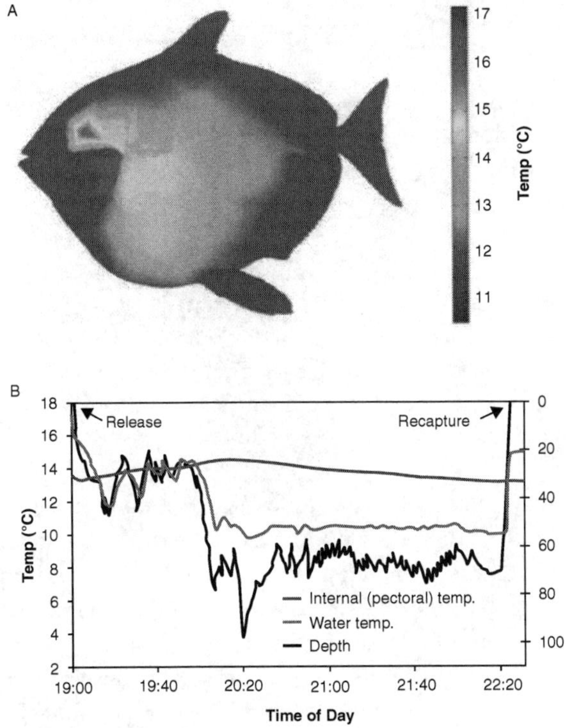

Fig. 1. Body temperature in the opah, *L. guttatus.* (A) In situ internal temperature profile (measurements taken ~4 to 5 cm below the skin) for a 98.0-cm fork length (40.0 kg) opah, with an ambient reference temperature of 10.5°C. (B) In vivo pectoral muscle temperature for a 96.4-cm (39.0 kg) opah swimming at depth.

Source: Wegner, Nicholas C., Owyn E. Snodgrass, Heidi Dewar, and John R. Hyde. 2015. "Whole-Body Endothermy in a Mesopelagic Fish, the Opah, Lampris guttatus." Science 348 (6236): 786–9.

FIGURE 12.11 Example of data presented visually and a descriptive caption from a *Science* report article. These visuals follow the guidelines for preparing and presenting data visually. In this example, the data from the first experiment (part A) leads to the next (part B). The descriptive caption provides an experimental overview and relevant criteria.

5. Maintain a consistent design between all visuals (e.g., all graphs should use symbols and color consistently to represent data from the same subjects).

Figure 12.11 shows data from a *Science* report article that adheres to these design principles. Furthermore, the descriptive caption is a good example of connecting the logic between two data sets, or data from different experiments, that constructs the presentation narrative. Your poster presentation will likely contain several visuals, like those shown in Figure 12.1, with the outcome of each visual leading to the question or issue explored in the next.

Box 12.3 Using Color Purposefully and Effectively

Purposeful use of color in graphs and diagrams effectively focuses the audience's attention on relevant information and contributes to the overall aesthetics of communications. Using color purposefully means using color meaningfully to facilitate the communication by:

- Drawing attention to (highlighting) key data with bright colors or deemphasizing less relevant data with muted colors,
- Representing values or intensity of data with standard color scales, and
- Grouping related objects or showing consistency of the same groups in multiple graphics by using the same colors to convey a relationship or using different colors to convey differences.

A secondary, less important, use of color is for design (or aesthetics). Colors used for aesthetic reasons should complement the use and selection of colors used for communicating information. Color perception is influenced by physiology (i.e., complete vision or partial color blindness), technology (i.e., printing and medium), and environment (i.e., lighting). Additionally, the meaning of colors is influenced by convention, culture, and personal taste. Although a comprehensive discussion of color selection is beyond the scope of this text, I have summarized the most relevant guidelines below:

- Use color sparingly (often 3 to 7 colors or shades of colors in total).
- Use warm-scale colors (red, orange, or yellow) to emphasize or highlight data.
- Use adequate contrast between colors for clear discrimination (to check for adequate contrast, print the image in black and white and make sure that you can easily see the contrast between the shades).
- Use color consistently in all images by using the same set of colors in all images and consistently using the same color for the same item or group in each image.
- Use conventional selection of colors and standard color output from instruments.
- Use analogous colors (see Figure 12.12) or monochromatic colors (different shades of the same color) for creating an overall professional, attractive visual (e.g., a different color for each bar or line on a graph or color scheme within a diagram).
- Use light color shades (often from the cool scale: blue, green, and violet) or white for the background.

- Use warm-scale and fully saturated colors sparingly.
- Use black for text and to show small details.

These recommendations are based on the research and experience of Horton (1991), Wong (2010), and Carter (2013).

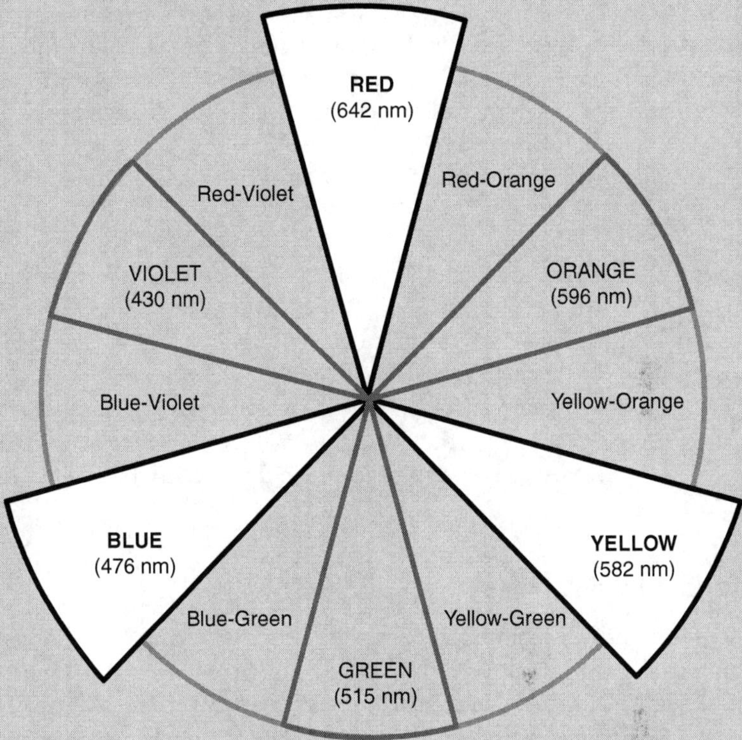

FIGURE 12.12 Color wheel with dominant wavelengths (nm) of primary (red, yellow, and blue) and secondary (orange, green, and violet) colors. Contrasting colors are opposite on the color wheel (e.g., blue and orange). Analogous colors (hues) are adjacent on the color wheel (e.g., blue, blue green, and blue violet).

Discussion (or Conclusions and Future Research)

Two subsections, the Conclusions and Future Research, often replace the Discussion section as these subheadings more accurately describe the key content being presented. Both sections should be exceedingly brief and are commonly presented as a bulleted list. In the Conclusions, you should summarize the Results and answer the research question or objective presented at the beginning of the poster (Shelledy 2004). You only need to list two to four key findings in the Conclusion (Hofmann 2010). In the Future Research section, you should present ideas for additional experimental plans. Both of these sections provide points for conversation between the presenter and attendees. Therefore, you should ensure that the Conclusions are carefully grounded or based in the data presented in the poster and avoid undue speculation or exaggerated claims of the importance or implications of the research because any unfounded scientific claims are likely to be challenged by attendees. Similarly, the Future Research section should be treated as points for discussion. Attendees often provide presenters with alternative approaches and insights to strengthen the research, which is a major purpose of scientific poster presentations (Woolsey 1989).

End Matter

At the end of your poster, you should include the References, Acknowledgments (if appropriate), and Contact Information (for key authors, such as the presenting author). For the References (and citation style), you should use the style that is most common in the field (Purrington 2015). If you do not know the most common citation style, you can look for the most used style of the journal articles you cited. As Figure 12.1 shows, the entire end-matter content occupies relatively little space and often uses a type size smaller than that used for the body text (see Table 12.1).

Concluding Points

The scientific poster is an important genre of scientific communication. It should be considered a unique genre, resembling the storyboard view of oral presentation slides rather than the journal article (Block 1996; MacIntosh-Murray 2007; Miller and Lewis 2008; Powell 2012; Woolsey 1989). In this chapter, we explored the format and content of scientific posters and how both contribute to an effective oral poster presentation. But unlike the typical oral presentation, the scientific poster presentation is a small, intimate presentation designed to facilitate dialogue between the presenter and one or more attendees at a time (Block 1996; Purrington 2015). The poster functions as the visual aid that the presenter, standing to the side, narrates for the audience (Carter 2013; Hofmann 2010; Lang 2010; Miller 2007). Since the interaction between the presenter and the audience is very close, you should also consider your professional appearance and hygiene during your presentation (Carter 2013; Keegan and Bannister 2003).

In addition to designing and printing the scientific poster, you want to ensure that you transport it carefully and have the necessary supplies to display it at a conference. To transport a scientific poster, most researchers use a "poster carrying tube," which is available from printing shops or can be ordered online.

Additionally, you should prepare a kit of desk supplies that can accommodate any last-minute need. Such kits should include pushpins, a small stapler (and staples), tape (single and double sided), correction fluid, ink pens that are similar in color and weight to the text (in order to make last-minute corrections), and the poster and images files (in case you need to reprint the poster at the conference location) (Carter 2013; Hofmann 2010; Lang 2010; Miller 2007; Purrington 2015). Many presenters also provide handouts of the entire poster scaled down and printed on US legal paper as well as business cards, which are made available to attendees by hanging envelopes below the poster itself (Block 1996; Carter 2013; Hamilton 2004; Hofmann 2010; Lang 2010; Lawson 2005; Miller 2007; Purrington 2015). You should also have materials available to take notes, as the discussions are likely to generate new research ideas, or you might need to record contact information in order to follow up with attendees on questions that you could not answer at the conference (Hamilton 2004; Lawson 2005). Sometimes presenters also post an empty envelope so that attendees can leave their business cards if they viewed the poster when the author was not present (Lang 2010).

SECTION IV

Communicating Research Findings with the Public and News Media

13

PUBLIC COMMUNICATION STRATEGY AND ETHICS

In this chapter, you will learn:

- Scientific literacy of general audiences
- Media relations strategy

As a researcher or professional writer involved in scientific and medical research, you may have the opportunity to communicate research directly to the public or be interviewed by news reporters. Yet, as a team of researchers cautioned in an article published in the *Journal of Clinical Epidemiology*, "Media disasters can result from us researchers getting ahead of ourselves (talking with the media when we shouldn't or in ways that we shouldn't), as well as from hasty, incompetent, or malicious reporting" (Sackett et al. 2006, 907).

In this chapter, we will examine the issues associated with communicating scientific and medical findings to the general public. Specifically, we will examine the scientific literacy of general audiences and media relations in science and medicine.

Scientific Literacy of General Audiences

A GENERAL (or LAY) AUDIENCE is defined as a group of individuals who lack specialized education or training in the subject area. Communicating scientific and medical research findings to the general public is called SCIENCE or HEALTH COMMUNICATION, although science communication encompasses both (Burns, O'Connor, and Stocklymayer 2003). Since 2010, the internet has been the primary medium for general audiences to learn about the latest science and technology news and is a major resource for detailed information about specific scientific issues (see Figure 13.1).

Effectively communicating science or medicine to a general audience is challenging because the education, knowledge, and attitudes toward science and medicine are markedly broad (National Science Board 2014). The public's understanding of scientific "facts" provides a measure for comparing scientific literacy across nations (Jon Miller 2004; Miller 1998; National Science Board 2014; Norris and Phillips 2003; Turner 2008). In Table 13.1, the percent of correct answers for each basic science question by country is provided for the most recent year of data collection. In Table 13.2, the percent of the US population who understand the scientific inquiry process (probability, experiment, and scientific study) is provided.

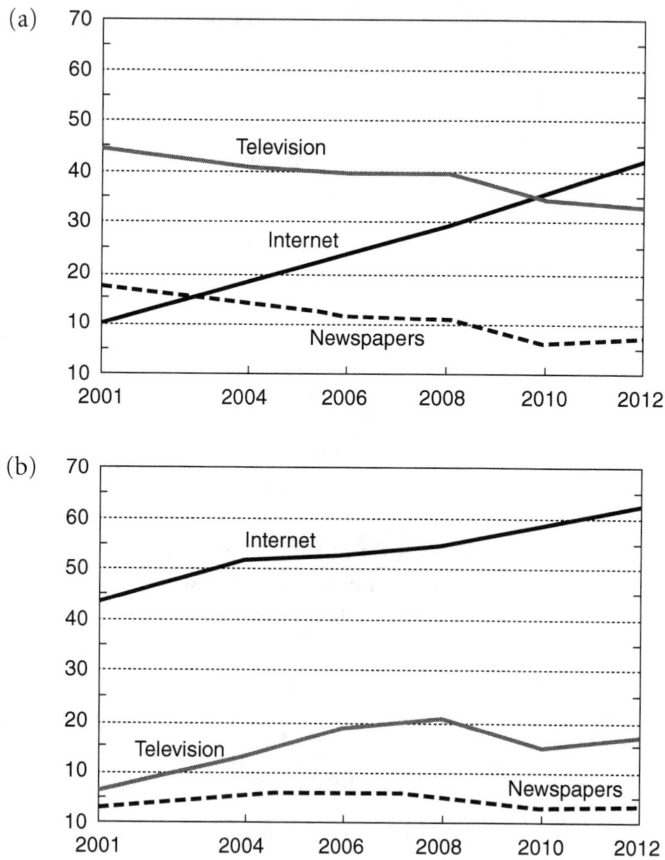

FIGURE 13.1 Primary sources of scientific information for the general public in the United States from 2001 to 2012 for (a) general science and technology news and (b) specific scientific issues.

Source: National Science Board (2014).

TABLE 13.1 Percent of correct answers to basic scientific questions by country or region for the most recent year surveyed.

Question (Answer)	United States (2012)	China (2010)	EU (2005)	India (2004)	Japan (2011)	Malaysia (2008)	Russia (2003)	South Korea (2004)
Physical science								
The center of the Earth is very hot. (True)	84	56	86	57	84	66	NA	87
The continents have been moving their location for millions of years and will continue to move. (True)	83	50	87	32	88	44	40	87
Does the Earth go around the Sun, or does the Sun go around the Earth? (Earth around Sun)	74	NA	66	70	NA	72	NA	86
All radioactivity is manmade. (False)	72	48	59	NA	69	14	35	48
Electrons are smaller than atoms. (True)	53	27	46	30	32	33	44	46
Lasers work by focusing sound waves. (False)	47	23	47	NA	32	16	24	31
The universe began with a huge explosion. (True)	39	NA	NA	34	NA	NA	35	67
Biological science								
It is the father's gene that decides whether the baby is a boy or a girl. (True)	63	58	64	38	29	40	22	59
Antibiotics kill viruses as well as bacteria. (False)	51	28	46	39	33	8	18	30
Human beings, as we know them today, developed from earlier species of animals. (True)	48	66	70	56	76	NA	44	64

NA = not available, question not asked.

EU = European Union; data are not available for Bulgaria and Romania.

Source: National Science Board (2014).

TABLE 13.2 Percent of correct answers (by year) showing an understanding of the scientific process by US adults.

Question	1999	2001	2004	2006	2008	2010	2012
Understanding of scientific inquiry scale	32	40	39	41	36	42	33
Components of understanding scientific inquiry scale							
Understanding of probability	64	67	64	69	64	66	65
Understanding of experiment	34	40	46	42	38	51	34
Understanding of scientific study	21	26	23	25	23	18	20

Source: National Science Board (2014).

TABLE 13.3 Percent of adults considered scientifically literate in the United States and European Union in 1995.

Country	Construct Vocabulary Score 67+	Understand Scientific Inquiry	Civic Scientific Literacy			*N*
			Literate	Partial	Not	
United States	27	21	12	25	63	2,006
European Union	20	12	5	22	73	12,147
Britain	27	20	10	26	64	1,000
Denmark	26	15	8	25	67	1,000
Netherlands	27	15	8	27	65	1,000
Italy	18	14	5	22	73	1,000
France	21	12	4	24	72	1,000
Germany	23	8	4	24	72	2,000
Spain	17	7	3	17	80	1,000
Belgium	16	10	3	20	77	1,000
Ireland	14	8	3	16	81	1,000
Greece	7	7	1	11	88	1,000
Portugal	8	2	1	8	91	1,000

Source: J. Miller (1998).

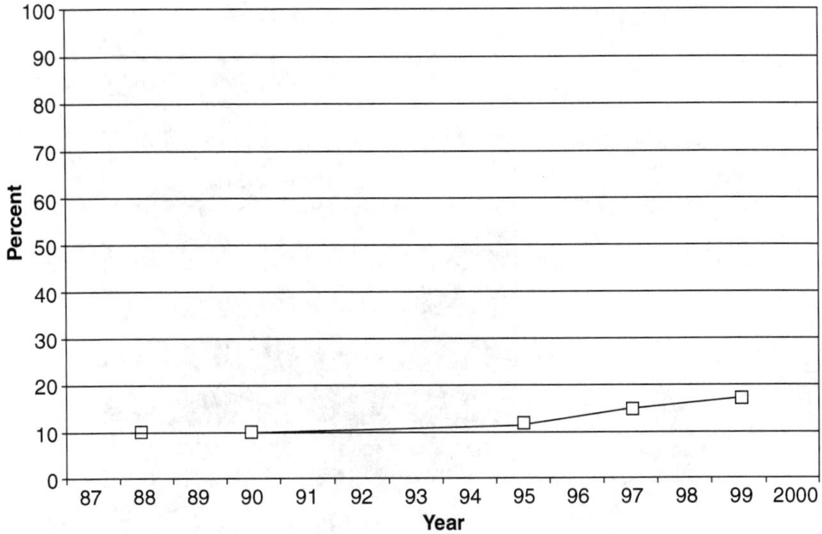

FIGURE 13.2 Percent of US adults who are scientifically literate.

Source: Jon Miller (2004).

Combined, results from these two measures indicate that only 12% of the United States and 5% of the European Union population had an adequate civic scientific literacy in 1995 (see Table 13.3). CIVIC SCIENTIFIC LITERACY is defined as an adequate understanding of scientific terms and concepts that an individual must have to read a daily newspaper or magazine and understand the essence of scientific public issues or controversies (J. Miller 1998; Shen 1975). Notably, the rate of civic scientific literacy is higher in the US than in the European Union or Japan (Jon Miller 2004). The higher civic scientific literacy in the US is attributed to the general education requirement at the university level, which includes science courses (Jon Miller 2004). Because the US is the only major country that requires general education courses for university degrees, these general education science courses increase the overall civic scientific literacy despite the lower performance of US high school students compared to those in other countries (Jon Miller 2004). As illustrated in Figure 13.2, scientific literacy has increased slightly in the United States from 1988–99, when the latest data were available (Jon Miller 2004).

Despite the low overall percentage of civic scientific literacy among adults, the general public (at least in the US) expresses a high degree of confidence in the scientific and medical communities (see Figure 13.3). This confidence affects the public's acceptance of the research findings and conclusions endorsed by scientific and medical leaders (National Science Board 2014). Thus, communicating with the general public and news media is an influential role and responsibility of those on research teams.

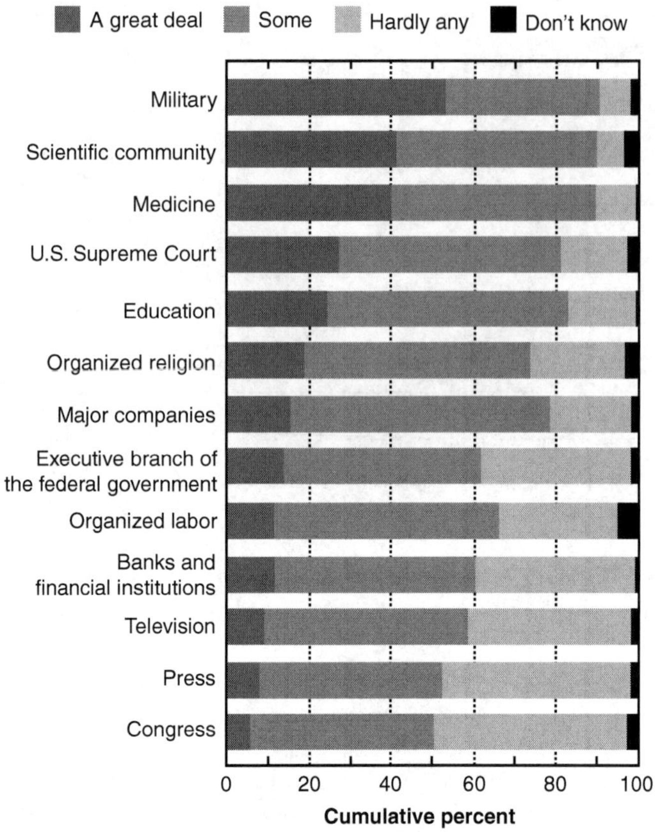

FIGURE 13.3 Public confidence in institution types in the United States.

Source: National Science Board (2014).

Media Relations Strategy

The publication of research in the news and popular media not only shares research with the public but also amplifies the transmission of research findings among the scientific and medical community (Phillips et al. 1991). However, journalists may not appropriately report important nuances of the research findings nor properly emphasize the newsworthiness of the research (Cohn 1989; Lenzer 2002).

Rather, the goal of journalists is to attract readers with dramatic headlines and news articles, which often results in the exaggeration of scientific findings (Cohn 1989; Lenzer 2002). This exaggeration, or "spin" (Lenzer 2002, 664), may be positive or negative. A positive bias is common for scientific research with human medical potential whereas a negative bias is common for research on the environment or risk of toxicity (Cohn 1989; Lenzer 2002). Victor Cohn (1989) of *The Washington Post* explained that:

> [reporters] tend to rely most on "authorities" who are either most quotable or quickly available or both, and they often tend to be those who get most carried away with their sketchy and unconfirmed but "exciting" data—or have big axes to grind however lofty their motives. The cautious, unbiased scientist who says, "Our results are inconclusive" or "We don't have enough data yet to make any strong statement" or "I don't know" tends to be omitted or buried someplace down in the story.
>
> *(6)*

As such, the reporter's objective does not complement the professional reputations of researchers or professional communicators (Lantz and Lanier 2003). The journal *Nature* (2002) recommends that researchers and communicators working with the news media be informed about media practices in order to represent the science clearly and avoid misrepresentation or spin.

Determining If the Research is Ready to be Reported

The primary issue to consider before communicating research findings to the news media is whether or not the research has been thoroughly vetted through the full peer-review process of journals. Announcement of research findings

"One thing I'll say for us, Meyer—we never stooped to popularizing science."

should coincide with publication of the research data in a journal article but not before communication to the scientific community (Committee on Publication Ethics 1999; International Committee of Medical Journal Editors 2013). Research presented at conferences should be carefully communicated only as preliminary findings in at all (Schwartz, Woloshin, and Baczek 2002). Moreover, preliminary announcement of research findings in the news media may be considered prior publication and, depending on interpretation by the editors of a journal, may violate the INGELFINGER RULE, which states that journals only publish research that has not been previously published (Langdon-Neuner 2009). The International Committee of Medical Journal Editors (2013) explains that:

> the general public has a legitimate interest in all journal content and is entitled to important information within a reasonable amount of time, and editors have a responsibility to facilitate that. However media reports of scientific research before it has been peer-reviewed and fully vetted may lead to dissemination of inaccurate or premature conclusions, and doctors in practice need to have research reports available in full detail before they can advise patients about the reports' conclusions.
>
> *(11)*

In the *Journal of Clinical Epidemiology*, researchers provided several questions that you should consider before responding to a request from the news media (Sackett et al. 2006):

1. Have your results and conclusions survived external peer review yet?
2. Might the top clinical journals reject your manuscript later if you give its details to a reporter sooner?
3. Are you sure that anybody needs to learn about your results by any route other than a peer-reviewed scientific journal?
4. If so, should this notification be left in the hands of a reporter you've never met, who may know little about the subject, and is extremely unlikely to let you see a draft of the news article?
5. Who should be the first target audience to learn about and apply your results?
6. What harm might be caused if others learned about the results first?
7. If you decide to respond to the request from the media, what strategies can you employ to protect the validity and appropriateness of your message?

Based on the answers to these notably cautious and discouraging questions, you will be able to decide if your research is ready to be communicated to the public.

Communicating with the Media

In interviews with the news media, you will often be perceived as representing your institution or organization rather than your own personal views. Therefore, you must confirm with your organization's media relations department that you are authorized to communicate with the media or be interviewed in a professional capacity (Selinger 2007). Many organizations have public relations specialists who deal directly with reporters and coordinate news reports and interviews with individuals inside of an organization. If you are contacted directly by a reporter, you should notify the media relations department or public relations specialist for permission and assistance.

If a reporter contacts you directly, you may have little time to respond to the request or prepare for an interview. Often reporters are working on short deadlines and have the constant pressure of being scooped by a rival news media outlet. When these pressures are coupled with the 24-hour news cycle, reporters often have limited time to thoroughly develop a science story (Cohn 1989; Lofstrom 2010; Selinger 2007). Thus, reporters often have limited knowledge or understanding of a topic before the interview. Such limitations will impact your dialogue with reporters.

When communicating to reporters, you should communicate the science simply and clearly. Reporters are looking to you, as a subject-matter expert, to provide quotes (or "sound bites") for a news story (Cohn 1989). Therefore, you have very limited opportunity to provide much information or background education for the audience. Rather you must communicate the essence of the issue in a manner that a general audience will understand. Specifically, you should develop three basic, but strong, information points that summarize the issue and represent your position on the topic (Lofstrom 2010).

Furthermore, some reporters may attempt to push interviewees into making dramatic comments based on a "worst-case" or "what-if" situation (Lofstrom 2010). This strategy relates to the reporters' attempt to sensationalize the research and attract an audience. To present scientific information accurately, then, you should do the following (Lofstrom 2010; Selinger 2007):

- Anticipate difficult and challenging questions on the topic,
- Comment only on facts and avoid speculation,
- Provide accurate information, which means you may research the answer and respond to a question later,
- Recognize that nothing is "off the record" and everything you say to a reporter is quotable, and
- Never use the phrase "no comment," which implies that you have something to hide.

Finally, when dealing with the news media, you should recognize that journalists may confirm facts and quotes, but rarely, if ever, do they share their stories with sources before the stories are published (Lofstrom 2010). Therefore, you must communicate accurately and clearly the first (and often only) time you speak with a journalist.

Concluding Points

Although communicating science to the public is an important role of those who are involved in scientific and medical research, you must ensure that you are communicating clearly, accurately, and appropriately. In the *IEEE Spectrum*, Selinger (2007) summarizes working with reporters as follows: "Reporters may be friendly, but they're never your friends. Assume nothing; they often have their own agendas" (59). Thus, you should communicate with the news media on your own time schedule when your research investigation is complete and being published in a journal article. The proper method for announcing research findings to the media is through press (or news) releases, which we will examine in the next chapter.

14

WRITING PRESS (NEWS) RELEASES

In this chapter, you will learn:

- The audiences and gatekeepers of press (news) releases
- The content, organization, style, and design of press releases announcing scientific and medical innovations to a general audience
- The process of distributing a press release to media outlets

A PRESS RELEASE (sometimes called a NEWS RELEASE or PRESS STATEMENT) is written and disseminated by universities, companies, government organizations, and other institutions to announce something of news value. Press releases are frequently used to communicate research innovations to the news media in order to encourage news editors to republish the content directly from the press release or encourage reporters to write a more in-depth news article (Dutch 2012). A press release may be distributed by the institution where researchers work or the journals that publish the research. As such, a press release is not an objective news article—it is designed to promote an organization's research, innovations, or activities. The resulting media attention may have important implications for funding from potential investors.

Scientists, physicians, and professional writers are often involved in writing (or contributing to writing) a press release because individuals in these positions understand the technology (Lofstrom 2010; Mogull 2008; Sackett et al. 2006; Selinger 2007). The challenge of writing an effective press release is that you, as the author, must communicate the value of a scientific innovation to a general audience who may have little or no prior knowledge of the topic. As we will discuss in this chapter, communicating to general audiences requires more than simply translating technical jargon; it also requires revising the content and style.

Audience of Press Releases

Press releases reach an audience that is different from that of other forms of scientific publication. The press release is a pseudo-news story that is sent directly to editors or journalists in news media. If these individuals believe that the information in the press release is of interest to the audience of their publication, then they may either write their own news story or directly republish the content provided in the press release. Therefore, the ultimate audience of a press release is the readers of the particular news publication or outlet. In a press release, you should emphasize the relevant information for the target audience of the news outlet, which may include other scientists and physicians (Lofstrom 2010). Even for technical publications and science news magazines, the audience is always less technical and more heterogeneous than that of journal articles.

Although the readers of a particular news outlet are the targeted audience, an editor, journalist, or both must be convinced of the value of the story to the audience. In this situation, the editor and journalist act as gatekeepers who must be convinced of the newsworthiness, importance, and relevance of the information in a press release in order to republish the information.

A common pitfall when writing a press release is that the author writes from an egocentric perspective—which is writing from one's own perspective rather than considering what is important to the journalist, editor, and readers of a particular news outlet. The journalist and editor are looking for something that fulfills their need for news. Moreover, they consider the perspective of their audience when determining content for their publication. If the readers of a particular news outlet would be interested in a story, then the editor and journalist are more likely to publish a particular story. When developing a press release, you should target the information to the audience by asking the following questions:

- Why would the readers of a particular news publication care about this research?
- How will the readers use or apply the information?

The answers to these questions will differ for different publications. For example, a business publication would emphasize the financial impact of a research innovation whereas a consumer health magazine would emphasize the impact of the therapeutic potential for an individual. Therefore, before writing a press release, you should review the publication to determine the information that is most important to the readers. This may mean that you write several different press releases that you distribute to different news media publications.

Writing a Press Release

A press release should focus on a single, specific topic that is positioned to be of interest to the readers of a particular news outlet. A press release is a short announcement, which usually contains:

1. A brief overview of the innovation,
2. A clear statement of the innovation's value or importance to the readers of the news outlet,
3. Quotes from individuals who were involved in the research and development of an innovation or who are affected by the innovations, and
4. A brief summary of the university, company, or institution where the research was performed.

A press release is a short 300 to 400 word genre that is often limited to one or two pages. In the following subsections, we will examine the content, organization, and style of a press release.

Content and Organization

Press releases consist of three sections: the header, body, and footer. We will examine each of these sections below.

Header

The header is the content that appears at the top of a press release. Example 14.1 illustrates the header of a typical printed press release. The header design varies slightly between organizations, as the examples in this chapter illustrate. Many organizations provide press releases on company letterhead or on a web page that contains the company logo. The header contains the following parts:

1. **Document title:** "PRESS RELEASE" or "NEWS RELEASE" (typically appears in bold and all capital letters).
2. **Release information:** "FOR IMMEDIATE RELEASE" or "UNDER EMBARGO UNTIL [date of release)]" (typically appears in bold and all capital letters). An EMBARGO delays publication of the press release until a future date, which may coincide with the publication of research in a peer-reviewed journal.
3. **News headline:** content that captures the readers' attention (which should be consistent with the lead, or the first sentence of a press release). The title should be short, declarative, and clear (Sackett et al. 2006). The title should summarize the content of the press release.
4. **Dateline:** the release date (which may be dated with the embargo date) and the location (city and state or country) where the news originates.

Body

The body of a press release provides the news content of the genre. The body should be approximately 350 words, although some press releases may be shorter (Sackett

EXAMPLE 14.1 HEADER OF A TYPICAL PRESS RELEASE.

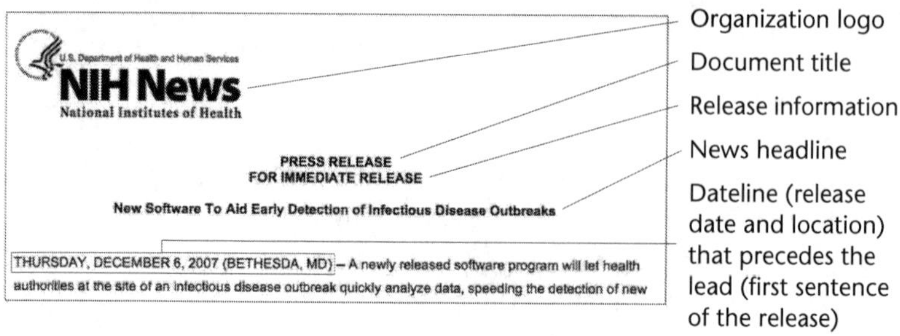

et al. 2006; Woloshin and Schwartz 2002). The news content of a press release begins with the LEAD (also written as "lede")—the first sentence of a press release or news article, which states the impact, novelty, timeliness, or usefulness of the information to the reader. A HARD LEAD, which is common in press releases, provides a comprehensive summary of the news. In contrast, a SOFT LEAD is more commonly used in long feature articles and consists of a creative, attention-grabbing introduction.

In a hard lead, you should strive to answer who? what? where? when? and why? (5 Ws) and occasionally how? But because lead sentences are often limited to 20 to 25 words in length, you may not be able to include all of the information in the first sentence. Therefore, your goal should be to provide the most important information first (remember to consider importance from the reader's perspective, which may change depending on the focus of the news publication). All of the 5 Ws should be summarized in the first few (one to three) sentences, which provides the key information about the research.

Following the lead, the remainder of the press release is used to provide information to support the thesis statement. The bulk of the press release is dedicated to supporting the claim(s) made in the lead. In the following sentences or paragraphs, you should provide details that summarize key information and support these statements with quotes from individuals who played key roles in the research or who are affected by it. When quoting individuals, you will use a news style, which includes the full name, position, and organizational affiliation in order to establish the relevance and credibility of the statements. Although the press release is not a place for overly technical details (which might confuse the audience), it should provide a source for more detailed information (such as a journal article).

Content in news articles is organized in respect to relative importance, which means you should present information that adheres to the INVERTED PYRAMID structure (Dutch 2012). In the inverted pyramid, the importance of the information decreases throughout the article. This organization is used by news writers because readers may stop reading before they reach the end of the article and because editors may cut content from the bottom of the article if they need space for other content. In either case, the most important content is maintained at the top of the article.

When announcing the publication of research findings, as in Example 14.2, press releases may contain a TECHNICAL SUMMARY in italics below the news headline and before the body of the press release. This technical summary is a short overview of the research findings announced in the journal article, which communicates the key information to the gatekeepers (or editors of news media). The technical summary will be removed before publication by a news media outlet.

Although I mentioned that press releases are biased toward the research findings, you should ensure that the press release does not introduce an inappropriate interpretation or analysis of the data beyond the claims established and reviewed in a journal article. Specifically, a press release should not (1) contain more direct or explicit medical advice than the corresponding journal article, (2) falsely show a deterministic or cause-and-effect relationship between variables that were correlational results, nor (3) inflate inference to humans from studies of animals, cells, or simulations (Sumner et al. 2014).

EXAMPLE 14.2 A PRESS RELEASE ANNOUNCING RESEARCH FINDINGS.

MAYO CLINIC
PROCEEDINGS

MAYO CLINIC

ELSEVIER

NEWS RELEASE
UNDER EMBARGO UNTIL FEBRUARY 27, 2014, 12:00 PM CT

Accompanying video available at: http://youtu.be/I7glVHaTo2c
(Video *under embargo until February 27, 2014, 12:00 PM CT*)

Contact:
Rachael Zaleski
Elsevier
Tel: 215–239–3658
mcpmedia@elsevier.com

Physicians' Stethoscopes More Contaminated Than Palms of Their Hands
Results of New Study Published in *Mayo Clinic Proceedings*

A comparative analysis shows that stethoscope diaphragms are more contaminated than the physician's own thenar eminence (group of muscles in the palm of the hand) following a physical examination.

Rochester, MN, February 27, 2014—Although healthcare workers' hands are the main source of bacterial transmission in hospitals, physicians' stethoscopes appear to play a role. To explore this question, investigators at the University of Geneva Hospitals assessed the level of bacterial contamination on physicians' hands and stethoscopes following a single physical examination. The study appears in the March issue of *Mayo Clinic Proceedings*.

"By considering that stethoscopes are used repeatedly over the course of a day, come directly into contact with patients' skin, and may harbor several thousands of bacteria (including MRSA) collected during a previous physical examination, we consider them as potentially significant vectors of transmission," commented lead investigator Didier Pittet, MD, MS, Director of the Infection Control Program and WHO Collaborating Centre on Patient Safety, University of Geneva Hospitals. "From infection control and patient safety perspectives, the stethoscope should be regarded as an extension of the physician's hands and be disinfected after every patient contact."

In this study, 71 patients were examined by one of three physicians using sterile gloves and a sterile stethoscope. After they completed the examination, two parts of the stethoscope (the tube and diaphragm) and four regions of the physician's hands (back, fingertips, and thenar and hypothenar eminences) were measured for the total number of bacteria present.

The stethoscope's diaphragm was more contaminated than all regions of the physician's hand except the fingertips. Further, the tube of the stethoscope was more heavily contaminated than the back of the physician's hand. Similar results were observed when contamination was due to methicillin-resistant *S. aureus* (MRSA) after examining MRSA-colonized patients.

This work is the first to compare directly the level of contamination of physicians' hands and stethoscopes. Stethoscope contamination is not trivial and is comparable to the contamination of healthcare workers' fingertips, the hand region most implicated in microbial cross-transmission. Physicians must be aware of the need to disinfect their stethoscopes after each use.

# # #

NOTES FOR EDITORS

"Contamination of Stethoscopes and Physicians' Hands After a Physical Examination," by Yves Longtin, MD; Alexis Schneider, MD; Clément Tschopp, MD; Gesuèle Renzi, MS; Angèle Gayet-Ageron, MD, PhD; Jacques Schrenzel, MD; and Didier Pittet, MD, MS (DOI:http://dx.doi.org/10.1016/j.mayocp.2013.11.016). *Mayo Clinic Proceedings*, Volume 89, Issue 3 (March 2014), published by Elsevier.

Full text of the article is available to credentialed journalists upon request. Contact Rachael Zaleski at 215-239-3658 or mcpmedia@elsevier.com to obtain copies. To view an accompanying video presented by Dr. Didier Pittet go to http://youtu.be/I7glVHaTo2c (under embargo until February 27, 2014, 12:00 PM CT).

Journalists wishing to interview the authors may contact Dr. Didier Pittet at +41 22 372 9833 or +41 22 372 9828 or via email at Didier.Pittet@hcuge.ch.

ABOUT MAYO CLINIC PROCEEDINGS

The flagship journal of Mayo and one of the premier peer-reviewed clinical journals in general medicine, *Mayo Clinic Proceedings* is among the most widely read and highly cited scientific publications for physicians, with a circulation of approximately 125,000. While the *Journal* is sponsored by Mayo Clinic, it welcomes submissions from authors worldwide, publishing articles that focus on clinical medicine and support the professional and educational needs of its readers. www.mayoclinicproceedings.org

ABOUT MAYO CLINIC

Mayo Clinic is a nonprofit worldwide leader in medical care, research, and education for people from all walks of life. For more information visit www.mayoclinic.org/about and www.mayoclinic.org/news.

ABOUT ELSEVIER

Elsevier is a world-leading provider of scientific, technical and medical information products and services. The company works in partnership with the global science and health communities to publish more than 2,000 journals, including *The Lancet* (www.thelancet.com) and *Cell* (www.cell.com), and close to 20,000 book titles, including major reference works from Mosby and Saunders. www.elsevier.com

Footer

The footer of a press release contains a summary of the institutions involved in the research (see Example 14.3). Every press release from an organization usually contains the same information in the footer, which includes these elements:

1. **Boilerplate:** a short paragraph that provides background information on the university, company, government organization, or institution. This section is often titled "About [name of company]" and includes five to six short sentences that describe the core business and mission of the organization. The bolierplate often provides a URL to the organization's home page.
2. **Close:** typically three number signs (###) placed at the bottom of a printed press release indicating the end of the press release.
3. **Media contact information:** the name, phone number, email address, mailing address, or other contact information of the individual at an institution serving as the primary contact for journalists. The media contact individual is usually a public relations professional who answers initial questions about the news item and arranges interviews with technical experts within the organization. (If you will be filling this role, see Chapter 13 for recommendations for communicating with the news media.)

EXAMPLE 14.3 FOOTER OF A TYPICAL PRESS RELEASE.

The National Institutes of Health (NIH) — The Nation's Medical Research Agency — includes 27 Institutes and Centers and is a component of the U.S. Department of Health and Human Services. It is the primary federal agency for conducting and supporting basic, clinical and translational medical research, and It investigates the causes, treatments. and cures for both common and rare diseases. For more Information about NIH and its programs, visit www.nlh.gov.	Boilerplate (Information about an organization)
###	Close (###)
Contact: Emily Carlson 301-555-5555	Media Contact Information
-2-	Page number (for press releases that are more than one page)

Writing Style and Issues for Communicating to a General Audience

The style of a press release, which is a science communication genre, is much different from that of the technical genres of SMC because the audience is a broad, general audience that often lacks a strong educational foundation in science or medicine (Dunwoody and Ryan 1985; Durant, Evans, and Thomas 1989; Millar and Wayne 1988). As Jeane Fahnestock (1986) of the University of Maryland cautioned, "Accommodating the scholarly piece for the nonscholarly magazine [or other publication] is not, therefore, simply a matter of translating technical jargon into nontechnical equivalents" (280). Rather, you should adapt the content to a general audience (Doak and Doak 2010; Fahnestock 1986; Rowan 1991) by:

- Removing unnecessary jargon and defining unfamiliar terms,
- Presenting general concepts and limiting technical details,
- Interpreting the meaning for the reader,
- Including visual diagrams (if possible), and
- Not requiring or expecting the reader to learn or remember technical details.

Furthermore, to facilitate communication with general audiences, you should adhere to the principles of plain language (Doak and Doak 2010; Locke 2003; Nancekivell 2010) by:

- Using common, everyday language in a conversational tone,
- Using the active voice,
- Keeping sentences short (averaging 15–20 words each),
- Keeping paragraphs short (limit to a few sentences each), and
- Using familiar, parallel, and predictable structure for sentences.

These guidelines will facilitate communication with general audiences, who are only casually reading science news articles and typically will only attend to or understand general concepts.

When communicating science and medicine to a general audience, you also need to attend to two other issues. First, the use of analogies and examples are helpful in relating abstract concepts to the everyday lives of the audience (Rowan 1991). Second, general audiences (as well as journalists) have difficulty understanding the inherent uncertainty in science (Cohn 1989). We will examine each of these issues in a bit more detail.

General audiences often rely on incorrect assumptions based on their familiar experiences that prevent understanding of the importance or relevance of complex issues. To properly communicate concepts, then, authors of science communication should use an analogy or metaphor that the general public can understand (Baake 2003; Rowan 1991). For example, Katherine Rowan (1991) of

Purdue University noted that a general population may not fully comprehend the issue of hitting a car windshield at 30 miles an hour. To properly communicate this concept, science writers should inform a general audience that hitting a car windshield at 30 miles an hour is like falling head first from a three-story building. In this example, the important information (e.g., harm to the individual) is communicated through an analogy that the audience can conceptualize.

The inherent uncertainty in science, which is qualified in scientific communication with hedging statements (see Chapter 8), is a confusing style for general audiences (Cohn 1989). Often, research conclusions or speculative applications of the research are presented to general audiences without extensive hedging statements, so the research appears to be reported with a high level of certainty (Fahnestock 1986). As a scientific writer, you must carefully monitor the degree of certainty to reflect the research accurately without misleading the audience. As Ken Baake (2003), a scholar of science communication and former journalist state, science writing is a "balancing act" (150) between presenting the research thoroughly and precisely at the level of detail and presenting the information clearly to the audience. In the generalization or translation of information from an expert to a general audience, the information loses precision and a slight degree of accuracy (see Example 14.4). Therefore, you must be extremely attentive to the claims and diction in order to communicate science properly to a general audience.

EXAMPLE 14.4 CONVERSION OF SCIENTIFIC INFORMATION FROM A JOURNAL ARTICLE TO A NEWS ARTICLE.

The process of generalizing from specific observations results in a loss of precision and a slight loss of accuracy.

Source	Audience	Excerpt[1]
Journal article	Subject-matter experts	"two conventionally studied groups of genes: 47 allozyme (allelic isozyme) loci and 155 soluble proteins resolved by two-dimensional gel electrophoresis. . . . The entire cheetah sample was invariant at each of the 47 loci."
Science magazine	General scientists	"all the cheetahs carried exactly the same form of every one of the 47 enzymes. . . . In another test of more than 150 proteins, 97% of them matched the cheetahs."

[1] *Source*: Fahnestock (1986).

Finally, when writing a press release, you should use the Associated Press (AP) style conventions for grammar, punctuation, and spelling so that journalists and editors do not have to do much (if any) revision. This will help facilitate your communication to the general public. For more information on AP style, visit http://www.apstylebook.com.

Distributing a Press Release

After you have written a press release, the final step is to distribute it to the news media. Press releases can be distributed to journalists and editors via email, fax, or sometimes post mail. Most large organizations have a public relations individual or department that coordinates the writing and distribution of press releases and acts as the liaison between the news media and others within the organization. In smaller organizations, the responsibility for dissemination of a press release may fall upon the writer (Lofstrom 2010). As a writer, you can create a list of contacts by finding contact information of local and topic-specific publications on the internet. If you don't have the time to research media contacts, other companies provide press release-distribution services for a fee. Such organizations include businesswire.com and prnewswire.com.

Typically a press release should be coordinated with the publication of a journal article (Committee on Publication Ethics 1999; Fontanarosa and DeAngelis 2002). A press release may be placed under an embargo, which restricts publication of the story until a specified date. An embargoed press release is provided to the media days (and sometimes a few weeks) in advance of the peer-reviewed publication. This early announcement to news organizations enables journalists to write and develop their own story—and emphasize the timeliness of research findings by coinciding with the publication of research in the peer-reviewed literature.

Concluding Points

Although press releases fall more into the field of marketing and public relations, we covered press releases in our discussion of scientific communication because you, as a technical expert, may be asked to help develop a press release. When writing a press release, the important thing to remember is that the audience is much different than scientific audiences and requires more than a superficial translation of technical jargon. For the press release, you have to take a step back from the research and look at the forest rather than the trees.

APPENDIX A: TYPESETTING GREEK LETTERS AND OTHER SYMBOLS IN MICROSOFT WORD

This table provides the Greek alphabet and instructions for inserting these letters in Microsoft Word. If the character you need does not appear in the table, then it is not a Greek letter. But you can still insert the character in a manuscript using Microsoft Word by selecting the following menus: **Insert** > **Symbol** > **Symbol Browser.** A menu will appear and you can select the desired symbol (which also includes accent, math, and currency symbols). Some journals specify that all symbols must be inserted through the Symbol Browser. Such requirements are stated in the Instructions to Authors.

Name	**Greek Letter**		**English Letter in Microsoft Word**[1]
	Capital	**Lowercase**	
Alpha	Α	α	A
Beta	Β	β	B
Gamma	Γ	γ	G
Delta	Δ	δ	D
Epsilon	Ε	ε	E
Zeta	Ζ	ζ	Z
Eta	Η	η	H
Theta	Θ	θ	Q

Name	Greek Letter		English Letter in Microsoft Word[1]
	Capital	Lowercase	
Iota	Ι	ι	I
Kappa	Κ	κ	K
Lambda	Λ	λ	L
Mu	Μ	μ	M
Nu	Ν	ν	N
Xi	Ξ	ξ	X
Omicron	Ο	ο	O
Pi	Π	π	P
Rho	Ρ	ρ	R
Sigma	Σ	σ	S
Tau	Τ	τ	T
Upsilon	ϒ	υ	U
Phi	Φ	φ	F
Chi	Χ	χ	C
Psi	Ψ	ψ	Y
Omega	Ω	ω	W

[1] For the Greek letter, insert the English letter in Microsoft Word and then change the typeface to "Symbol." For Greek capital letters (e.g., Σ), use uppercase English letters (e.g., S); for lowercase Greek letters (e.g., σ), use lowercase English (e.g., s).

APPENDIX B: GENERAL FORMATTING REQUIREMENTS FOR RESEARCH MANUSCRIPTS

These general formatting requirements for manuscripts should be used to supplement the Instructions to Authors provided by journals and the guidelines provided by the style manuals of professional organizations. In other words, the following list of typical specifications serves to fill in the gaps of formatting requirements that may be omitted in a particular Instructions to Authors document.

This list of formatting requirements provides a relative consensus of prominent styles, assembled from the following style manuals and journals: *AMA Manual of Style: A Guide for Authors and Editors (10th ed.); The ACS Style Guide: Effective Communication of Scientific Information (3rd ed.)*; *Cell*; *Chest*; *Nature*; *Journal of Bacteriology*; *Publication Manual of the American Psychological Association (6th ed.)*; *Recommendations for the Conduct, Reporting, Editing and Publication of Scholarly Work in Medical Journals* (previously the *Uniform Requirements for Manuscripts*); *Science*; and *Scientific Style and Format: The CSE Manual for Authors, Editors, and Publishers.*

Refer to the tips (shaded boxes) for instructions to avoid common manuscript errors and access special features in Microsoft Word, and for notes regarding less common conventions.

	Typical Specifications
Page setup	8.5 by 11 inches (US Letter)
Margins	1 to 1.5 inches (default settings in Microsoft Word are usually acceptable)
Typeface	12-point Times (or Symbol for special characters). Some journals may accept Unicode characters for author's names in non-Latin languages **TIP**: For more information on the use of special characters, refer to the section Scientific Nomenclature, Abbreviations, and Acronyms in Chapter 5.

	Typical Specifications
Line spacing	Double spaced (all sections, including title, headings, subheadings, footnotes, quotations, tables, figure captions, and references)
Text alignment	Left justify all text (ragged right). Possible exceptions include the following: title, headings/subheadings, and alignment of text or data in tables.
Headings and subheadings	In full-length journal articles, the IMRAD section typically establishes the first-level headings (Introduction, Methods, Results, and Discussion). The Abstract, Acknowledgments, and References are added as first-level headings. Some journals add a Conclusions section following the Discussion. Also, some journals exclude the Introduction heading (since the Introduction of the text usually begins on a separate page of the manuscript). Shorter articles are usually not divided by headings or subheadings. To ensure proper division of first-level headings, refer to the Instructions to Authors and recent articles from the journal. Use the exact name of the sections in the target journal (such as "Methods" versus "Materials and Methods"). Also note singular or plural use of names (such as "Conclusion" versus "Conclusions"). Many journals encourage use of subheadings to organize material in the Methods, Results, and Discussion sections. Authors are permitted to define and name subheadings below the first-level headings. Often journals require that sections contain at least two subheadings to be divided. The number of subheadings within a manuscript should be limited to two or three levels so that readers can easily interpret the hierarchical structure (although some style guides permit up to five levels to accommodate unusual cases). Each heading level should be formatted consistently and distinctly from the other levels, as shown below:

Heading Level	Example[1]
1	**Centered, Boldfaced, Uppercase and Lowercase Letters**
2	**Left Justified, Boldfaced, Uppercase and Lowercase Letters**
3	**Indent, boldfaced, lowercase paragraph heading that ends with a period.** Following the title, the paragraph text begins without bold text (just like this).
4	***Indent as third level but also italicized.***
5	*Indent as fourth level but not in boldfaced font.*

[1] *Source*: American Psychological Association (2010).

Continued

	Typical Specifications
Indentation	Indent the first line of each paragraph and footnote by approximately 0.5 inches. Sections and elements that are typically not indented include the Abstract, block quotes, titles, headings, and captions. **TIP:** Use the paragraph indent option located in the format menu in Microsoft Word (indented 0.5 inches and the first line selected under options). In contrast, the Reference section is usually formatted with hanging indentation, in which the first line extends to the left-hand margin and the remainder of the reference is indented. **TIP:** To create hanging indentation in Microsoft Word, select Paragraph from the Format menu. A dialogue box will appear and you can select Hanging from the Special option under the Indentation section.
Page numbers and running heads	Often required: Typically the entire manuscript is numbered consecutively beginning with the title page. Page numbers are usually integrated into the running head (which is an abbreviated title of the article and often limited to approximately 50 characters). **TIP:** To create a running head in Microsoft Word, select Header and Footer from the View menu. (The document must be in Print Layout, which also can be selected from the View menu.) When the header is visible, you can type the running head. You can insert page numbers by selecting Page Numbers from the Insert menu.
Footnotes	Often discouraged by publishers but may be used sparingly. Refer to the Instructions to Authors. **TIP:** To insert a footnote in Microsoft Word, place the cursor at the location to insert the note in the text. Then select Footnote from the Insert menu. A dialogue box will appear and allow you to insert footnotes. Journals that accept footnotes often prefer the note to be placed at the bottom of the page rather than creating an endnote that appears at the end of the document, but refer to the Instructions to Authors.
Line numbers	Occasionally requested **TIP:** To insert line numbers in Microsoft Word, select the Layout tab from the toolbar, then locate Line Numbers in the Text Layout box and select Continuous. (Document must be in Print Layout, which can be selected from the View menu.)

	Typical Specifications
Tables and figures (and figure captions)	Number tables and figures consecutively in order of first citation (using Arabic numbers). Tables and figures are numbered independently, so you may have both a Table 1 and a Figure 1. Any artwork that is not a table is considered a figure. The submission format and placement of tables, figures, and figure captions varies by journal: Some journals request that tables and figures be inserted at the end of the document (after the References section), some journals request that tables and figures be uploaded as separate files, and a few journals permit integration of tables and figures in the manuscript following the paragraph with the initial reference to the respective table or figure. The figure caption usually accompanies the figure or is entered into an online content-management system during submission of the manuscript (see Chapter 10). **TIP**: Do not create figures in Microsoft Word or PowerPoint. Most journals do not accept images created in those programs.
File names	Usually not specified since file names are changed during the upload process to content management systems (along with removal of identifying information in file metadata). But, some journals do specify naming conventions. In the absence of a naming convention, you may decide to use the convention specified by the journal *Science* in the Instructions to Authors: All files should begin with the last name of the first author, then identify the content of the file (such as "text" for the primary text file, "fig 1" for the first figure, or "supp" for a supplemental file). All file titles should end with the appropriate file extension (such as ".doc"). In this case, a primary manuscript file with Smith as the first author would be "smithtext.doc" and the first figure would be "smithfig1.eps."
File size limits	Varies; may be limited to 8–100 MB for manuscript files within the content management system. Most journals can accommodate large files, which is occasionally necessary for supplemental data files.

APPENDIX C: GLOSSARY OF TERMS TO DESCRIBE MEDICAL RESEARCH STUDIES

Source: Ad Hoc Working Group for Critical Appraisal of the Medical Literature (1987).

BEFORE-AFTER TRIAL investigation of therapeutic alternatives in which patients of one time period and under one treatment are compared with patients at a subsequent time who are treated in a different fashion. If the disorder is not fatal and the "before" treatment is not curative, the same patients may be studied in the before and after periods, strengthening the design through increased group comparability for the two periods.

BLIND or BLINDED masked, unaware. The term may be modified according to the purpose of the blinding. For example, clinicians or patients can be blinded to the treatments that patients are receiving, or observers can be blinded to each other's assessments, making their observations uninfluenced by one another (see also DOUBLE BLIND). To avoid confusion, the term MASKED is preferred in studies in which vision loss of patients is an outcome of interest.

CASE-CONTROL STUDY (CASE-REFERENT or CASE-COMPARISON STUDY) investigation generally used to test possible causes of a disease or disorder, in which persons with and without a designated disorder are compared with respect to prior or current exposure to a putative causal factor. For example, persons with hepatic cancer (cases) are compared with persons without hepatic cancer (controls), and the history of hepatitis B is determined for the two groups. A case-control study is often referred to as a RETROSPECTIVE STUDY (even if patients are recruited prospectively) because the logic of the design leads from effect to cause.

CASE SERIES a series of patients with a defined disorder. Usually, the term is used to describe a study reporting on a consecutive collection of patients treated in a similar manner, without a concurrent control group. For example, a surgeon might describe the characteristics of, and outcomes for, 100 consecutive patients with cerebral ischemia who received a revascularization procedure. See also CONSECUTIVE SAMPLE.

COHORT a group of persons with a common characteristic or set of characteristics. Typically, the group is followed for a specified period to determine the incidence of a disorder or complications of an established disorder (that is, prognosis), as in cohort study (prospective study). See also INCEPTION COHORT.

COHORT ANALYTIC STUDY prospective investigation of the factors that might cause a disorder. A cohort of persons who do not have evidence of an outcome of interest but who are exposed to the putative cause are compared with a concurrent cohort who are also free of the outcome but not exposed to the putative cause. Both cohorts are then followed to compare the incidence of the outcome of interest.

CONSECUTIVE SAMPLE sample in which the units are chosen on a strict "first come, first chosen" basis. All persons who are eligible should be included as they are seen.

CONVENIENCE SAMPLE persons or groups selected at the convenience of the investigator or primarily because they were available at a convenient time or place.

COST-BENEFIT ANALYSIS a form of economic assessment, usually from society's perspective, in which the costs of medical care are compared with the economic benefits of the care, with both costs and benefits expressed in units of currency. The benefits typically include reductions in future healthcare costs and increased earnings due to the improved health of those receiving the care.

COST-EFFECTIVENESS ANALYSIS an economic evaluation in which alternative programs, services, or interventions are compared in terms of the cost per unit of clinical effect: for example, cost per life saved, cost per millimeter of mercury of blood pressure lowering, or cost per quality-adjusted life-year gained. The cost per quality-adjusted life-years (and equivalents such as "healthy days of life gained") gives rise to what is also referred to as COST-UTILITY ANALYSIS.

COST-UTILITY ANALYSIS see COST-EFFECTIVENESS ANALYSIS.

CRITERION STANDARD a method having established or widely accepted accuracy for determining a diagnosis, providing a standard to which a new screening or diagnostic test can be compared. The method need not be a single or simple procedure but could include follow-up of patients to observe the evolution of their conditions or the consensus of an expert panel of clinicians, as is frequently used in the study of psychiatric conditions. Criterion standard can also be used in studies of the quality of care to indicate a level of performance, agreed to by experts or peers, to which the performance of individual practitioners or institutions can be compared.

DOUBLE BLIND or DOUBLE MASK neither the subject nor the study staff (those responsible for patient treatment and data collection) is aware of the group or intervention to which the subject has been assigned. Also, the term can indicate any condition in which two different groups of persons are purposely denied access to information in order to keep that information from influencing some measurement, observation, or process.

Economic Evaluation comparative analysis of alternative courses of action in terms of both their costs and consequences.

Gold Standard see CRITERION STANDARD.

Inception Cohort a designated group of persons, assembled at a common time early in the development of a specific clinical disorder (for example, at the time of first exposure to the putative cause or at the time of initial diagnosis) who are followed thereafter. See also COHORT.

Likelihood Ratio for a screening or diagnostic test (including clinical signs or symptoms), expresses the relative odds that a given test result would be expected in a patient with (as opposed to one without) a disorder of interest.

Masked see BLIND.

Nonrandomized Controlled Trial experiment in which assignment of patients to the intervention groups is at the convenience of the investigator or according to a preset plan that does not conform to the definition of random. See also RANDOMIZED CONTROLLED TRIAL.

Primary Care medical care provided by the clinician of first contact for the patient. Typically, the primary care physician is a general practitioner, family practitioner, internist, or pediatrician. Primary care may also be administered by health professionals other than physicians, notably, specially trained nurses (nurse practitioners) and paramedics. Usually, a general practitioner, family practitioner, nurse practitioner, or paramedic provides only primary care services, but a person with specialty qualifications may provide primary care, alone or in combination with referral services (see also REFERRED CARE). Thus, it is the nature of the contact (first as opposed to referred) that determines the care designation, not the qualifications of the practitioner.

Primary Care Center, Primary Care Setting medical care facility that offers first-contact healthcare only. Patients requiring specialized medical care are referred elsewhere. Some primary care centers provide a mixture of primary and referred care. Thus it is the nature of the service provided (first contact), rather than the setting per se, that distinguishes primary from more advanced levels of care. See also PRIMARY CARE, REFERRED CARE, and TERTIARY CARE CENTER.

Prospective Study see COHORT and COHORT ANALYTIC STUDY.

Random governed by a formal chance process in which the occurrence of previous events is of no value in predicting future events. The probability of assignment of, for example, a given person to a specified treatment group is fixed and constant (typically 0.50), but the person's actual assignment cannot be known until it occurs.

Random Sample a sample derived by selecting sampling units (for example, individual patients) such that each unit has an independent and fixed (generally equal) chance of selection. Whether or not a given unit is selected is determined by chance, for example, by a table of randomly ordered numbers.

Randomization, Random Allocation allocation of persons to groups by chance, usually done with the aid of a table of random numbers. Not to be confused with systematic allocation (for example, on even and odd days of the month) or allocation at the convenience or discretion of the investigator.

Randomized Trial (Randomized Controlled Trial, Randomized Clinical Trial) experiment in which subjects are randomly allocated to receive or not receive an experimental preventive, therapeutic, or diagnostic procedure and then are followed to determine the effect of the intervention.

Referred Care medical care provided to a patient when referred by one health professional to another with more specialized qualifications or interests. The two levels of referred care are secondary and tertiary. Secondary care is usually provided by a broadly skilled specialist, such as a general surgeon, general internist, or obstetrician. Tertiary care is provided on referral of a patient to a subspecialist, such as an orthopedic surgeon, neurologist, or neonatologist. See also TERTIARY CARE CENTER.

Retrospective Study see CASE-CONTROL STUDY.

Secondary Care see REFERRED CARE.

Sensitivity for a diagnostic or screening test, the proportion of people who truly have a designated disorder who are so identified by the test. The test may consist of, or include, clinical observations.

Sequential Sample see CONSECUTIVE SAMPLE.

Specificity for a diagnostic or screening test, the proportion of people who are truly free of a designated disorder who are so identified by the test. The test may consist of, or include, clinical observations.

Survey observational or descriptive, nonexperimental study in which persons are systematically examined for the absence or presence (or degree of presence) of characteristics of interest.

Tertiary Care see REFERRED CARE.

Tertiary Care Center a medical facility that receives referrals from both primary and secondary care levels and usually offers tests, treatments, and procedures not available elsewhere. Most tertiary care centers offer a mixture of primary, secondary, and tertiary care services so that it is the specific level of service rendered, rather than the facility, that determines the designation of care in a given study. See also REFERRED CARE.

APPENDIX D: CREATING TABLES IN MICROSOFT WORD

This appendix, provides an overview to creating publication-quality tables in Microsoft Word using the table editor.

Stage 1: Determining the Table Size

Before you begin creating a table in Microsoft Word (or any other program), you need to determine the number of rows and columns that the final table will contain. Although you can modify the number of rows and columns after creating the table, such changes will require you to spend a bit of time readjusting the table. Therefore, you may want to hand sketch the table first. The table in our example has 6 columns (Example 7.13, see Chapter 7):

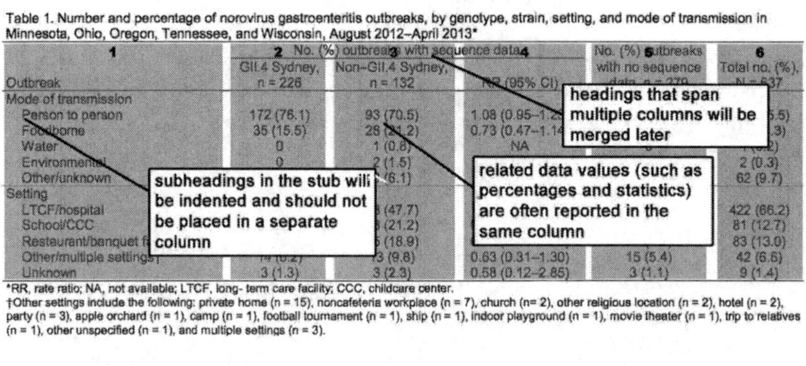

The table in this example has 14 rows:

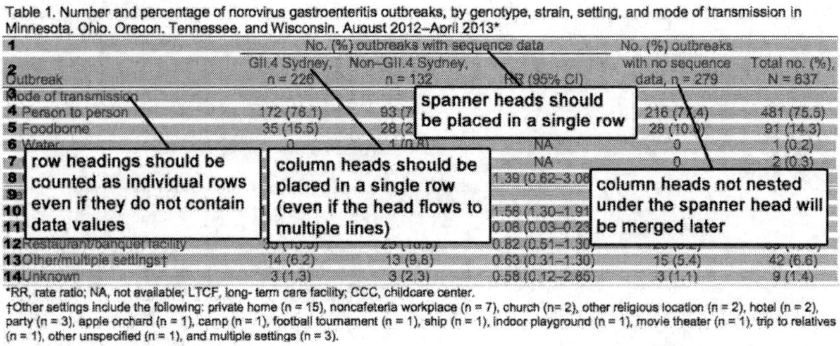

Stage 2: Making a Basic Table

To create a table, place your cursor in the document where you want the table to be inserted.

1. From the **Table** menu, select **Insert** > **Table.**

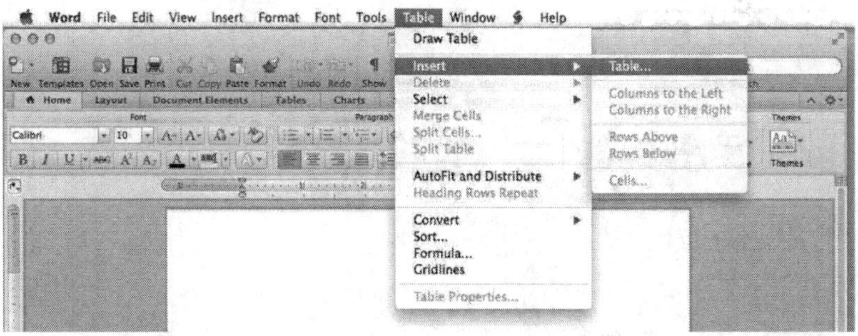

2. In the **Insert Table** dialogue box, enter the **number of columns** and **number of rows.**

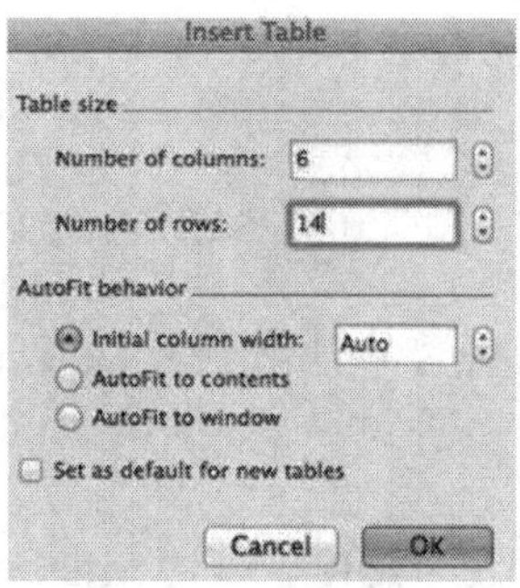

3. Click **OK.**

Stage 3: Creating the Boxhead

To nest columns under a spanning head and adjust the columns on either side of the spanning head, you need to merge cells in the top two rows (see the following illustration).

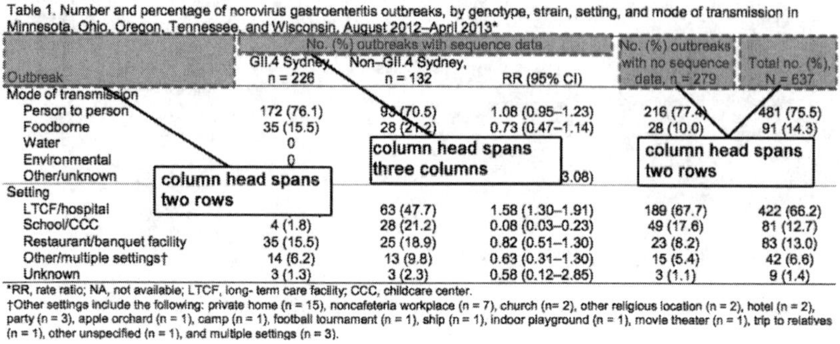

Table 1. Number and percentage of norovirus gastroenteritis outbreaks, by genotype, strain, setting, and mode of transmission in Minnesota, Ohio, Oregon, Tennessee, and Wisconsin, August 2012–April 2013*

Outbreak	No. (%) outbreaks with sequence data			No. (%) outbreaks with no sequence data, n = 279	Total no. (%), N = 637
	GII.4 Sydney, n = 226	Non–GII.4 Sydney, n = 132	RR (95% CI)		
Mode of transmission					
Person to person	172 (76.1)	93 (70.5)	1.08 (0.95–1.23)	216 (77.4)	481 (75.5)
Foodborne	35 (15.5)	28 (21.2)	0.73 (0.47–1.14)	28 (10.0)	91 (14.3)
Water	0				
Environmental	0				
Other/unknown			3.08)		
Setting					
LTCF/hospital		63 (47.7)	1.58 (1.30–1.91)	189 (67.7)	422 (66.2)
School/CCC	4 (1.8)	28 (21.2)	0.08 (0.03–0.23)	49 (17.6)	81 (12.7)
Restaurant/banquet facility	35 (15.5)	25 (18.9)	0.82 (0.51–1.30)	23 (8.2)	83 (13.0)
Other/multiple settings†	14 (6.2)	13 (9.8)	0.63 (0.31–1.30)	15 (5.4)	42 (6.6)
Unknown	3 (1.3)	3 (2.3)	0.58 (0.12–2.85)	3 (1.1)	9 (1.4)

*RR, rate ratio; NA, not available; LTCF, long- term care facility; CCC, childcare center.

†Other settings include the following: private home (n = 15), noncafeteria workplace (n = 7), church (n= 2), other religious location (n = 2), hotel (n = 2), party (n = 3), apple orchard (n = 1), camp (n = 1), football tournament (n = 1), ship (n = 1), indoor playground (n = 1), movie theater (n = 1), trip to relatives (n = 1), other unspecified (n = 1), and multiple settings (n = 3).

To merge cells, do the following:

1. Select, or highlight, the cells to be merged (see below).
2. From the **Table** menu, select **Merge Cells.**

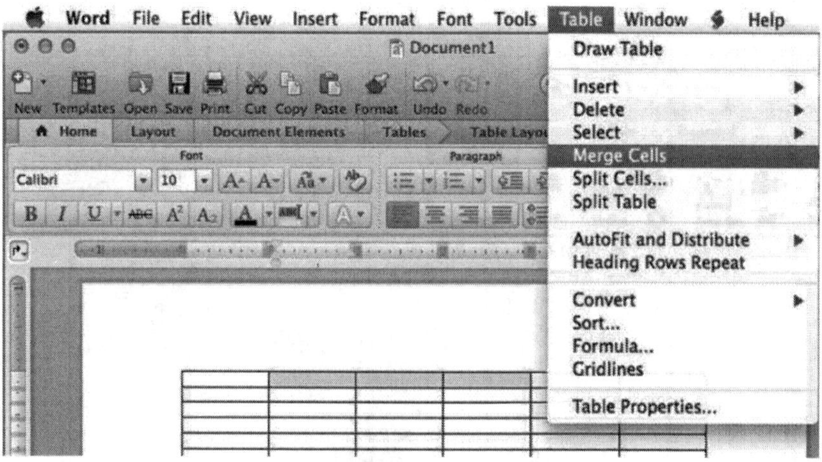

The cells you have highlighted will be merged into a single cell. This will allow you to input content that spans multiple columns or rows.

Stage 4: Adjusting the Table to Fit the Content

After you have entered data into the table, you can adjust the width of the columns to best fit the content. Although you may do this manually, the easiest way to adjust the margins is to set them to automatically fit to the content.

1. Select the entire table (see below).
2. From the **Table** menu, select **AutoFit and Distribute** > **AutoFit to Contents.**

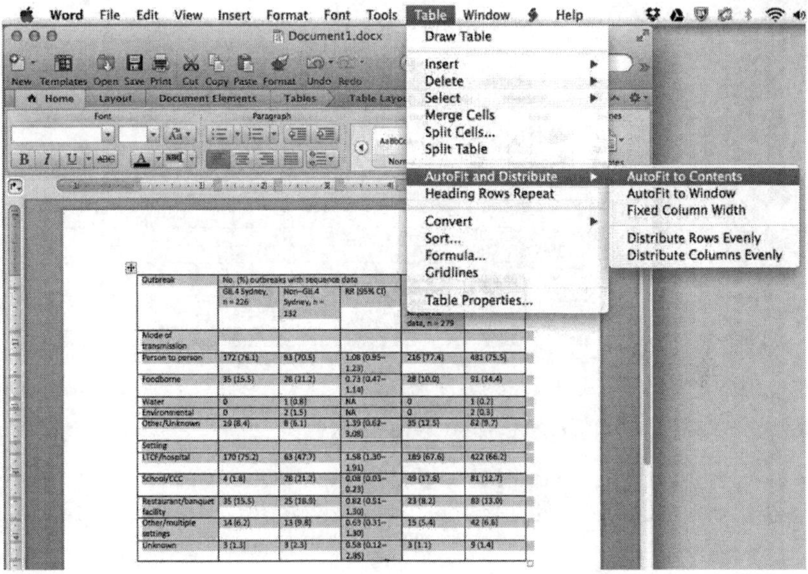

Microsoft Word will automatically adjust the width of the columns to best fit the content. You can further adjust any column width by selecting the column and then dragging the vertical border to the left or right.

Stage 5: Indenting Subheadings in Rows

If your table contains subheadings, you will need to indent these subheadings in the rows. Since the tab key advances the cursor to the next cell, you cannot use the tab key. The easiest way to indent subheadings, however, is to adjust the left-hand margin from the ruler, as summarized below:

1. Select the cells that you want to adjust (see following page).
2. From the **Ruler,** select the **left margin icon**. While holding the mouse button, slide the margin to the desired position.
3. Release the button on the mouse to set the indentation.

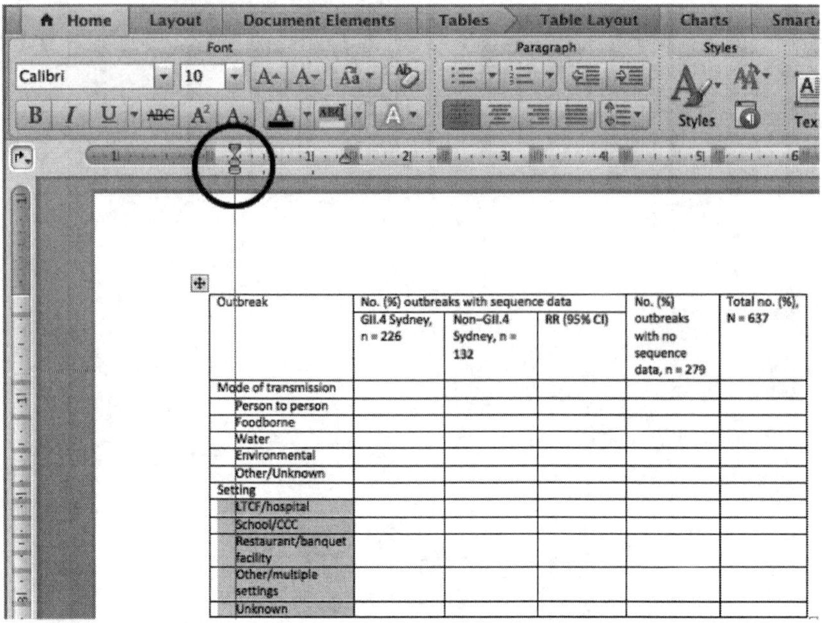

Stage 6: Aligning Content in the Cells

Content in each cell can be aligned horizontally by selecting the cell and using the justification buttons (left, center, or right) in the Home toolbar (see below).

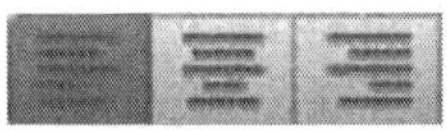

Alternatively, to adjust the content both horizontally and vertically, you can use the text alignment feature in the Table Layout toolbar.

1. Select the cell that you want to align. (You can also select multiple cells, columns, or rows.)
2. From the **Table Layout** toolbar, click on the **Align** button.
3. From the drop-down menu, select the desired alignment.

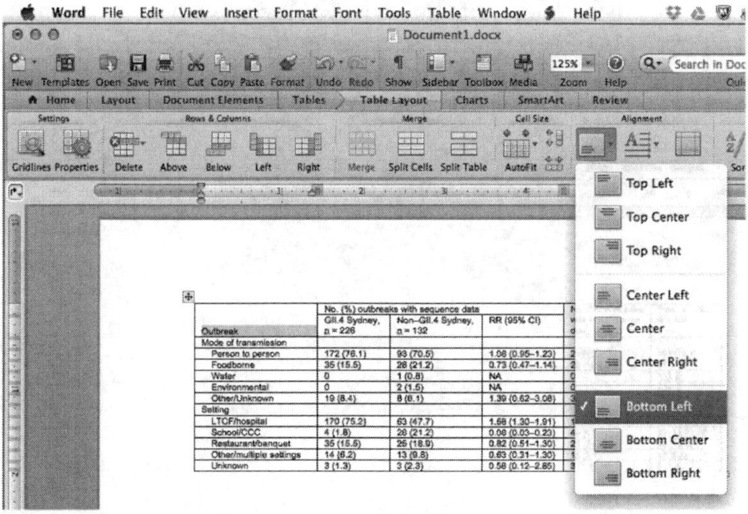

Stage 7: Formatting the Rule Lines of the Table

You can modify the visible rule lines of the table by using the **Borders** menu in the **Table** toolbar. Since professional tables typically have only a few rule lines, the easiest way to begin is to erase all the visible rule lines and then selectively add lines.

1. Begin by selecting all of the cells (the entire table, as in the following illustration).
2. From the **Table** toolbar, click on the **Borders** button.
3. Change the **Borders** to **None**. (This removes all lines from the table.)

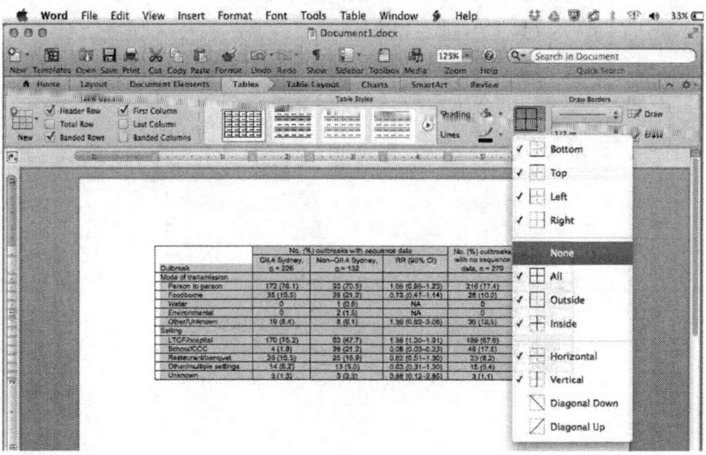

4. Select the cell (or row or column) to add a border. (In the following illustration, the bottom row of the table was selected to add a horizontal rule line.)
5. From the **Table** toolbar, click on the **Borders** button.
6. Select the **Border** that you want to add (in this case, **Bottom**).

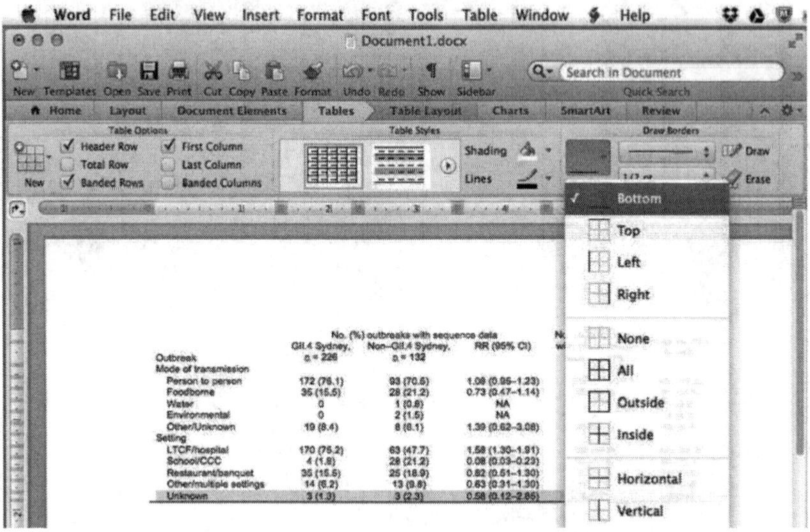

7. Continue adding horizontal rule lines to complete your table.

APPENDIX E: COMMON MEASUREMENTS AND INTERNATIONAL SYSTEM UNITS

TABLE E.1 Common measurements, or base units, for various quantities.

Quantity	Unit	Abbreviation
Mass	Kilogram*	Kg
Length	Meter**	M
Time	Second	S
Temperature	Kelvin	K
Amount of Substance	Mole	Mol
Volume	Liter	L

* Kilogram is the base unit for mass, but attach a prefix (see Table E.2) to "gram" to make larger or smaller units.

** "Metre" is the internationally accepted spelling, except in the United States, where "meter" is the common spelling.

TABLE E.2 International System (SI) Units. Metric units are the standard units of measurement for scientific data. A standard set of prefixes in multiples of 10 may be used to derive larger and smaller units; however, prefixes for multiples of 1,000 are the most commonly used prefixes.

10^n	Prefix	Symbol	Decimal Equivalent (in SI format)
10^{24}	Yotta-	Y	1 000 000 000 000 000 000 000 000
10^{21}	Zetta-	Z	1 000 000 000 000 000 000 000
10^{18}	Exa-	E	1 000 000 000 000 000 000
10^{15}	Peta-	P	1 000 000 000 000 000
10^{12}	Tera-	T	1 000 000 000 000
10^{9}	Giga-	G	1 000 000 000
10^{6}	Mega-	M	1 000 000
10^{3}	Kilo-	k	1 000
10^{2}	Hecto-	h	100
10^{1}	Deca-	da	10
10^{0}	(None)	(none)	1
10^{-1}	Deci-	d	0.1
10^{-2}	Centi-	c	0.01
10^{-3}	Milli-	m	0.001
10^{-6}	Micro-	μ	0.000 001
10^{-9}	Nano-	n	0.000 000 001
10^{-12}	Pico-	p	0.000 000 000 001
10^{-15}	Femto-	f	0.000 000 000 000 001
10^{-18}	Atto-	a	0.000 000 000 000 000 001
10^{-21}	Zepto-	z	0.000 000 000 000 000 000 001
10^{-24}	Yocto-	y	0.000 000 000 000 000 000 000 001

REFERENCES

Ad Hoc Working Group for Critical Appraisal of the Medical Literature. 1987. "A Proposal for More Informative Abstracts of Clinical Articles." *Annals of Internal Medicine* 106 (4): 598–604.

Adams Smith, Diana E. 1983. "Style in Medical Journals." *BMJ* 287 (6399): 1122–4.

Al-Benna, Sammy, Prachi Rajgarhia, Safraz Ahmed, and Zeeshan Sheikh. 2009. "Accuracy of References in Burns Journals." *BURNS* 35 (5): 677–80.

Albert, Karen M. 2006. "Open Access: Implications for Scholarly Publishing and Medical Libraries." *Journal of the Medical Library Association* 94 (3): 253–62.

Albert, Tim. 2000. "Writing for Journals: A Paradigm Lost?" *Journal of Epidemiology and Community Health* 54 (9): 642–3.

Alexander, Lori L., and Sue Hudson. 2008. "Ghostbusting." *AMWA Journal* 23 (3): 54–5.

Allen, Elena A., Erik B. Erhardt, and Vince D. Calhoun. 2012. "Data Visualization in the Neurosciences: Overcoming the Curse of Dimensionality." *Neuron* 74 (4): 603–8.

Alred, Gerald J., Walter E. Oliu, and Charles T. Brusaw. 1992. *The Professional Writer: A Guide for Advanced Technical Writing*. New York: St. Martin's Press.

Altman, Douglas G., Steven N. Goodman, and Sara Schroter. 2002. "How Statistical Expertise Is Used in Medical Research." *JAMA* 287 (21): 2817–20.

Altman, Douglas G., and David Moher. 2014. "Importance of Transparent Reporting of Health Research." In *Guidelines for Reporting Health Research: A User's Manual*, edited by David Moher, Douglas G. Altman, Kenneth F. Schulz, Iveta Simera, and Elizabeth Wager, 3–13. Oxford, UK: Wiley & Sons.

Amann, K., and K. Knorr-Cetina. 1990. "The Fixation of (Visual) Evidence." In *Representation in Scientific Practice*, edited by Michael Lynch and Steve Woolgar, 85–121. Cambridge, MA: MIT Press.

Amdur, Robert J., Jessica Kirwan, and Christopher G. Morris. 2010. "Use of the Passive Voice in Medical Journal Articles." *American Medical Writers Association Journal* 25 (3): 98–104.

American Chemical Society. 2015. *Newsworthy Highlights from 2015 Spring American Chemical Society National Meeting*. Accessed June 16, 2015. http://www.acs.org

/content/acs/en/pressroom/newsreleases/2015/march/newsworthy-highlights-from-2015-spring-acs-national-meeting.html.

American Psychological Association. 2010. *Publication Manual of the American Psychological Association*. 6th ed. Washington, DC: American Psychological Association.

American Society for Microbiology. 2014. *Instructions to Authors*. Washington, DC: American Society for Microbiology, 2014. http://jb.asm.org/site/misc/2014JBITA.pdf

———. 2015. "About ASM." Accessed June 15, 2015. http://gm.asm.org/index.php/abstract-submission/abstract-faq.

Angell, Marcia, and Jerome P. Kassirer. 1991. "The Ingelfinger Rule Revisited." *The New England Journal of Medicine* 325 (19): 1371–3.

Annesley, Thomas M. 2010. "The Discussion Section: Your Closing Argument." *Clinical Chemistry* 56 (11): 1671–4.

Anscombe, F. J. 1973. "Graphs in Statistical Analysis." *The American Statistician* 27 (1): 17–21.

Aoki, Naomi J., Joanne C. Enticott, and Louise E. Phillips. 2013. "Searching the Literature: Four Simple Steps." *Transfusion* 53 (1): 14–17.

Applewhite, Lottie. 1979. "Examination of the Medical/Scientific Manuscript." *Journal of Technical Writing and Communication* 9 (1): 17–25.

Arrington, Phillip, and Shirley K. Rose. 1987. "Prologues to What Is Possible: Introductions as Metadiscourse." *College Composition and Communication* 38 (3): 306–18.

Arsenault, Darin J., Laurence D. Smith, and Edith A. Beauchamp. 2006. "Visual Inscriptions in the Scientific Hierarchy: Mapping the 'Treasures of Science.'" *Science Communication* 27 (3): 376–428.

Audisio, Riccardo A., Rolf A. Stahel, Matti S. Aapro, Alberto Costa, Manoj Pandey, and Nicholas Pavlidis. 2009. "Successful Publishing: How to Get Your Paper Accepted." *Surgical Oncology* 18 (4): 350–6.

Avery, Oswald T., Colin M. MacLeod, and Maclyn McCarty. 1944. "Studies on the Chemical Nature of the Substance Inducing Transformation of Pneumococcal Types: Induction of Transformation by a Desoxyribonucleic Acid Fraction Isolated from Pneumococcus Type III." *Journal of Experimental Medicine* 79 (2): 137–58.

Awrey, Julianne, Kenji Inaba, Galinos Barmparas, Gustavo Recinos, Pedro G. R. Teixeira, Linda S. Chan, Peep Talving, and Demetrios Demetriades. 2011. "Reference Accuracy in the General Surgery Literature." *World Journal of Surgery* 35 (3): 475–9.

Baake, Ken. 2003. *Metaphor and Knowledge: The Challenges of Writing Science*. Albany, NY: SUNY Press.

Baethge, Christopher. 2012. "Impact Factor—A Useful Tool, but Not for All Purposes." *Deutsches Ärzteblatt International* 109 (15): 267–72.

Bailar, John C. III, and Frederick Mosteller. 1988. "Guidelines for Statistical Reporting in Articles for Medical Journals." *Annals of Internal Medicine* 108 (2): 266–73.

Bailey, Byron J. 2002. "Duplicate Publication in the Field of Otolaryngology-Head and Neck Surgery." *Otolaryngology—Head and Neck Surgery* 126 (3): 211–16.

Baker, John R. 1955. "English Style in Scientific Papers." *Nature* 176 (4488): 851–2.

Ball, Tamara. 2006. "Starting Out as a Medical Writer: How to Generate Writing Samples." *AMWA Journal* 21 (1): 26–7.

Bargiela-Chiappini, F., and C. Nickerson. 1999. "Business Writing as Social Action." In *Writing Business: Genres, Media and Discourses*, edited by F. Bargiela-Chiappini and C. Nickerson, 1–32. Harlow, UK: Pearson.

Bartels, Else Marie. 2009. “How to Keep Up with Medical Literature.” *Best Practice & Research Clinical Rheumatology* 23 (2): 281–90.

Bastide, Franciose. 1990. “The Iconography of Scientific Texts: Principles of Analysis.” In *Representation in Scientific Practice*, edited by Michael Lynch and Steve Woolgar, 187–229. Cambridge, MA: MIT Press.

Bates, Marcia J. 1989. “The Design of Browsing and Berrypicking Techniques for the Online Search Interface.” *Online Information Review* 13 (5): 407–24.

Bayer, Nancy L. 1991. “Scientific Visualization: New Opportunities for Technical Communicators.” *Technical Communication* 38 (2): 223–7.

Bazerman, Charles. 1983. “Scientific Writing as a Social Act: A Review of the Literature of the Sociology of Science.” In *New Essays in Technical and Scientific Communication: Research, Theory, Practice*, edited by Paul V. Anderson, John Brockmann, and Carolyn R. Miller, 156–84. Farmingdale, NY: Baywood.

———. 1984. “Modern Evolution of the Experimental Report in Physics: Spectroscopic Articles in *Physical Review*, 1893–980.” *Social Studies in Science* 14 (2): 163–96.

———. 1985. “Physicists Reading Physics: Schema-Laden Purposes and Purpose-Laden Schema.” *Written Communication* 2 (1): 3–23.

———. 1998. *Shaping Written Knowledge: The Genre and Activity of the Experimental Article in Science*. Madison: University of Wisconsin Press.

Ben-Chaim, Michael. 1996. “How Do Facts Speak for Themselves? The Doctrine and Practice of Classical Empiricism.” *Journal of Technical Writing and Communication* 26 (1): 3–19.

Benson, Philippa Jane. 1985. “‘Writing Visually’ Design Considerations in Technical Publications.” *Technical Communication* 32 (4): 35–9.

———. 1998. “Changing Moorings in Scientific Writing: Suggestions to Authors, Allusions for Teachers.” In *Essays in the Study of Scientific Discourse: Methods, Practice, and Pedagogy*, edited by John T. Battalio, 209–25. Stamford, CT: Ablex.

Benson, Philippa Jane, and Susan C. Silver. 2013. *What Editors Want: An Author’s Guide to Scientific Journal Publishing*. Chicago: University of Chicago Press.

Berkenkotter, Carol, and Thomas N. Huckin. 1995. *Genre Knowledge in Disciplinary Communication: Cognition/Culture/Power*. Hillsdale, NJ: Lawrence Erlbaum.

Bernhardt, Stephen A. 1985. “The Writer, the Reader, and the Scientific Text.” *Journal of Technical Writing and Communication* 15 (2): 163–74.

Bernstam, Elmer V., Jorge R. Herskovic, Yindalon Aphinyanaphongs, Constantin F. Aliferis, Madurai G. Sriram, and William R. Hersh. 2006. “Using Citation Data to Improve Retrieval from MEDLINE.” *Journal of the American Medical Informatics Association* 13 (1): 96–105.

Bertrand, Carol A. 2001. “Color by Number: Imaging Large Data.” *Science* 293 (5533): 1335–6.

Betterton-Lewis, Deana G. 2007. “Career Pathways in Medical Communication.” *AMWA Journal* 22 (1): 29–32.

Bhatia, Vijay. 1993. *Analysing Genre: Language Use in Professional Settings*. Harlow, UK: Addison Wesley Longman.

Billings, John Shaw. 1881. “Our Medical Literature.” *Transactions of the International Medical Congress, Seventh Session, Held in London, August 2nd to 9th, 1881*, Vol.1, 54–70.

Bishop, M. J. 2007. “Editorial.” *Briefings in Bioinformatics* 8 (6): 381.

Björk, Bo-Christer, Annikki Roos, and Mare Lauri. 2009. “Scientific Journal Publishing—Yearly Volume and Open Access Availability.” *Information Research* 14 (1): 391.

Björk, Bo-Christer, and David Solomon. 2012. "Open Access Versus Subscription Journals: A Comparison of Scientific Impact." *BMC Medicine* 10 (1): 73.

Blakeslee, Ann M. 1994. "The Rhetorical Construction of Novelty: Presenting Claims in a Letters Forum." *Science, Technology, & Human Values* 19 (1): 88–100.

Blanton, Jesse D., Nadine Y. Bowden, Millicent Eidson, Jeffrey D. Wyatt, and Cathleen A. Hanlon. 2005. "Rabies Postexposure Prophylaxis, New York, 1995–2000." *Emerging Infectious Diseases* 11 (12): 1921–7.

Block, A. Jay. 1998. "Blinded Reviews." *CHEST* 114 (6): 1501–2.

Block, Steven M. 1996. "Do's and Don'ts of Poster Presentation." *Biophysical Journal* 71 (6): 3527–29.

Bloemenkamp, D. G., H. C. Walvoort, W. Hart, and A. J. Overbeke. 1999. "Duplicate publication of articles in the Dutch Journal of Medicine in 1996." *Nederlands Tijdschrift voor Geneeskunde* 143 (43): 2150–3.

BMJ. n.d. "Fast Track Publication." Accessed March 20, 2014. http://www.bmj.com/about-bmj/resources-authors/fast-track-publication.

———. 2007. "BMJ Online Questionnaire." Accessed October 5, 2012. http://www.bmj.com/about-bmj/visitor-statistics/bmj-online-questionnaire-july-2007.

Boellstorff, Tom. 2011. "Submission and Acceptance: Where, Why, and How to Publish Your Article." *American Anthropologist* 113 (3): 383–8.

Boers, Maarten. 2004. "Null Bar and Null Zone Are Better Than the Error Bar to Compare Group Means in Graphs." *Journal of Clinical Epidemiology* 57 (7): 712–5.

Boss, Jeremy M., and Susan H. Eckert. 2003. "Academic Scientists at Work: Publishing at the Top of the Heap." *Science Careers*, July 11. Accessed July 20, 2013. http://sciencecareers.sciencemag.org/career_magazine/previous_issues/articles/2003_07_11/nodoi.5314906010509081373.

Boutron, Isabelle, Susan Dutton, Philippe Ravaud, and Douglas G. Altman. 2010. "Reporting and Interpretation of Randomized Controlled Trials with Statistically Nonsignificant Results for Primary Outcomes." *JAMA* 303 (20): 2058–64.

Brand, Richard A., and Rik Huiskes. 2001. "Structural Outline of an Archival Paper for the *Journal of Biomechanics*." *Journal of Biomechanics* 34 (11): 1371–4.

Bredan, Amin S., and Frans Van Roy. 2006. "Writing Readable Prose." *EMBO Reports* 7 (9): 846–9.

Bryan, John. 1995. "Seven Types of Distortion: A Taxonomy of Manipulative Techniques Used in Charts and Graphs." *Journal of Technical Writing and Communication* 25 (2): 127–79.

Buchan, John C., John Norris, and Hannah Kuper. 2005. "Accuracy of Referencing in the Ophthalmic Literature." *American Journal of Ophthalmology* 140 (6): 1146–8.

Buehler, Mary Fran. 1977. "Report Construction: Tables." *IEEE Transactions on Professional Communication* 20 (1): 29–32.

Burns, T. W., D. J. O'Connor, and S. M. Stocklymayer. 2003. "Science Communication: A Contemporary Definition." *Public Understanding of Science* 12 (2): 183–202.

Burrough-Boenisch, Joy. 1999. "International Reading Strategies for IMRD Articles." *Written Communication* 16 (3): 296–316.

Bussard, Alain E. 2005. "A Scientific Revolution? The Prion Anomaly May Challenge the Central Dogma of Molecular Biology." *EMBO Reports* 6 (8): 691–4.

Butz, Arlene M., Lisa Kohr, and Dolores Jones. 2004. "Developing a Successful Poster Presentation." *Journal of Pediatric Health Care* 18 (1): 45–8.

Byrne, Daniel W. 2000. "Common Reasons for Rejecting Manuscripts at Medical Journals: A Survey of Editors and Peer Reviewers." *Science Editor* 23 (2): 39–44.

Cain Project in Engineering and Professional Communication. 2008. "Common Errors to Avoid in Scientific Writing." *OpenStax CNX*. Accessed April 23. http://cnx.org/contents/4f8a53b4-8f24-491a-bef5-c16cc33f5b63@1@1.

Cals, Jochen W. L., and Daniel Kotz. 2013a. "Effective Writing and Publishing Scientific Papers, Part III: Introduction." *Journal of Clinical Epidemiology* 66 (7): 702.

———. 2013b. "Effective Writing and Publishing Scientific Papers, Part VI: Discussion." *Journal of Clinical Epidemiology* 66 (10): 1064.

Campanario, Juan Miguel. 1998a. "Peer Review for Journals as It Stands Today—Part 1." *Science Communication* 19 (3): 181–211.

———. 1998b. "Peer Review for Journals as It Stands Today—Part 2." *Science Communication* 19 (4): 277–306.

Campbell, Eric G., Brian R. Clarridge, Manjusha Gokhale, Lauren Birenbaum, Stephen Hilgartner, Neil A. Holtzman, and David Blumenthal. 2002. "Data Withholding in Academic Genetics: Evidence from a National Survey." *JAMA* 287 (4): 473–80.

Cargill, Margaret, and Patrick O' Connor. 2013. *Writing Scientific Research Articles: Strategies and Steps*. Singapore: Wiley-Blackwell.

Carroll, Aaron E., Colin M. Sox, Beth A. Tarini, Sarah Ringold, and Dimitri A. Christakis. 2003. "Does Presentation Format at the Pediatric Academic Societies' Annual Meeting Predict Subsequent Publication?" *Pediatrics* 112 (6): 1238–41.

Carter, Matt. 2013. *Designing Science Presentations: A Visual Guide to Figures, Papers, Slides, Posters, and More*. London, UK: Academic Press.

Caulfield, R. H., A. Maleki-Tabrizi, J. M. Pleat, and M. P. H. Tyler. 2008. "The Factors Considered by Editors of Plastic Surgery Journals in Evaluating Submitted Manuscripts." *Anesthetic Plastic Surgery* 32 (2): 353–8.

Cetin, Selma, and David J. Hackam. 2005. "An Approach to the Writing of a Scientific Manuscript." *Journal of Surgical Research* 128 (2): 165–7.

Chalmers, Iain. 1990. "Underreporting Research is Scientific Misconduct." *JAMA* 263 (10): 1405–8.

Chapman, Andrea L., Laura C. Morgan, and Gerald Gartlehner. 2009. "Semi-Automating the Manual Literature Search for System Reviews Increases Efficiency." *Health Information and Libraries Journal* 27 (1): 22–7.

Charney, Davida. 1993. "A Study in Rhetorical Reading: How Evolutionists Read 'The Spandrels of San Marco.'" In *Understanding Scientific Prose*, edited by Jack Selzer, 203–31. Madison: University of Wisconsin Press.

Chipperfield, Leighton, Leslie Citrome, Juli Clark, Frank S. David, Robert Enck, Michelle Evangelista, John Gonzalez, Trish Groves, Jay Magrann, Bernadette Mansi, Charles Miller, LaVerne A. Mooney, Ann Murphy, John Shelton, Philip D. Walson, and Al Weigel. 2010. "Authors' Submission Toolkit: A Practical Guide to Getting Your Research Published." *Current Medical Research & Opinion* 26 (8): 1968–82.

Chiswick, Malcolm. 2002. "Writing a Research Paper." *Current Paediatrics* 12 (5): 414–8.

Cho, Mildred K., Amy C. Justice, Margaret A. Winker, Jesse A. Berlin, Joseph F. Waeckerle, Michael L. Callaham, Drummond Rennie, and peer investigators. 1998. "Masking author identity in peer review: What factors influence masking success?" *JAMA* 280, no. 3, 243–245.

Cicutto, Lisa. 2008. "Plagiarism: Avoiding the Peril in Scientific Writing." *CHEST* 133 (2): 579–81.

Citrome, Leslie, Stuart V. Moss, and Chris Graf. 2009. "How to Search and Harvest the Medical Literature: Let the Citations Come to You, and How to Proceed When They Do." *The International Journal of Clinical Practice* 63 (11): 1565–70.

Clapham, Phil. 2005. "Publish or Perish." *BioScience* 55 (5): 390–1.

Cleveland, William S. 1984a. "Graphical Methods for Data Presentation: Full Scale Breaks, Dot Charts, and Multibased Logging." *The American Statistician* 38 (4): 270–80.

———. 1984b. "Graphs in Scientific Publications." *American Statistician* 38 (4): 261–9.

Cleveland, William S., Persi Diaconis, and Robert McGill. 1982. "Variables on Scatterplots Look More Highly Correlated When the Scales Are Increased." *Science* 216 (4550): 1138–41.

Cleveland, William S., and Robert McGill. 1985. "Graphical Perception and Graphical Methods for Analyzing Scientific Data." *Science* 229 (4716): 828–33.

Coghill, Anne M., and Lorrin R. Garson. 2006. *The ACS Style Guide: Effective Communication of Scientific Information*. 3rd ed. New York: Oxford University Press.

Cohn, Victor. 1989. *News & Numbers: A Guide to Reporting Statistical Claims and Controversies in Health and Other Fields*. Ames: Iowa State University Press.

Committee on Publication Ethics. 1999. "The COPE Report 1999: Guidelines on Good Publication Practice." *Family Practice* 17 (3): 218–21.

Committee on Responsibilities of Authorship in the Biological Sciences (National Research Council). 2003. *Sharing Publication-Related Data and Materials: Responsibilities of Authorship in the Life Sciences*. Washington, DC: National Academies Press.

Committee on Science, Engineering, and Public Policy (National Academy of Sciences, National Academy of Engineering, and Institute of Medicine of the National Academies). 2009. *On Being a Scientist: A Guide to Responsible Conduct in Research*, 3rd ed. Washington, DC: The National Academies Press.

Coney, Mary B. 1992. "*Terministic Screens*: A Burkean Reading of the Experimental Paper." *Journal of Technical Writing and Communication* 22 (2): 149–58.

Connor, Jason T. 2009. "Statistical Graphics in *AJG*: Save the Ink for the Information." *The American Journal of Gastroenterology* 104 (7): 1624–30.

Cooper, Richelle J., David L. Schriger, and Reb J. H. Close. 2002. "Graphical Literacy: The Quality of Graphs in a Large-Circulation Journal." *Annals of Emergency Medicine* 40 (3): 317–22.

Cooper, Richelle J., David L. Schriger, and David A. Tashman. 2001. "An Evaluation of the Graphical Literacy of Annals of Emergency Medicine." *Annals of Emergency Medicine* 37 (1): 13–19.

Coultas, David. 2007. "Ethical Considerations in the Interpretation and Communication of Clinical Trial Results." *Proceedings of the American Thoracic Society* 4 (2): 194–9.

Council of Science Editors. 2006. *Scientific Style and Format: The CSE Manual for Authors, Editors, and Publishers*. 7th ed. Reston, VA: Council.

Council of Science Editors Editorial Policy Committee. 2012. *CSE's White Paper on Promoting Integrity in Scientific Journal Publications, 2012 Update*. Accessed May 1, 2014. http://www.councilscienceeditors.org/resource-library/editorial-policies/white-paper-on-publication-ethics/.

Creswell, John W. 2012. *Educational Research*. 4th ed. Boston: Pearson.

Cummings, Louise. 1998. "The Scientific Reductionism of Relevance Theory: The Lesson from Logical Positivism." *Journal of Pragmatics* 29 (1): 1–12.

Davidoff, Frank. 1998. "Masking, Blinding, and Peer Review: The Blind Leading the Blinded." *Annals of Internal Medicine* 128 (1): 66–8.

Davids, Jon R., Daniel M. Weigl, Joye P. Edmonds, and Dawn W. Blackhurst. 2010. "Reference Accuracy in Peer-reviewed Pediatric Orthopaedic Literature." *The Journal of Bone & Joint Surgery* 92 (5): 1155–61.

Day, Robert A., and Barbara Gastel. 2006. *How to Write and Publish a Scientific Paper*. 6th ed. Westport, CT: Greenwood Press.

De Amici, D., C. Klersy, and C. Tinelli. 1997. "Graphic Data Representation in Anaesthesiological Journals: A Proposed Methodology for Assessment of Appropriateness." *Anaesthesia and Intensive Care* 25 (6): 659–64.

De Angelis, Catherine D., Jeffrey M. Drazen, Frank A. Frizelle, Charlotte Haug, John Hoey, Richard Horton, Sheldon Kotzin, Christine Laine, Ana Marusic, John P. M. Overbeke, Torben V. Schroeder, Hal C. Sox, and Martin B. Van Der Weyden. 2004. "Clinical Trial Registration: A Statement from the International Committee of Medical Journal Editors." *JAMA* 292 (11): 1363–4.

———. 2005. "Is This Clinical Trial Fully Registered?—A Statement from the International Committee of Medical Journal Editors." *The New England Journal of Medicine* 352 (23): 2436–8.

De Angelis, Catherine D., and Phil B. Fontanarosa. 2008. "Impugning the Integrity of Medical Science: The Adverse Effects of Industry Influence." *JAMA* 299 (15): 1833–5.

de Lacey, Gerald, Carol Record, and Jenny Wade. 1985. "How Accurate Are Quotations and References in Medical Journals?" *BMJ* 291 (6499): 884–6.

De Muth, James E. 2008. "Preparing for the First Meeting with a Statistician." *American Journal of Health-System Pharmacy* 65 (24): 2358–66.

Derish, Pamela A., John Maa, Nancy L. Ascher, and Hobart W. Harris. 2007. "Enhancing the Mission of Academic Surgery by Promoting Scientific Writing Skills." *Journal of Surgical Research* 140 (2): 177–83.

Desnoyers, Luc. 2011. "Toward a Taxonomy of Visuals in Science Communication." *Technical Communication* 58 (2): 119–34.

Ding, Daniel D. 2002. "The Passive Voice and Social Values in Science." *Journal of Technical Writing and Communication* 32 (2): 137–54.

Doak, Leonard G., and Cecilia C. Doak. 2010. "Writing for Readers with a Wide Range of Reading Skills." *AMWA Journal* 25 (4): 149–54.

Docherty, Michael, and Richard Smith. 1999. "The Case for Structuring the Discussion of Scientific Papers: Much the Same as That for Structuring Abstracts." *BMJ* 318 (7193): 1224–5.

Doumont, Jean-Luc, and Philippe Vandenbroeck. 2002. "Choosing the Right Graph." *IEEE Transactions on Professional Communication* 45 (1): 1–6.

Drake, D. C., B. Maritz, S. M. Jacobs, C. J. Crous, A. Engelbrecht, A. Etale, M. J. Fourie, D. G. Furniss, S. L. Scott, S. Parusnath, and D. R. Tye. 2013. "The Propagation and Dispersal of Misinformation in Ecology: Is There a Relationship Between Citation Accuracy and Journal Impact Factor?" *Hydrobiologia* 702 (1): 1–4.

Driscoll, Peter. 1997. "How to Write a Paper." *Journal of Accident & Emergency Medicine* 14 (2): 65–9.

Drotar, Dennis. 2009. "How to Write an Effective Results and Discussion for the Journal of Pediatric Psychology." *Journal of Pediatric Psychology* 34 (4): 339–43.

Dunwoody, Sharon, and Michael Ryan. 1985. "Scientific Barriers to the Popularization of Science in the Mass Media." *Journal of Communication* 35 (1): 26–42.

Durant, John R., Geoffrey A. Evans, and Geoffrey P. Thomas. 1989. "The Public Understanding of Science." *Nature* 340 (6228): 11–14.

Durbin, Charles G. 2004. "Effective Use of Tables and Figures in Abstracts, Presentations, and Papers." *Respiratory Care* 49 (10): 1233–7.

Dutch, Alyson. 2012. "The Art of a Press Release: How to Write One." *Investors Digest*, September.

Eichorn, Philip, and Alfred Yankauer. 1987. "Do Authors Check Their References? A Survey of Accuracy of References in Three Public Health Journals." *American Journal of Public Health* 77 (8): 1011–12.

Ettenberg, Eugene M. 1980. "Typographic Legibility." In *Graphic Arts Manual*, edited by Janet N. Field, 116–18. New York: Arno Press.

Evans, James T., Howard Nadjari, and Sherry A. Burchell. 1990. "Quotational and Reference Accuracy in Surgical Journals." *JAMA* 263 (10): 1353–4.

Eysenbach, Gunther. 2006. "Citation Advantage of Open Access Articles." *PLoS Biology* 4 (5): e157.

Fahnestock, Jeanne. 1986. "Accommodating Science: The Rhetorical Life of Scientific Facts." *Written Communication* 3 (3): 275–96.

———. 2005. "Rhetoric of Science: Enriching the Discipline." *Technical Communication Quarterly* 14 (3): 277–86.

Fahy, Kathleen. 2008. "Writing for Publication: Argument and Evidence." *Women and Birth* 21 (3): 113–17.

Falagas, Matthew, and Pashalis L. Vergidis. 2004. "Addressing the Limitations of Structured Abstracts." *Annals of Internal Medicine* 141 (7): 576–7.

Fanelli, Daniele. 2009. "How Many Scientists Fabricate and Falsify Research? Systematic Review and Meta-Analysis of Survey Data." *PLoS ONE* 4 (5): e5738.

Fenton, J. E., H. Brazier, A. Souza, J. P. Hughes, and D. P. McShane. 2000. "The Accuracy of Citation and Quotation in Otolaryngology/Head and Neck Surgery Journals." *Clinical Otolaryngology & Allied Sciences* 25 (1): 40–4.

Fleming, Alexander. 2013. "Nobel Lecture: Penicillin." Accessed February 7, 2013. http://www.nobelprize.org/nobel_prizes/medicine/laureates/1945/fleming-lecture.html.

Flower, Linda. 1981. *Problem-Solving Strategies for Writing*. Orlando, FL: Harcourt.

Fontanarosa, Phil B., and Catherine D. DeAngelis. 2002. "The Importance of the Journal Embargo." *JAMA* 288 (6): 748–50.

Fontelo, Paul, Alex Gavino, and Raymond Francis Sarmiento. 2013. "Comparing Data Accuracy Between Structured Abstracts and Full-Text Journal Articles: Implications in Their Use for Informing Clinical Decisions." *Evidence Based Medicine* 18 (6): 207–11.

Frank, Erica. 1994. "Authors' Criteria for Selecting Journals." *JAMA* 272 (2): 163–4.

Frankel, Felice C., and Angela H. DePace. 2012. *Visual Strategies: A Practical Guide to Graphics for Scientists and Engineers*. New Haven, CT: Yale University Press.

Franzblau, Lauren E., and Kevin C. Chung. 2012. "Graphs, Tables, and Figures in Scientific Publications: The Good, the Bad, and How Not to Be the Latter." *Journal of Hand Surgery* 37 (3): 591–6.

Frels, Rebecca K., Anthony J. Onwuegbuzie, and John R. Slate. 2010. "Editorial: A Step-by-Step Guide for Creating Tables." *Research in the Schools* 17 (2): xxxviii–lix.

Friedman, Barbara A. 1981. "Copyright from a Permissions Person's Point of View." *Journal of Chemical Information & Computer Sciences* 22 (2): 70–2.

Friedman, Paul J. 1990. "Correcting the Literature Following Fraudulent Publication." *JAMA* 263 (10): 1416–19.

Furler, John, Parker Magin, Marie Pirotta, and Mieke van Driel. 2012. "Participant Demographics Reported in 'Table 1' of Randomised Controlled Trials: A Case of 'Inverse Evidence?'" *International Journal for Equity in Health* 11 (1): 1–4.

Gardner, M. J., D. Machin, and M. J. Campbell. 1986. "Use of Check Lists in Assessing the Statistical Content of Medical Studies." *BMJ* 292 (6523): 810–12.

Garfield, Eugene. 1955. "Citation Indexes for Science." *Science* 122 (3159): 108–11.

———. 1980. "Has Scientific Communication Changed in 300 Years?" *Essays of an Information Scientist* 4 (8): 394–400.

———. 1986. "Which Medical Journals Have the Greatest Impact?" *Annals of Internal Medicine* 105 (2): 313–20.

———. 2006. "The History and Meaning of the Journal Impact Factor." *JAMA* 295 (1): 90–3.

Garrett, Lori K., and Judy M. Penn. 2008. *Get Ready for Microbiology*. San Francisco, CA: Benjamin Cummings.

Gasparyan, Armen Yuri. 2013. "Choosing the Target Journal: Do Authors Need a Comprehensive Approach?" *Journal of Korean Medical Science* 28 (8): 1117–19.

Gavras, Haralambos. 2002. "Inappropriate Attribution: The 'Lazy Author Syndrome.'" *American Journal of Hypertension* 15 (9): 831.

Gerstein, Hertzel C. 2014. "Writing Science that Your Colleagues Can Read." *Journal of Diabetes and Its Complications* 28 (1): 4–5.

Ghai, Babita, Akshay Kumar Saxena, and Jeetinder Kaur Makkar. 2007. "A Guide to Reducing Citation Errors in Bibliographies." *Emergency Medical Journal* 24 (3): 232–3.

Gillen, Christopher M. 2006. "Criticism and Interpretation: Teaching the Persuasive Act of Research Articles." *CBE—Life Sciences Education* 5 (1): 34–8.

Goldbeck-Wood, Sandra. 1999. "Evidence on Peer Review—Scientific Quality Control or Smokescreen?" *BMJ* 318 (7175): 445.

Goldberg, Richard, Edward Newton, Julie Cameron, Raymond Jacobson, Linda Chan, W. Richard Bukata, and Amine Rakab. 1993. "Reference Accuracy in the Emergency Medicine Literature." *Annals of Emergency Medicine* 22 (9): 1450–4.

Goldbort, Robert C. 1998. "Scientific Writing: Three Neglected Aspects." *Journal of Environmental Health* 60 (8): 26.

———. 2001. "Scientific Writing as an Art and as a Science." *Journal of Environmental Health* 63 (7): 22–5.

Goodrich, June E., and Charles G. Roland. 1977. "Accuracy of Published Medical Reference Citations." *Journal of Technical Writing and Communication* 7 (1): 15–19.

Gopen, George D., and Judith A. Swan. 1990. "The Science of Scientific Writing." *American Scientist* 78 (6): 550–8.

Grant, Maria M. 2001. "Reading Critically: Getting What You Need from Scientific Papers." *Canadian Journal of Medical Laboratory Science* 63 (1): 18–24.

Greenberg, Steven A. 2009. "How Citation Distortions Create Unfounded Authority: Analysis of a Citation Network." *BMJ* 339 (7714): 210–13.

Greenhalgh, Trisha, and Richard Peacock. 2005. "Effectiveness and Efficiency of Search Methods in Systematic Reviews of Complex Evidence: Audit of Primary Sources." *BMJ* 331 (7524): 1064–5.

Gregory, Martin W. 1992. "The Infectiousness of Pompous Prose." *Nature* 360 (5): 11–12.

Gross, Alan G. 1983. "A Primer on Tables and Figures." *Journal of Technical Writing and Communication* 13 (1): 33–55.

———. 1985. "The Form of the Experimental Paper: A Realization of the Myth of Induction." *Journal of Technical Writing and Communication* 15 (1): 15–26.

———. 1990. "Extending the Expressive Power of Language: Tables, Graphs, and Diagrams." *Journal of Technical Writing and Communication* 20 (3): 221–35.

———. 1991. "Does Rhetoric of Science Matter? The Case of the Floppy-Eared Rabbits." *College English* 53 (8): 933–43.

———. 2007. "Medical Tables, Graphics and Photographs: How They Work." *Journal of Technical Writing and Communication* 37 (4): 419–33.

Gross, Alan G., Joseph E. Harmon, and Michael S. Reidy. 2009. *Communicating Science: The Scientific Article from the 17th Century to the Present.* Anderson, SC: Parlor Press.

Hagge, John. 1997. "Disciplinary Style Manuals as Reliable Guides to Scientific Discourse Norms." *Technical Communication* 44 (2): 129–41.

Hamilton, Cindy W. 2004. "At a Glance: A Stepwise Approach to Successful Poster Presentations." *CHEST* 134 (2): 457–9.

———. 2009. "On the Table: Form and Function." *CHEST* 135 (4): 1087–9.

Hansen, Margaret E., and Donald D. McIntire. 1994. "Reference Citations in Radiology: Accuracy and Appropriateness of Use in Two Major Journals." *AJR. American Journal of Roentgenology* 163 (3): 719–23.

Harmon, Joseph E. 1989. "Development of the Modern Technical Article." *Technical Communication* 36 (1): 33–8.

Harmon, Joseph E., and Alan G. Gross. 1996. "The Scientific Style Manual: A Reliable Guide to Practice?" *Technical Communication* 43 (1): 61–72.

———. 2009. "The Structure of Scientific Titles." *Journal of Technical Writing and Communication* 39 (4): 455–65.

———. 2010. *The Craft of Scientific Communication.* Chicago: University of Chicago Press.

Harris, Donald J. 2011. "Sharing of Research Data." *The Lancet* 378 (9808): 1194–5.

Hartley, James. 1991. "Tabling Information." *American Psychologist* 46 (6): 655–6.

———. 1998. "Is It Appropriate to Use Structured Abstracts in Non-Medical Science Journals?" *Journal of Information Science* 24 (5): 359–64.

———. 2005. "To Attract Or to Inform: What Are Titles For?" *Journal of Technical Writing and Communication* 35 (2): 203–13.

———. 2007. "There's More to the Title Than Meets the Eye: Exploring the Possibilities." *Journal of Technical Writing and Communication* 37 (1): 95–101.

Hartley, James, and Michele Benjamin. 1998. "An Evaluation of Structured Abstracts in Journals Published by the British Psychological Society." *British Journal of Educational Psychology* 68 (3): 443–56.

Hartley, James, and Matthew Sydes. 1997. "Are Structured Abstracts Easier to Read Than Traditional Ones?" *Journal of Research in Reading* 20 (2): 122–36.

Hartley, James, Matthew Sydes, and Anthony Blurton. 1996. "Obtaining Information Accurately and Quickly: Are Structured Abstracts More Efficient?" *Journal of Information Science* 22 (5): 349–56.

Harzing, Anne-Wil. 2002. "Are Our Referencing Errors Undermining Our Scholarship and Credibility? The Case of Expatriate Failure Rates." *Journal of Organizational Behavior* 23 (1): 127–48.

Haussmann, Natalie Suzette, Trevor McIntyre, Adam John Bumby, and Michael John Loubser. 2013. "Referencing Practices in Physical Geography: How Well Do We Cite What We Write?" *Progress in Physical Geography* 37 (4): 543–9.

Hawes, Thomas, and Sarah Thomas. 1997. "Tense Choices in Citations." *Research in the Teaching of English* 31 (3): 393–414.

Healthcare Convention and Exhibitors Association. 2015. "HCEA Top 50 by Total Attendance." Accessed June 16, 2015. http://c.ymcdn.com/sites/www.hcea.org/resource/resmgr/docs/june_2014_top_50_total_atten.pdf.

Herbison, Peter. 2007. "How to Make Your Article More Acceptable for the Statistical Reviewer." *Neurourology and Urodynamics* 26 (3): 318–22.

Hess, Dean R. 2004. "How to Write an Effective Discussion." *Respiratory Care* 49 (10): 1238–41.

Hess, George R., Kathryn W. Tosney, and Leon H. Liegel. 2009. "Creating Effective Poster Presentations: AMEE Guide No. 40." *Medical Teacher* 31 (4): 319–21.

Hill, Kate. 2006. "Consent, Confidentiality and Record Keeping for the Recording and Usage of Medical Images." *Journal of Visual Communication in Medicine* 29 (2): 76–9.

Hill, Susan S., Betty F. Soppelsa, and Gregory K. West. 1982. "Teaching ESL Students to Read and Write Experimental-Research Papers." *TESOL Quarterly* 16 (3): 333–47.

Hofmann, Angelika H. 2010. *Scientific Writing and Communication: Papers, Proposals, and Presentations*. New York, NY: Oxford University Press.

———. 2014. *Scientific Writing and Communication: Papers, Proposals, and Presentations*. 2nd ed. New York: Oxford University Press.

Holmes, Frederic L. 1987. "Scientific Writing and Scientific Discovery." *The History of Science Society* 78 (2): 220–35.

Honeyman-Buck, Janice C. 2003a. "Reviewing Manuscripts with Manuscript Central." *Journal of Digital Imaging* 16 (4): 319–23.

———. 2003b. "Submitting Manuscripts for Publication through Manuscript Central." *Journal of Digital Imaging* 16 (3): 245–50.

Horton, Richard. 1995. "The Rhetoric of Research." *BMJ* 310 (6985): 985–7.

Horton, Richard, and Trisha Greenhalgh. 1995. "The Rhetoric of Research." *BMJ* 310 (6985): 985–8.

Horton, William. 1991. Illustrating *Computer Documentation: The Art of Presenting Information Graphically on Paper and Online*. New York: Wiley & Sons.

Howarth, Richard J., and Mark St. J. Turner. 1987. "Statistical Graphics in Geochemical Journals." *Mathematical Geology* 19 (1): 1–24.

Hu, Xiaojun, Ronald Rousseau, and Jin Chen. 2011. "On the Definition of Forward and Backward Citation Generations." *Journal of Informetrics* 5 (1): 27–36.

Huckin, Thomas N. 1983. "A Cognitive Approach to Readability." In *New Essays in Technical and Scientific Communication Research: Research, Theory, Practice*, edited by Paul V. Anderson, John R. Brockman, and Carolyn R. Miller, 90–108. Farmingdale, NY: Baywood.

Huler, Scott. 1990. "How to Get Your Research Published: Editors' Thoughts." *The Scientist*, November 12. Accessed September 9, 2012. http://www.the-scientist.com/?articles.view/articleNo/11467/title/How-To-Get-Your-Research-Published—Editors—Thoughts/.

Huston, Patricia, and David Moher. 1996. "Redundancy, Disaggregation, and the Integrity of Medical Research." *The Lancet* 347 (9007): 1024–6.

Huth, Edward J. 1982. *How to Write and Publish Papers in the Medical Sciences*. Philadelphia, PA: ISI Press.

———. 1986. "Irresponsible Authorship and Wasteful Publication." *Annals of Internal Medicine* 104 (2): 257–9.

———. 1987. "Structured Abstracts for Papers Reporting Clinical Trials." *Annals of Internal Medicine* 106 (4): 626–7.

Hyland, Ken. 1995. "The Author in the Text: Hedging Scientific Writing." *Hong Kong Papers in Linguistics and Language* 18: 33–42.

———. 1996. "Talking to the Academy: Forms of Hedging in Scientific Research Articles." *Written Communication* 13 (2): 251–81.

———. 2001. "Humble Servants of the Discipline? Self-Mention in Research Articles." *English for Specific Purposes* 20 (3): 207–26.

Infection and Immunity. 2015. "About this Journal." Accessed July 1, 2015. http://iai.asm.org/site/misc/about.xhtml.

International Committee of Medical Journal Editors. 2013. *Recommendations for the Conduct, Reporting, Editing, and Publication of Scholarly Work in Medical Journals*. Accessed February 2, 2015. http://www.icmje.org/recommendations.

———. 2014. *Uniform Requirements for Manuscripts Submitted to Biomedical Journals: Writing and Editing for Biomedical Publication*. Accessed May 21, 2015. http://www.icmje.org/icmje-recommendations.pdf.

Ioannidis, John P. A. 2007. "Limitations Are Not Properly Acknowledged in the Scientific Literature." *Journal of Clinical Epidemiology* 60 (4): 324–9.

———. 2008. "Why Most Discovered True Associations Are Inflated." *Epidemiology* 19 (5): 640–8.

Ioannidis, John P. A., Stephen J. W. Evans, Peter C. Gøtzsche, Robert T. O'Neill, Douglas G. Altman, Kenneth Schulz, and David Moher. 2004. "Better Reporting of Harms in Randomized Trials: An Extension of the CONSORT Statement." *Annals of Internal Medicine* 141: 781–8.

Irwin, Richard S. 2007. "Clinical Trial Registration Promotes Patient Protection and Benefit, Advances the Trust of Everyone, and Is Required." *CHEST* 131 (3): 639–41.

———. 2009. "The Role of Conflict of Interest in Reporting Scientific Information." *CHEST* 136 (1): 253–9.

Iverson, Cheryl, Stacy Christiansen, Annette Flanagin, Phil B. Fontanarosa, Richard M. Glass, Brenda Gregoline, Stephen J. Lurie, Harriet S. Meyer, Margaret A. Winker, and Roxanne K. Young. 2007. *AMA Manual of Style: A Guide for Authors and Editors*. 10th ed. New York: Oxford University Press.

Jacobs, Adam. 2004. "Transparency and Trust: In Defence of Medical Writers." *BMJ* 329 (7478): 1345–6.

Jacobs, Adam, and Elizabeth Wager. 2005. "European Medical Writers Association (EMWA) Guidelines on the Role of Medical Writers in Developing Peer-Reviewed Publications." *Current Medical Research and Opinion* 21 (2): 317–21.

JAMA. n.d. "JAMA Instructions for Authors." Accessed July 19, 2013. http://jama.jamanetwork.com/public/instructionsForAuthors.aspx.

Jefferson, Tom, Elizabeth Wager, and Frank Davidoff. 2002. "Measuring the Quality of Editorial Peer Review." *JAMA* 287 (21): 2786–9.

John, Michael. 2009. "The Target Journal: Choosing the Right Place to Submit Your Paper." *HSR Proceedings in Intensive Care and Cardiovascular Anesthesia* 1 (3): 60–2.

Johnson-Sheehan, Richard, and Scott Rode. 1999. "On Scientific Narrative: Stories of Light by Newton and Einstein." *Journal of Business and Technical Communication* 13 (3): 336–58.

Jones, Dan. 1998. *Technical Writing Style*. Boston, MA: Allyn and Bacon.

Journal of Neuroscience. n.d. "The Neuroscience Peer Review Consortium." Accessed July 19, 2013. http://www.jneurosci.org/content/29/5/1255.full.

Kanoksilapatham, Budsaba. 2005. "Rhetorical Structure of Biochemistry Research Articles." *English for Specific Purposes* 24 (3): 269–92.

———. 2012. "Structure of Research Article Introductions in Three Engineering Subdisciplines." *IEEE Transactions on Professional Communication* 55 (4): 294–309.

Kaplan, Judith B., and Trude Bennett. 2003. "Use of Race and Ethnicity in Biomedical Publication." *JAMA* 289 (20): 2709–16.

Katz, Michael Jay. 2009. *From Research to Manuscript: A Guide to Scientific Writing*. 2nd ed. New York: Springer.

Keegan, David A., and Susan L. Bannister. 2003. "Effect of Colour Coordination of Attire with Poster Presentation on Poster Popularity." *Canadian Medical Association Journal* 169 (12): 1291–2.

Kelleher, Christa, and Thorsten Wagner. 2011. "Ten Guidelines for Effective Data Visualization in Scientific Publications." *Environment Modeling & Software* 26 (6): 822–7.

Kelly, Kevin. 2010. "Evolving the Scientific Method: Technology Is Changing the Way We Conduct Science." *The Scientist*, December 1.

Kirby, Kate, and Frances A. Houle. 2004. "Ethics and the Welfare of the Physics Profession." *Physics Today* 57 (11): 42–6.

Knight, Jonathan. 2003. "Scientific Literacy: Clear as Mud." *Nature* 423 (6938): 376–8.

Knottnerus, J. André, and Peter Tugwell. 2007. "Communicating Research to the Peers." *Journal of Clinical Epidemiology* 60 (7): 645–7.

Korner, Ann M. 2008. *Guide to Publishing a Scientific Paper*. New York: Routledge.

Kosslyn, Stephen M. 1989. "Understanding Charts and Graphs." *Applied Cognitive Psychology* 3 (3): 185–226.

Kostelnick, Charles. 1998. "Conflicting Standards for Designing Data Displays: Following, Flouting, and Reconciling Them." *Technical Communication* 45 (4): 473–82.

———. 2007. "The Visual Rhetoric of Data Displays: The Conundrum of Clarity." *IEEE Transactions on Professional Communication* 50 (4): 280–94.

Kotz, Daniel, and Jochen W. L. Cals. 2013a. "Effective Writing and Publishing Scientific Papers, Part V: Results." *Journal of Clinical Epidemiology* 66 (9): 945.

———. 2013b. "Effective Writing and Publishing Scientific Papers, Part VII: Tables and Figures." *Journal of Clinical Epidemiology* 66 (11): 1197.

———. 2014. "Effective Writing and Publishing Scientific Papers, Part XII: Responding to Reviewers." *Journal of Clinical Epidemiology* 67 (3): 243.

Kozak, Marcin, and James Hartley. 2012. "Presenting Numerical Values within Sentences and Text Tables." *Journal of the American Society for Information Science and Technology* 63 (1): 108–13.

Kraft, Michelle A. 2014. "Making the Most of Google Scholar." *AMWA Journal* 29 (2): 80–2.

Krebs, Charles J., Emma Bredesen, Andrea Coombs, Rachelle Daniel, Ramona de Graf, Anna Elz, Anna Hall, Laura Hooper, Trent Hoover, Alida Janmaat, Ladan Mehranvar, Emily Rubidge, Greta Slooten, Patrick Tamkee, and Kym Welstead. 2001. "Graphical Presentation in the Journal Ecology." *Bulletin of the Ecological Society of America* 82 (4): 247–8.

Krohn, Roger. 1991. "Why Are Graphs So Central in Science?" *Biology and Philosophy* 6 (2): 181–203.

Kuhn, Thomas S. 2000. *The Road Since Structure: Philosophical Essays, 1970–93*. Chicago: University of Chicago Press.

Kuo, Chih-Hua. 1998. "The Use of Personal Pronouns: Role Relationships in Scientific Journal Articles." *English for Specific Purposes* 18 (2): 121–38.

Kuo, Yen-Hong. 2002. "Extrapolation of Correlation Between 2 Variables in 4 General Medical Journals." *JAMA* 287 (21): 2815–17.

Kurata, Keiko, Tomoko Morioka, Keiko Yokoi, and Mamiko Matsubayashi. 2013. "Remarkable Growth of Open Access in the Biomedical Field Articles from 2006 to 2010." *PLoS One* 8 (5): e60925.

Kurmis, A. P. 2003. "Contributing to Research: The Basic Elements of a Scientific Manuscript." *Radiography* 9 (4): 277–82.

Kurup, Viji, and Denise Hersey. 2010. "Finding Information by 'Design': Search Strategies for Cardiothoracic and Vascular Anesthesia Literature." *Journal of Cardiothoracic and Vascular Anesthesia* 24 (5): 845–58.

Laakso, Mikael, and Bo-Christer Bjork. 2012. "Anatomy of Open Access Publishing: A Study of Longitudinal Development and Internal Structure." *BMC Medicine* 10 (1): 124.

LaFollette, Marcel C. 2000. "The Evolution of the 'Scientific Misconduct' Issue: An Historical Overview." *Experimental Biology and Medicine* 224 (4): 211–15.

Lagowski, J. J. 1981. "Journal Copyright Problems: An Editor's View." *Journal of Chemical Information & Computer Sciences* 22 (2): 72–3.

Lancet. n.d. "Fast-Track Publication." Accessed March 20, 2014. http://www.thelancet.com/lancet-infectious-diseases-information-for-authors/fast-track.

Lane, David M., and Aniko Sandor. 2009. "Designing Better Graphs by Including Distributional Information and Integrating Words, Numbers, and Images." *Psychology Methods* 14 (3): 239–57.

Lang, Thomas A. 2009. "Just Who Are We and What Are We Doing, Anyway? Needed Research in Medical Writing." *AMWA Journal* 24 (3): 106–12.

———. 2010. *How to Write, Publish, & Present in the Health Sciences: A Guide for Clinicians & Laboratory Researchers*. Philadelphia: American College of Physicians.

Lang, Thomas A., and Douglas G. Altman. 2014. "Statistical Analysis and Methods in the Published Literature: The SAMPL Guidelines." In *Guidelines for Reporting Heath Research: A User's Manual*, edited by David Moher, Douglas G. Altman, Kenneth F. Schulz, Iveta Simera, and Elizabeth Wager, 265–74. Oxford, UK: Wiley & Sons.

Langdon-Neuner, Elise. 2009. "When Does Previous Disclosure Become a 'Prior Publication' Problem?" *CHEST* 135 (1): 233–7.

Lantz, Jane C., and William L. Lanier. 2003. "Observations from the Mayo Clinic National Conference on Medicine and the Media." *Science Editor* 26 (3): 85.

Latour, Bruno. 1990. "Drawing Things Together." In *Representation in Scientific Practice*, edited by Michael Lynch and Steve Woolgar, 19–68. Cambridge, MA: MIT Press.

Latour, Bruno, and Steve Woolgar. 1986. *Laboratory Life: The Construction of Scientific Facts*. Princeton, NJ: Princeton University Press.

Lawson, Glenda. 2005. "The Poster Presentation: An Exercise in Effective Communication." *Journal of Vascular Nursing* 23 (4): 157–8.

Lenzer, Jeanne. 2002. "The Operation Was a Success (But the Patients Died): How Media Spin Distorted Outcomes of a Study Comparing Radical Prostatectomy with Watchful Waiting." *BMJ* 325 (7365): 664.

Levine, Mitchell, Ahman I. El-Nahas, and Benjamin Asa. 2010. "Relative Risk and Odds Ratio Data Are Still Portrayed with Inappropriate Scales in the Medical Literature." *Journal of Clinical Epidemiology* 63 (9): 1045–7.

Lim, Jason Miin-Hwa. 2012. "How Do Writers Establish Research Niches? A Genre-Based Investigation into Management Researchers' Rhetorical Steps and Linguistic Mechanisms." *Journal of English for Academic Purposes* 11 (3): 229–45.

Lindeborg, Richard A. 1990. "Faster Manuscript Publishing: A Case Study." *IEEE Transactions on Professional Communication* 33 (1): 7–11.

Lipson, Carol. 1988. "A Social View of Technical Writing." *Journal of Business and Technical Communication* 2 (1): 7–20.

Liumbruno, Giancarlo Maria, Claudio Velati, Patrizio Pasqualetti, and Massimo Franchini. 2013. "How to Write a Scientific Manuscript for Publication." *Blood Transfusion* 11 (2): 217–26.

Lock, Stephen. 1988. "Structured Abstracts for Papers Reporting Clinical Trials." *BMJ* 297 (6642): 156.

Locke, Simon. 2003. "Golem Science and the Public Understanding of Science: From Deficit to Dilemma." *Public Understanding of Science* 8 (2): 75–92.

Lofstrom, Joyce. 2010. "A Collaborative Approach for Media Training between Technical Communication and Public Relations: Tutorial." *IEEE Transactions on Professional Communication* 53 (2): 164–73.

Loke, Yoon K., and Sheena Derry. 2003. "Does Anybody Read 'Evidence-Based' Articles?" *BMC Medical Research Methodology* 3 (1): 14–18.

Lukic, Ivan Kresimir, Anita Lukic, Vico Gluncic, Vedran Katavic, Vladimira Vucenik, and Ana Marusic. 2004. "Citation and Quotation Accuracy in Three Anatomy Journals." *Clinical Anatomy* 17 (7): 534–9.

Luo, Ma, Charles Chuan, Domingo Molina IV, Clark R. Andersen, and Vinod K. Panchbhavi. 2013. "Accuracy of Citation and Quotation in Foot and Ankle Surgery Journals." *Foot & Ankle International* 34 (7): 949–55.

Luzon, Maria Jose. 2005. "Genre Analysis in Technical Communication." *IEEE Transactions on Professional Communication* 48 (3): 285–95.

———. 2007. "The Added Value Features of Online Scholarly Journals." *Journal of Technical Writing and Communication* 37 (1): 59–73.

Lynch, Michael. 1990. "The Externalized Retina: Selection and Mathematization in the Visual Documentation of Objects in the Life Sciences." In *Representation in Scientific Practice*, edited by Michael Lynch and Steve Woolgar, 153–86. Cambridge, MA: MIT Press.

Lynch, Michael, and Steve Woolgar. 1990. "Introduction: Sociological Orientations to Representational Practice in Science." In *Representation in Scientific Practice*, edited by Michael Lynch and Steve Woolgar, 1–18. Cambridge, MA: MIT Press.

Ma, Irene W. Y., Nadia A. Khan, Anna Kang, Nadia Zalunardo, and Anita Palepu. 2007. "Systematic Review Identified Suboptimal Reporting and Use of Race/Ethnicity in General Medical Journals." *Journal of Clinical Epidemiology* 60 (6): 572–8.

MacGregor, A. Jean. 1982. "Selecting the Appropriate Chart." *IEEE Transactions on Professional Communication* 25 (2): 102–4.

MacIntosh-Murray, Anu. 2007. "Poster Presentations as a Genre in Knowledge Communication: A Case Study of Forms, Norms, and Values." *Science Communication* 28 (3): 347–76.

Maddox, John. 1983. "Must Science Be Impenetrable?" *Nature* 305 (5934): 477–8.

Markel, Mike. 1993. "Induction, Social Constructionism, and the Form of the Scientific Paper." *Journal of Technical Writing and Communication* 23 (1): 7–22.

Martinson, Brian C., Melissa S. Anderson, and Raymond de Vries. 2005. "Scientists Behaving Badly." *Nature* 435 (7043): 737–8.

Masic, Izet. 2011. "How to Search, Write, Prepare and Publish the Scientific Papers in the Biomedical Journals." *AIM* 19 (2): 68–79.

———. 2012. "Ethical Aspects and Dilemmas of Preparing, Writing and Publishing of the Scientific Papers in the Biomedical Journals." *Acta Informatica Medica* 20 (3): 141–8.

Massoudi, Mehrdad. 2003. "Can Scientific Writing Be Creative?" *Journal of Science Education and Technology* 12 (2): 115–28.

Matsubayashi, Mamiko, Keiko Kurata, Yukiko Sakai, Tomoko Morioka, Shinya Kato, Shinji Mine, and Shuichi Ueda. 2009. "Status of Open Access in the Biomedical Field in 2005." *Journal of the Medical Library Association* 97 (1): 4–11.

Matsuda, Paul Kei, and Christine M. Tardy. 2007. "Voice in Academic Writing: The Rhetorical Construction of Author Identity in Blind Manuscript Review." *English for Specific Purposes* 26 (2): 235–49.

Matthews, Diane L. 1990. "The Scientific Poster: Guidelines for Effective Visual Communication." *Technical Communication* 37 (3): 225–32.

Matthews, Janice R., John M. Bowen, and Robert W. Matthews. 2000. *Successful Scientific Writing*. 2nd ed. Cambridge, UK: Cambridge University Press.

Matthews, John N. S., and Douglas G. Altman. 1996. "Interaction 2: Compare Effect Sizes not *p* Values." *BMJ* 313 (7060): 808.

McKenzie, Kwame, and Natasha Crowcroft. 1996. "Ethnicity, Race, and Culture: Guidelines for Research, Audit, and Publication." *BMJ* 312 (7038): 1094.

McMillan, Victoria E. 2012. *Writing Papers in the Biological Sciences*. 5th ed. Boston: Bedford/St. Martin's.

McNamara, D. A., M. Grannell, R. G. K. Watson, and D. J. Bouchier-Hayes. 2001. "The Research Abstract: Worth Getting It Right." *Irish Journal of Medical Science* 170 (1): 38–40.

McNutt, Marcia. 2015. "It Starts with a Poster." *Science* 347 (6226): 1047.

Meyer, Joachim, Kuskin Shamo, and Daniel Gopher. 1999. "Information Structure and the Relative Efficacy of Tables and Graphs." *Human Factors* 41 (4): 570–87.

Mike, Valerie, and Robert A. Good. 1977. "Medical Research: Statistics and Ethics. Old Problems, New Challenges." *Science* 198 (4318): 677–8.

Millar, Robin, and Brian Wayne. 1988. "Public Understanding of Science: From Contents to Processes." *International Journal of Science Education* 10 (4): 388–98.

Miller, Carolyn R. 1979. "A Humanistic Rationale for Technical Writing." *College English* 40 (6): 610–17.

———. 1992. "*Kairos* in the Rhetoric of Science." In *A Rhetoric of Doing: Essays on Written Discourse in Honor of James L. Kinneavy*, edited by Stephen P. Whitte, Neil Nakadate, and Roger D. Cherry, 310–27. Carbondale: Southern Illinois University Press.

Miller, Elaine T., and Katrina D. Lewis. 2008. "Bridging the Gap Between Poster and Manuscript." *Rehabilitation Nursing* 33 (2): 52–4.

Miller, Franklin G., and Howard Brody. 2005. Professional Integrity in Industry-Sponsored Clinical Trials. *Academic Medicine* 80 (10): 899–904.

Miller, Jane E. 2004. *The Chicago Guide to Writing about Numbers*. Chicago: University of Chicago Press.

———. 2007. "Preparing and Presenting Effective Research Posters." *Health Services Research* 42 (1): 311–28.

Miller, Jon D. 1998. "The Measurement of Civic Scientific Literacy." *Public Understanding of Science* 7 (3): 203–23.

———. 2004. "Public Understanding of, and Attitudes Toward, Scientific Research: What We Know and What We Need to Know." *Public Understanding of Science* 13 (3): 273–94.

Miller, Thomas. 1998. "Visual Persuasion: A Comparison of Visuals in Academic Texts and the Popular Press." *English for Specific Purposes* 17 (1): 29–46.

Mishra, Punyasloke. 2004. "The Role of Abstraction in Scientific Illustration: Implications for Pedagogy." In *Visual Rhetoric in a Digital World: A Critical Sourcebook*, edited by Carolyn Handa, 177–94. Boston: Bedford/St. Martin's.

Mogull, Scott A. 2008. "A Call for New Courses to Train Scientists as Effective Communicators in Contemporary Government and Business Settings." *Technical Communication* 55 (4): 357–69.

Mogull, Scott A., and Candice T. Stanfield. 2015. "Current Use of Visuals in Scientific Communication." 2015 IEEE *International Professional Communication Conference (IPCC)*, 1–6. DOI: 10.1109/IPCC.2015.7235818

Mogull, Scott A., and Steven B. Katz. 2012. "Scientific Communication Ethics vs. Ethics in Science." Paper presented at the annual conference of the Association of Teachers of Technical Writing (ATTW), St. Louis, MO.

Moher, David, Sally Hopewell, Kenneth F. Schulz, Victor Montori, Peter C. Gotzsche, P. J. Devereaux, Diana Elbourne, Matthias Egger, and Douglas G. Altman. 2010. "CONSORT 2010 Explanation and Elaboration: Updated Guidelines for Reporting Parallel Group Randomized Trials." *BMJ* 340: c869.

Montgomery, Scott L. 1996. *The Scientific Voice*. New York: Guilford Press.

Moore, Randy. 1991. "How We Write about Biology." *American Biology Teacher* 53 (7): 388–9.

Morton, James P. 2009. "Reviewing Scientific Manuscripts: How Much Statistical Knowledge Should a Reviewer Really Know?" *Advances in Physiological Education* 33 (1): 7–9.

Motsitsi, N. S. 2008. "Writing a Convincing and Effective Discussion Section of a Biomedical Research Paper." *East and Central African Journal of Surgery* 13 (1): 4–6.

Murray, Rowena, Morag Thow, and Rosanne Strachan. 1998. "Visual Literacy: Designing and Presenting a Poster." *Physiotherapy* 84 (7): 319–27.

Myers, Greg. 1985. "Texts as Knowledge Claims: The Social Construction of Two Biology Articles." *Social Studies of Science* 15 (4): 593–630.

———. 1992. "'In This Paper We Report . . .': Speech Acts and Scientific Facts." *Journal of Pragmatics* 17 (4): 295–313.

Nair, P. K. R. 2005. "How (Not) to Write Research Papers in Agroforestry." *Agroforestry Systems* 64 (1): v–xvi.

Nakaoka, Hiroshi, Lovett Lawson, S. Bertel Squire, Brian Coulter, Pernille Ravn, Inger Brock, C. Anthony Hart, and Luis E. Cuevas. 2006. "Risk for Tuberculosis among Children." *Emerging Infectious Diseases* 12: 1383–88.

Nancekivell, Sharon. 2010. "Meeting the Challenges of Health Literacy: The Medical Communicator's Role." *AMWA Journal* 25 (4): 146–8.

Nastasee, Susan A. 2010. "Acknowledgment of Medical Writers in Medical Journal Articles: A Comparison from the Years 2000 and 2007." *AMWA Journal* 25 (1): 2–6.

National Science Board. 2014. "Science and Engineering Indicators 2014." Accessed May 24, 2015. http://www.nsf.gov/statistics/seind14.

Nature. n.d. "How to Anonymize Your Manuscript for Double Blind Peer Review." Accessed July 7, 2015. http://www.nature.com/nclimate/pdf/double-blind-checklist.pdf.

———. 2002. "Media Studies for Scientists." *Nature* 416 (6880): 461.

———. 2014. "English Communication for Scientists." Accessed November 17, 2014. http://www.nature.com/scitable/ebooks/english-communication-for-scientists-14053993/126083980.

Nature Cell Biology. 2009. "How Robust Are Your Data? New Rules for the Presentation of Statistics." *Nature Cell Biology* 11 (6): 667.

Nature Neuroscience. 2007. "Got Data?" *Nature Neuroscience* 10 (8): 931.

Naylor, W. Patrick, and Carlos A. Muñoz-Viveros. 2005. "The Art of Scientific Writing: How to Get Your Research Published!" *Journal of Contemporary Dental Practice* 6 (2): 164–80.

Ng, K. H., and W. C. G. Peh. 2009a. "Preparing Effective Illustrations: Part 1: Graphs." *Singapore Medical Journal* 50 (3): 245–9.

———. 2009b. "Preparing Effective Illustrations: Part 2: Photographs, Images, and Diagrams." *Singapore Medical Journal* 50 (4): 330–5.

———. 2009c. "Preparing Effective Tables." *Singapore Medical Journal* 50 (2): 117–19.

Norris, Stephen P., and Linda M. Phillips. 2003. "How Literacy in its Fundamental Sense is Central to Scientific Literacy." *Science Education* 87 (2): 224–40.

Northcut, Kathryn M. 2011. "Insights from Illustrators: The Rhetorical Invention of Palentology Representations." *Technical Communication Quarterly* 20 (3): 303–26.

Nourbakhsh, Eva, Rebecca Nugent, Helen Wang, Cihan Cevik, and Kenneth Nugent. 2012. "Medical Literature Searches: A Comparison of PubMed and Google Scholar." *Health Information and Libraries Journal* 29 (3): 214–22.

Nuzzo, Regina. 2014. "Statistical Errors." *Nature* 506 (13): 150–2.

Nwogu, Kevin Ngozi. 1997. "The Medical Research Paper: Structure and Functions." *English for Specific Purposes* 16 (2): 119–38.

O'Connor, Timothy R., and Gerald P. Holmquist. 2009. "Algorithm for Writing a Scientific Manuscript." *Biochemistry and Molecular Biology Education* 37 (6): 344–8.

Ortinau, David J. 2011. "Writing and Publishing Important Scientific Articles: A Reviewer's Perspective." *Journal of Business Research* 64 (2): 150–6.

Overholser, Brian R., and Kevin M. Sowinski. 2007. "Biostatistics Primer: Part 1." *Nutrition in Clinical Practice* 22 (6): 629–35.

———. 2008. "Biostatistics Primer: Part 2." *Nutrition in Clinical Practice* 23 (1): 76–84.

Overstreet, Kristen. 2010. "Why Manuscript Formatting Can Reduce You to a Whiny 4-Year-Old." *Journal for Specialists in Pediatric Nursing* 15 (1): 4–5.

Oxford University Press. n.d. "How to Blind a Manuscript for Online Submission." Accessed July 7, 2015. http://www.oxfordjournals.org/our_journals/poq/for_authors/blinding.pdf.

Özçakar, L., F. Franchignoni, M. Kara, and S. Muñoz Lasa. 2012. "Choosing a Scholarly Journal During Manuscript Submission: The Way How It Rings True for Physiatrists." *European Journal of Physical and Rehabilitation Medicine* 48 (4): 643–7.

Pagel, Walter J., Frances E. Kendall, and Harry R. Gibbs. 2002. "Self-Identified Publishing Needs of Nonnative English-Speaking Faculty and Fellows at an Academic Medical Institution." *Science Editor* 25 (4): 111–14.

Paniello, Randal C., J. Gail Neely, Jason T. Rich, Eric L. Slattery, and Courtney C. J. Voelker. 2011. "Practical Guide to Choosing an Appropriate Data Display." *Otolaryngology—Head and Neck Surgery* 145 (6): 886–94.

Parkinson, Jean. 2011. "The Discussion Section as Argument: The Language Used to Prove Knowledge Claims." *English for Specific Purposes* 30 (3): 164–75.

Patterson, T. T. 1973. "Why You Should Publish." *Journal of Technical Writing and Communication* 3 (4): 285–90.

Peacock, Matthew. 2002. "Communicative Moves in the Discussion Section of Research Articles." *System* 30 (4): 479–97.

Penrose, Ann M., and Steven B. Katz. 2010. *Writing in the Sciences: Exploring the Conventions of Scientific Discourse*. 3rd ed. Upper Saddle River, NJ: Pearson/Longman.

Peterson, Martin S. 1961. *Scientific Thinking and Scientific Writing*. New York: Reinhold.

Phillips, David P., Elliot J. Kanter, Bridget Bednarczyk, and Particia L. Tastad. 1991. "Importance of the Lay Press in the Transmission of Medical Knowledge to the Scientific Community." *The New England Journal of Medicine* 325 (16): 1180–3.

Phillips, Stephanie G. 2009. "Authorship and Writing Practices in the Health Care Industry." *AMWA Journal* 24 (1): 4–8.

Pierson, David J. 2004. "How to Write an Abstract That Will Be Accepted for Presentation at a National Meeting." *Respiratory Care* 49 (10): 1206–12.

Pipes, Alan. 2009. *Production for Graphic Designers*. 5th ed. Upper Saddle River, NJ: Prentice Hall.

Piqué-Angordans, Jordi, and Santiago Posteguillo. 2006. "Medical Discourse and Academic Genres." In *Encyclopedia of Language & Linguistics*, 2nd ed., edited by Keith Brown, 649–57. Waltham, MA: Elsevier.

Pirolli, Peter, and Stuart Card. 1999. "Information Foraging." *Psychological Review* 106 (4): 643–75.

Pitkin, Roy M., and Mary Ann Branagan. 1998. "Can the Accuracy of Abstracts be Improved by Providing Specific Instructions? A Randomized Controlled Trial." *JAMA* 280 (3): 267–9.

Pitkin, Roy M., Mary Ann Branagan, and Leon F. Burmeister. 1999. "Accuracy of Data in Abstracts of Published Research Articles." *JAMA* 281 (12): 1110–11.

Plaxco, Kevin W. 2010. "The Art of Writing Science." *Protein Science* 19 (12): 2261–6.

PLoS Medicine Editors. 2006. "The Impact Factor Game." *PLoS Medicine* 3 (6): e291.

Pololi, Linda, Sharon Knight, and Kathleen Dunn. 2004. "Facilitating Scholarly Writing in Academic Medicine: Lessons Learned from a Collaborative Peer Mentoring Program." *Journal of General Internal Medicine* 19 (1): 64–8.

Popper, Karl. 2002. *The Logic of Scientific Discovery*. 2nd ed. New York: Routledge.

Powell, Kendall. 2012. "Presentations: Billboard Science." *Nature* 483 (7387): 113–15.

Price, Derek J. de Solla. 1989. *Little Science, Big Science . . . and Beyond*. New York: Columbia University Press.

Proto, Anthony V. 2007. "*Radiology* 2007: Evaluating and Processing Your Manuscript for Publication." *Radiology* 244 (1): 3–6.

Puhan, Milo A., Elie A. Akl, Dianne Bryant, Feng Xie, Giovanni Apolone, and Gerben ter Riet. 2012. "Discussing Study Limitations in Reports of Biomedical Studies—The Need for More Transparency." *Health and Quality of Life Outcomes* 10: 23.

Puhan, Milo A., Gerben Ter Riet, Klaus Eichler, Johann Steurer, and Lucas M. Bachmann. 2006. "More Medical Journals Should Inform Their Contributors About Three Key Principles of Graph Construction." *Journal of Clinical Epidemiology* 59 (10): 1017–22.

Purrington, Colin B. 2015. "Designing Conference Posters." Accessed June 11, 2015. http://colinpurrington.com/tips/poster-design.

Radford, D. R., L. Smillie, R. F. Wilson, and A. M. Grace. 1999. "The Criteria Used by Editors of Scientific Dental Journals in the Assessment of Manuscripts Submitted for Publication." *British Dental Journal* 187 (7): 376–9.

Ramji, Siddarth. 2011. "Effective Online Search for Medical Literature." *Indian Journal of Medical Specialities* 2 (2): 189–91.

Rastegar, Darius A., and Leah Wolfe. 2002. "Experience, Expertise, or Specialty? Uses and Misuses of a Reference." *Journal of Family Practice* 51 (2): 168.

Ratnoff, Oscar D. 1981. "How to Read a Paper." In *Coping with the Biomedical Literature: A Primer for the Scientist and Clinician*, edited by Kenneth S. Warren, 95–101. New York: Praeger.

Reddy, M. S., S. Srinivas, N. Sabanayagam, and S. P. Balasubramanian. 2008. "Accuracy of References in General Surgical Journals—An Old Problem Revisited." *The Surgeon* 6(2): 71–5.

Reese, David M. 1999. "Restoring the Literary to Medical Writing." *Lancet* 353 (9152): 585–6.

Reeves, Carol. 2005. "'I Knew There Was Something Wrong with That Paper': Scientific Rhetorical Styles and Scientific Misunderstandings." *Technical Communication Quarterly* 14 (3): 267–75.

Reidenbach, Faith. 2008. "ISMPP Recommends New Procedure for Disclosing Medical Writers' Contributions to Journal Manuscripts." *AMWA Journal* 23 (1): 70–1.

Rennie, Drummond. 2014. "Forward: Guide to Guidelines." In *Guidelines for Reporting Heath Research: A User's Manual*, edited by David Moher, Douglas G. Altman, Kenneth F. Schulz, Iveta Simera, and Elizabeth Wager, xiii–xvii. Oxford, UK: Wiley & Sons.

Rennie, Drummond, and Richard M. Glass. 1991. "Structuring Abstracts to Make Them More Informative." *JAMA* 266 (1): 116–17.

Rhodes, Susan. 1997. "The Active and Passive Voice are Equally Comprehensible in Scientific Writing." PhD diss., University of Washington. https://digital.lib.washington.edu/researchworks/handle/1773/9033.

Richards, Anne R. 2003. "Argument and Authority in the Visual Representations of Science." *Technical Communication Quarterly* 12 (2): 183–206.

Riley, Kathryn. 1991. "Passive Voice and Rhetorical Role in Scientific Writing." *Journal of Technical Writing and Communication* 21 (3): 239–57.

Ripple, Anna M., James G. Mork, Lou S. Knecht, and Betsy L. Humphreys. 2011. "A Retrospective Cohort Study of Structured Abstracts in MEDLINE, 1992–2006." *Journal of the Medical Library Association* 99 (2): 160–3.

Rivara, Frederick P., and Laurence Finberg. 2001. "Use of the Terms Race and Ethnicity." *Archives of Pediatrics & Adolescent Medicine* 155 (2): 119.

Rodman, Lilita. 1981. "The Passive in Technical and Scientific Writing." *Journal of Advanced Composition* 2 (1/2): 165–72.

———. 1994. "The Active Voice in Scientific Articles: Frequency and Discourse Functions." *Journal of Technical Writing and Communication* 24 (3): 309–31.

Roederer, Mary, Macary Weck Marciniak, Shanna K. O'Connor, and Stephen F. Eckel. 2013. "An Integrated Approach to Research Manuscript Development." *American Journal of Health-System Pharmacy* 70 (14): 1211–18.

Rossner, Mike, and Kenneth M. Yamada. 2004. "What's in a Picture? The Temptation of Image Manipulation." *The Journal of Cell Biology* 166 (1): 11–15.

Rothwell, Peter M., and Christopher N. Martyn. 2000. "Reproducibility of Peer Review in Clinical Neuroscience: Is Agreement Between Reviewers Any Greater Than Would Be Expected by Chance Alone?" *Brain* 123 (9): 1964–9.

Rowan, Katherine E. 1991. "When Simple Language Fails: Presenting Difficult Science to the Public." *Journal of Technical Writing and Communication* 21 (4): 369–82.

Rowe, Nicholas, and Dragan Ilic. 2011. "Poster Presentation—A Visual Medium for Academic and Scientific Meetings." *Paediatric Respiratory Reviews* 11 (3): 208–13.

Royal Society. 2012. "History." Accessed September 14, 2012. http://royalsociety.org/about-us/history/.

Rude, Carolyn D. 1992. "The Rhetoric of Scientific Inquiry." *IEEE Transactions on Professional Communication* 35 (2): 88–90.

Sackett, David L., R. Brian Haynes, Gordon H. Guyatt, and Peter Tugwell. 2006. "Dealing with the Media." *Journal of Clinical Epidemiology* 59 (9): 907–13.

Salager-Meyer, Francoise. 1990. "Discoursal Flaws in Medical English Abstracts: A Genre Analysis Per Research- and Text-type." *Text-Interdisciplinary Journal for the Study of Discourse* 10 (4): 365–84.

———. 1994. "Reading Medical English Abstracts: A Genre Study of the Interaction Between Structural Variables and the Reader's Linguistico-Conceptual Competence." *Journal of Research in Reading* 17 (2): 120–46.

———. 2006. "Medical Discourse: Structured Abstracts." *Encyclopedia of Language and Linguistics* 14-Volume Set: 717–20.

Samraj, Betty. 2002. "Introductions in Research Articles: Variations Across Disciplines." *English for Specific Purposes* 21 (1): 1–17.

———. 2005. "An Exploration of a Genre Set: Research Article Abstracts and Introductions in Two Disciplines." *English for Specific Purposes* 24 (2): 141–56.

Sanford, Clive C. 1991. "Augmenting the Gatekeeper's Role: A Decision Support System for a Journal Editor." *IEEE Transactions on Professional Communication* 34 (3): 140–6.

Saracino, Giovanna, Linda W. Jennings, and Jeanette M. Hasse. 2013. "Basic Statistical Concepts in Nutrition Research." *Nutrition in Clinical Practice* 28 (2): 182–93.

Schein, Moshe, and Ramesh Paladugu. 2001. "Redundant Surgical Publications: Tip of the Iceberg?" *Surgery* 129 (6): 655–61.

Scherer, Roberta W., Cesar Ugarte-Gil, Christine Schmucker, and Joerg J. Meerpohl. 2015. "Authors Report Lack of Time as Main Reason for Unpublished Research Presented at Biomedical Conferences: A Systematic Review." *Journal of Clinical Epidemiology* 68 (7): 803–10.

Schriger, David L. 2005. "Suggestions for Improving the Reporting of Clinical Research: The Role of Narrative." *Annals of Emergency Medicine* 45 (4): 437–43.

Schriger, David L., Sanjay Arora, and Douglas G. Altman. 2006." The Content of Medical Journal Instructions for Authors." *Annals of Emergency Medicine* 48 (6): 743–9.

Schriger, David L., and Richelle J. Cooper. 2001. "Achieving Graphical Excellence: Suggestions and Methods for Creating High-Quality Visual Displays of Experimental Data." *Annals of Emergency Medicine* 37 (1): 75–87.

Schriger, David L., Reshmi Sinha, Sara Schroter, Pamela Y. Liu, and Douglas G. Altman. 2006. "From Submission to Publication: A Retrospective Review of the Tables and Figures in a Cohort of Randomized Controlled Trials Submitted to the British Medical Journal." *Annals of Emergency Medicine* 48 (6): 750–6.

Schroter, Sara, Leanne Tite, and Richard Smith. 2005. "Perceptions of Open Access Publishing: Interviews with Journal Authors." *BMJ* 330 (7494): 756.

Schulte, Bruce A. 2003. "Scientific Writing & the Scientific Method: Parallel 'Hourglass' Structure in Form & Content." *The American Biology Teacher* 65 (8): 591–4.

Schwartz, Lisa M., Steven Woloshin, and Linda Baczek. 2002. "Media Coverage of Scientific Meetings: Too Much, Too Soon?" *JAMA* 287 (21): 2859–63.

Science. n.d. "The Science Contributors FAQ."Accessed July 19, 2013. http://www.sciencemag.org/site/feature/contribinfo/faq/#pct_faq.

Selinger, Carl. 2007. "Dealing with the Media." *IEEE Spectrum*. August.

Shah, Jatin, Anand Shah, and Ricardo Pietrobon. 2009. "Scientific Writing of Novice Researchers: What Difficulties and Encouragements Do They Encounter?" *Academic Medicine* 84 (4): 511–16.

Shea, Elizabeth Parthenia 2008. *How the Gene Got Its Groove: Figurative Language, Science, and the Rhetoric of the Real*. Albany, NY: State University of New York Press.

Shehzad, Wasima. 2011. "Outlining Purpose, Stating the Nature of the Present Research, and Listing Research Questions or Hypotheses in Academic Papers." *Journal of Technical Writing and Communication* 41 (2): 136–60.

Shelledy, David C. 2004. "How to Make an Effective Poster." *Respiratory Care* 49 (10): 1213–16.

Shen, Benjamin S. P. 1975. "Scientific Literacy and the Public Understanding of Science." In *Communication of Scientific Information*, edited by Stacey B. Day, 44–52. New York: Basel Karger.

Shultz, Mary. 2007. "Comparing Test Searches in PubMed and Google Scholar." *Journal of the Medical Library Association* 95 (4): 442–5.

Siebers, Robert, and Shaun Holt. 2000. "Accuracy of References in Five Leading Medical Journals." *The Lancet* 356 (9239): 1445.

Simkin, Mikhail, and Vwani P. Roychowdhury. 2003. "Read Before You Cite!" *Complex Systems* 14: 269–74.

———. 2005. "Stochastic Modeling of Citation Slips." *Scientometrics* 62 (3): 367–84.

———. 2007. "A Mathematical Theory of Citing." *Journal of the American Society for Information Science and Technology* 58 (11): 1661–73.

Siontis, George C. M., Nikolaos A. Patsopoulos, Antonios P. Vlahos, and John P. A. Ioannidis. 2010. "Selection and Presentation of Imaging Figures in the Medical Literature." *PLoS One* 5 (5): e10888.

Skelton, John R. 1994. "Analysis of the Structure of Original Research Papers: An Aid to Writing Original Papers for Publication." *British Journal of General Practice* 44 (387): 455–9.

Skelton, John R., and Sarah J. L. Edwards. 2000. "The Function of the Discussion Section in Academic Medical Writing." *BMJ* 320 (7244): 1269–70.

Smith, Laurence D., Lisa A. Best, D. Alan Stubbs, Andrea Bastiani Archibald, and Roxann Roberson-Nay. 2002. "Constructing Knowledge: The Role of Graphs and Tables in Hard and Soft Psychology." *American Psychologist* 57 (10): 749–61.

Sonnad, Seema S. 2002. "Describing Data: Statistical and Graphical Methods." *Radiology* 225 (3): 622–8.

Søreide, Kjetil, and Desmond C. Winter. 2010. "Global Survey of Factors Influencing Choice of Surgical Journal for Manuscript Submission." *Surgery* 147 (4): 475–80.

Spigt, Mark, and Ilja C. W. Arts. 2010. "How to Review a Manuscript." *Journal of Clinical Epidemiology* 63 (12): 1385–90.

Sprague, Sheila, Mohit Bhandari, P. J. Devereaux, Marc F. Swiontkowski, Paul Tornetta, Deborah J. Cook, Douglas Dirschl, Emil H. Schemitsch, and Gordon H. Guyatt. 2003. "Barriers to Full-Text Publication Following Presentation of Abstracts at Annual Orthopaedic Meetings." *Journal of Bone & Joint Surgery* 85 (1): 158–63.

Spriestersbach, Albert, Bernd Rohrig, Jean-Baptist du Prel, Aslihan Gerhold-Ay, and Maria Blettner. 2009. "Descriptive Statistics: The Specification of Statistical Measures and Their Presentation in Tables and Graphs." *Deutsches Arzteblatt International* 106 (36): 578–83.

St. Leger, A. S. 1985. "How Accurate Are Quotations and References in Medical Journals?" *BMJ* 291 (6506): 1420.

Steel, C. M. 1996. "Read Before You Cite." *The Lancet* 348 (9021): 144.

Steen, R. Grant. 2011. "Retractions in the Medical Literature: Who Is Responsible for Scientific Integrity?" *AMWA Journal* 26 (1): 2–7.

Steinbrook, Robert. 2006. "Searching for the Right Search—Researching the Medical Literature." *The New England Journal of Medicine* 354 (1): 4–7.

Stengel, Dirk, Georgio M. Calori, and Peter V. Giannoudis. 2008. "Graphical Data Presentation." *Injury, International Journal of the Care of the Injured* 39 (6): 659–65.

Sternberg, Robert J., and Tamara Gordeeva. 1996. "The Anatomy of Impact: What Makes an Article Influential?" *Psychology Science* 7 (2): 69–75.

Strunk, William. (1918) 1999. *The Elements of Style*. Facsimile of the first edition. Ithaca, NY: Priv. print. Geneva: Press of W.P. Humphrey, Bartleby.com.

Suls, Jerry, and Rene Martin. 2009. "The Air We Breathe: A Critical Look at Practices and Alternatives in the Peer-Review Process." *Perspectives on Psychological Science* 4 (1): 40–50.

Sumner, Petroc, Solveiga Vivian-Griffiths, Jacky Boivin, Andy Williams, Christos A. Venetis, Aimée Davies, Jack Ogden, Leanne Whelan, Bethan Hughes, Bethan Dalton, Fred Boy, and Christopher D. Chambers. 2014. "The Association Between Exaggeration in Health Related Science News and Academic Press Releases: Retrospective Observational Study." *BMJ* 349: g7015.

Swales, John. 1990. *Genre Analysis: English in Academic and Research Settings*. Cambridge, UK: Cambridge University Press.

———. 2004. *Research Genre: Explorations and Applications*. Cambridge, UK: Cambridge University Press.

Swales, John, and Christine B. Feak. 2004. *Academic Writing for Graduate Students: Essential Tasks and Skills*. 2nd ed. Ann Arbor, MI: University of Michigan Press.

Swales, John, and Hazem Najjar. 1987. "The Writing of Research Article Introductions." *Written Communication* 4 (2): 175–91.

Swan, Alma P., and Sheridan N. Brown. 2004. *JISC/OSI Journal Authors Survey*. Truro, UK: Key Perspectives.

Swazey, Judith P., Melissa S. Anderson, and Karen Seashore Louis. 1993. "Ethical Problems in Academic Research: A Survey of Doctoral Candidates and Faculty Raises Important Questions About the Environment of Graduate Education and Research." *American Scientist*, November/December. Accessed November 4, 2014. http://www.americanscientist.org/issues/pub/ethical-problems-in-academic-research.

Szklo, Moyses. 2006. "Quality of Scientific Articles." *Rev Saúde Pública* 40: 3–35.

Szoka, Kathryn. 1982. "A Guide to Choosing the Right Chart Type." *IEEE Transactions on Professional Communication* 25 (2): 98–101.

Taddio, Anna, T. Pain, F. F. Fassos, H. Boon, A. Lane Ilersich, and T. R. Einarson. 1994. "Quality of Nonstructured and Structured Abstracts of Original Research Articles in the British Medical Journal, the Canadian Medical Association Journal and the Journal of the American Medical Association." *CMAJ: Canadian Medical Association Journal* 150 (10): 1611–15.

Taylor, David. 2002. "The Appropriate Use of References in a Scientific Research Paper." *Emergency Medicine* 14 (2): 166–70.

Teijlingen, Edwin Van, and Vanora Hundley. 2002. "Getting Your Paper to the Right Journal: A Case Study of an Academic Paper." *Journal of Advanced Nursing* 37 (6): 506–11.

Teixeira, Mariana C., Sidinei M. Thomaz, Thaisa S. Michelan, Roger P. Mormul, Thamis Meurer, Jose Vitor B. Fasolli, and Marcio J. Silveira. 2013. "Incorrect Citations Give Unfair Credit to Review Authors in Ecology Journals." *PLoS One* 8 (12): e81871.

Thompson, Brian J. 1995. "What Is a Reference?" *Optical Engineering* 34 (7): 1861.

Thomson Reuters. n.d. "ScholarOne Manuscripts." Accessed June 2, 2014. http://scholarone.com/products/manuscript/.

Thornton, Jim. 2003. "The Discussion Section." *BJOG: an International Journal of Obstetrics and Gynaecology* 110 (5). 419.

Thrower, Peter A. 2006. "Abbreviate!" *Carbon* 44 (5): 825–7.

———. 2007. "Writing a Scientific Paper: I. Titles and Abstracts." *Carbon* 45 (11): 2143–4.

———. 2008a. "Writing a Scientific Paper: II. Introduction and References." *Carbon* 46 (2): 183–4.

———. 2008b. "Writing a Scientific Paper: III. Experimental." *Carbon* 46 (8): 1113–14.

———. 2010. "Writing a Scientific Paper: IV. Results and Discussion." *Carbon* 48 (10): 2675–6.

Todd, Peter A., James R. Guest, Junxiu Lu, and Loke Ming Chou. 2010. "One in Four Citations in Marine Biology Papers Is Inappropriate." *Marine Ecology Progress Series* 408: 299–303.

Todd, Peter A., Darren C. J. Yeo, Daiqin Li, and Richard J. Ladle. 2007. "Citing Practices in Ecology: Can We Believe Our Own Words?" *Oikos* 116 (9): 1599–601.

Triadafilopoulos, George. 2006. "The Manuscript Review Process." *Gastrointestinal Endoscopy* 64 (6): S23–5.

Tse, Tony, Rebecca J. Williams, and Deborah A. Zarin. 2009. "Reporting 'Basic Results' in ClinicalTrials.gov." *CHEST* 136 (1): 295–303.

Tufte, Edward. 2001. *The Visual Display of Quantitative Information*. Cheshire: Graphics Press.

Tukey, John W. 1977. "Some Thoughts on Clinical Trials, Especially Problems of Multiplicity." *Science* 198 (4318): 679–84.

Tulsky, Asher A., and Ruth W. Kouides. 1998. "Abstract Presentations: What Do SGIM Presenters Prefer?" *Journal of General Internal Medicine* 13 (6): 417–18.

Turner, Steven. 2008. "School Science and Its Controversies: Or, Whatever Happened to Scientific Literacy?" *Public Understanding of Science* 17 (1): 55–72.

U.S. National Institutes of Health. 2008. *NIH Public Access Policy Details* accessed February 25, 2014, http://publicaccess.nih.gov/policy.htm.

U.S. National Library of Medicine. 2013. "Summary Indexing Statistics: 1965–2010." Accessed February 25, 2014. http://www.nlm.nih.gov/bsd/index_stats.html.

U.S. National Library of Medicine, "Summary Indexing Statistics: 1965–2010." Accessed February 25, 2014. http://www.nlm.nih.gov/bsd/index_stats.html.

Van Raan, Anthony F. J. 2004. "Sleeping Beauties in Science." *Scientometrics* 59 (3): 467–72.

Van Rooyen, Susan, Nick Black, and Fiona Godlee. 1999. "Development of the Review Quality Instrument (RQI) for Assessing Peer Reviews of Manuscripts." *Journal of Clinical Epidemiology* 52 (7): 625–9.

Van Rooyen, Susan, Fiona Godlee, Stephen Evans, Richard Smith, and Nick Black. 1999. "Effect of Blinding and Unmasking on the Quality of Peer Review." *Journal of General Intern Medicine* 14 (10): 622–4.

Van Way, Charles W. 2007. "Writing a Scientific Paper." *Nutrition in Clinical Practice* 22 (6): 636–40.

Vande Kopple, William J. 1985. "Some Exploratory Discourse on Metadiscourse." *College Composition and Communication* 36 (1): 82–93.

Vandenbroucke, Jan P., Erik Von Elm, Douglas G. Altman, Peter C. Gøtzsche, Cynthia D. Mulrow, Stuart J. Pocock, Charles Poole, James J. Schlesselman, and Matthias Egger. 2007. "Strengthening the Reporting of Observational Studies in Epidemiology (STROBE): Explanation and Elaboration." *PLoS Medicine* 4 (10): 1628–54.

Vanhecke, Thomas E., Michael A. Barnes, Janet Zimmerman, Sandor Shoichet. 2007. "PubMed Vs. HighWire Press: A Head-to-Head Comparison of Two Medical Literature Search Engines." *Computers in Biology and Medicine* 37 (9): 1252–8.

Velickovic, Vladica M. 2015. "What Everyone Should Know about Statistical Correlation." *American Scientist* 103 (1): 26–9.

Veness, Michael J. 2010. "Strategies to Successfully Publish Your First Manuscript." *Journal of Medical Imaging and Radiation Oncology* 54 (4): 395–400.

Vines, Timothy H., Arianne Y. K. Albert, Rose L. Andrew, Florence Debarre, Dan G. Bock, Michelle T. Franklin, Kimberly J. Gilbert, Jean-Sebastien Moore, Sebastien Renaut, and Diana J. Rennison. 2014. "The Availability of Research Data Declines Rapidly with Article Age." *Current Biology* 24 (1): 94–7.

Von Elm, Erik, Greta Poglia, Bernhard Walder, and Martin R. Tramer. 2004. "Different Patterns of Duplicate Publication: An Analysis of Articles Used in Systematic Reviews." *JAMA* 291 (8): 974–80.

Wachs, Joy E. 1996. "From Idea to Publication: The Secrets of Publishing." *American Association of Occupational Health Nurses Journal* 44 (6): 273–7.

Wainer, Howard. 1997. "Improving Tabular Displays, with NAEP Tables as Examples and Inspirations." *Journal of Educational and Behavioral Statistics* 22 (1): 1–30.

Walkowski, Debbie. 1991. "Working Successfully with Technical Experts—From Their Perspective." *Technical Communication* 38 (1): 65–7.

Wang, Min-Fen, and Lori L. Bakken. 2004. "An Academic Writing Needs Assessment of English-as-a-Second-Language Clinical Investigators." *The Journal of Continuing Education in the Health Professions* 24 (3): 181–9.

Warlick, Stefanie E., and K. T. L. Vaughan. 2007. "Factors Influencing Publication Choice: Why Faculty Choose Open Access." *Biomedical Digital Libraries* 4 (1): 1.

Warman, P. C. 1903. "A Plea for Better English in Science." *Science* 18 (461): 563–8.

Watson, James D., and Francis H. C. Crick. 1953. "Molecular Structure of Nucleic Acids: A Structure for Deoxyribose Nucleic Acid." *Nature* 171 (4356): 737–8.

Webster, Angela C., Nicholas B. Cross, Ruth Mitchell, and Jonathan C. Craig. 2010. "How to Get the Most from the Medical Literature: Searching the Medical Literature Effectively." *Nephrology* 15 (1): 12–19.

Weeks, William B., and Amy E. Wallace. 2002. "Readability of British and American Medical Prose at the Start of the 21st Century." *BMJ* 325 (7378): 1451–2.

Weil, Ben H. 1981. "Why Should Chemists Care about Copyright?" *Journal of Chemical Information & Computer Sciences* 22 (2): 61–3.

Weinert, Clarann. 2010. "Are All Abstracts Created Equal?" *Applied Nursing Research* 23 (2): 106–9.

Welch, Stephen J. 2006. "Preparing Manuscripts for Online Submission: Basic Information and Avoidance of Common Pitfalls." *CHEST* 129 (3): 822–5.

———. 2007. "Avoiding Common Problems During Online Submission of Manuscripts." *CHEST* 131 (5): 1591–4.

Wenger, Neil S., Stanley G. Korenman, Richard Berk, and Sandy Berry. 1997. "The Ethics of Scientific Research: An Analysis of Focus Groups of Scientists and Institutional Representatives." *Journal of Investigative Medicine* 45 (6): 371–80.

Wickman, Chad. 2013. "Observing Inscriptions at Work: Visualization and Text Production in Experimental Physics Research." *Technical Communication Quarterly* 22 (2): 150–71.

Wilkinson, A. M. 1992. "Jargon and the Passive Voice: Prescriptions and Proscriptions for Scientific Writing." *Journal of Technical Writing and Communication* 22 (3): 319–25.

Williams, Joseph M. 2011. *Problems into PROBLEMS: A Rhetoric of Motivation*. Fort Collins, CO: WAC Clearinghouse.

Winker, Margaret A. 1999. "The Need for Concrete Improvement in Abstract Quality." *JAMA* 281 (12): 1129–30.

———. 2004. "Measuring Race and Ethnicity: Why and How?" *JAMA* 292 (13): 1612–14.

Wolff, Wirt M. 1970. "A Study of Criteria for Journal Manuscripts." *American Psychologist* 25 (7): 636–9.

Wolka, Anne M., and Gordon B. Cutler. 2008. "Writing a Manuscript Discussion in Clinical Medicine." *AMWA Journal* 23 (2): 74–5.

Woloshin, Steven, and Lisa M. Schwartz. 2002. "Press Releases: Translating Research into News." *JAMA* 287 (21): 2856–8.

Wong, Donna M. 2010. *The Wall Street Journal Guide to Information Graphics: The Dos and Don'ts of Presenting Data, Facts, and Figures*. New York: Norton.

Wood, Gordon J., and R. Sean Morrison. 2011. "Writing Abstracts and Developing Posters for National Meetings." *Journal of Palliative Medicine* 14 (3): 353–9.

Woodford, F. Peter. 1967. "Sounder Thinking Through Clearer Writing." *Science* 156 (3776): 743–5.

———. 1997. "Keynote Address: From Vigorous Writing to Scientific Integrity." *CBE Views* 20 (5): 159–62.

Woolgar, Steve. 1981. "Discovery: Logic and Sequence in a Scientific Text." In *The Social Process of Scientific Investigation*, edited by W. R. Knorr, R. Krohn, and Richard P. Whitley, 239–68. New York: Springer.

Woolley, Karen L. 2006. "Goodbye Ghostwriters! How to Work Ethically and Efficiently with Professional Medical Writers." *CHEST* 130 (3): 921–3.

Woolsey, John D. 1989. "Combating Poster Fatigue: How to Use Visual Grammar and Analysis to Effect Better Visual Communications." *Trends in Neurosciences* 12 (9): 325–32.

World Health Organization. n.d. *Top 10 Reasons for Rejecting a Manuscript*. Accessed May 26, 2014. http://www.who.int/bulletin/contributors/rejection/en/.

Worsham, Lynn. 2008. "What Editors Want." *Chronicle of Higher Education*, September 12. Accessed September 9, 2015. http://chronicle.com/article/What-Editors-Want/45909.

Wright, Patricia, and Kathryn Fox. 1970. "Presenting Information in Tables." *Applied Ergonomics* 1 (4): 234–42.

Yankauer, Alfred. 1991. "How Blind is Blind Review?" *American Journal of Public Health* 81 (7): 843–5.

Yore, Larry D., Marilyn K. Florence, Terry W. Pearson, and Andrew J. Weaver. 2006. "Written Discourse in Scientific Communities: A Conversation with Two Scientists about Their Views of Science, Use of Language, Role of Writing in Doing Science, and Compatibility between Their Epistemic Views and Language." *International Journal of Science Education* 28 (2–3): 109–41.

Youmans, Hubert L. 1974. "How to Read a Chemical Research Report." *Journal of Technical Writing and Communication* 4 (3): 251–9.

Zappen, James P. 1989. "The Discourse Community in Scientific and Technical Communication: Institutional and Social Views." *Journal of Technical Writing and Communication* 19 (1): 1–11.

Zarin, Deborah A., Tony Tse, Rebecca J. Williams, Robert M. Califf, and Nicholas C. Ide. 2011. "The ClinicalTrials.gov Results Database—Update and Key Issues." *The New England Journal of Medicine* 364 (9): 852–60.

Zeiger, Mimi. 2000. *Essentials of Writing Biomedical Research Papers*. 2nd ed. New York: McGraw-Hill.

Zhang, Zaixin. 1989. "Is Scientific Research Part of Prewriting in the Scientific Writing Process?" *Journal of Technical Writing and Communication* 19 (3): 285–96.

Zylke, Jody W. 2013. "Research Letters in *JAMA*: Small but Mighty." *JAMA* 310 (6): 589–90.

INDEX

Note: Page numbers in italic indicate a figure or table on the corresponding page.